T0272001

DISCRETE CHAOS

Second Edition

WITH APPLICATIONS IN SCIENCE AND ENGINEERING

DISCRETE CHAOS
Second Edition

WITH APPLICATIONS IN
SCIENCE AND ENGINEERING

Saber N. Elaydi

Trinity University
San Antonio, Texas, U.S.A.

Chapman & Hall/CRC
Taylor & Francis Group
Boca Raton   London   New York

Chapman & Hall/CRC is an imprint of the
Taylor & Francis Group, an **informa** business

Chapman & Hall/CRC
Taylor & Francis Group
6000 Broken Sound Parkway NW, Suite 300
Boca Raton, FL 33487-2742

© 2008 by Taylor & Francis Group, LLC
Chapman & Hall/CRC is an imprint of Taylor & Francis Group, an Informa business

No claim to original U.S. Government works
Printed in the United States of America on acid-free paper
10 9 8 7 6 5 4 3 2 1

International Standard Book Number-13: 978-1-58488-592-4 (Hardcover)

Library of Congress Cataloging-in-Publication Data

Elaydi, Saber, 1943-
    Discrete chaos : with applications in science and engineering / Saber N. Elaydi. -- 2nd ed.
        p. cm.
    Includes bibliographical references and index.
    ISBN 978-1-58488-592-4 (alk. paper)
    1. Differentiable dynamical systems. 2. Difference equations. 3. Chaotic behavior in systems. I. Title.

QA614.8.E53 2007
515'.39--dc22   2007034217

Contents

Preface

Preface to the Second Edition

The second edition maintains the lucidity of the first edition. Its main feature is the inclusion of many recent results on global stability, bifurcation, chaos, and fractals.

The first five chapters of this book include the most comprehensive exposition on discrete dynamical systems at the level of advanced undergraduates and beginning graduate students. Notable additions in this book are the L-systems, the trace-determinant stability and bifurcation analysis, the periodic structure of the bulbs in the Mandelbrot set, the detailed analysis of the center manifold theory, and new results on global stability. Moreover, new applications to biology, chemistry, and physics were added.

The biggest improvement, however, is in technology. A CD of an adapted version of PHASER by Jason Glick and Huseyin Kocac is attached to the back cover of the book. It contains all the material in Chapters 1 and 4. The material for the remaining chapters may be downloaded from

$$http://www.math.miami.edu/{\sim}hk/elaydi/.$$

You can immediately download the CD and start experimenting with PHASER. You may be able to generate all the graphs in the book and use it to solve many problems in the exercises. If you are a Maple or Mathematica user, programs were written by Henrique Oliveira and Rafael Luis and are posted on the Taylor and Francis website. They may be downloaded from "Electronic Products" and then "Downloads and Updates" found in the right pane of the website at *http://www.crcpress.com/*.

I am greatly indebted to Darnum Huang who reviewed thoroughly the first edition and made a plethora of insightful comments and suggestions. He caught numerous typos in the first edition. His contribution to this book is immeasurable. Sincere appreciation goes to Richard Neidinger whose critique led to improvements in several parts of this book. I would like to thank Greg Morrow who read the first edition and made several useful comments and to his student Dennis Duncan who caught several typos and misprints. I am grateful to Henrique Oliveira and our joint student Rafael Luis for developing Maple and Mathematica programs for this book and for proofreading the first draft. My sincere gratitude goes to Huseyin Kocac for his tremendous work on adapting PHASER to the contents of this book. I would like to thank Ronald Mickens for his many contributions to this book, AbiTUMath students and its director Andreas Ruffing who made many suggestions that improved this book, Xinjuan Chen for his comments about period-doubling bifurcation and to Fozi Dannan for his suggestions that improved this book. My sincere thanks go to Sunil Nair whose unwavering support and encouragement made a difference in the completion of this book. It is my great pleasure to acknowledge the support of Lord Robert M. May who graciously accepted and wrote a foreword for this book. Finally, I have to express my deep appreciation to Denise Wilson who not only typed the manuscript but also created many of its graphs. Without her help this book would have never seen the light.

Saber Elaydi

San Antonio, April 2007.

- Technology

    1. PHASER: A CD is attached on the inside back cover. You may download it and start using it immediately. It takes less than half an hour to learn it and no experience in programming is necessary. PHASER can generate all the graphs in the book and may be used in solving problems in the text as well as for research in discrete dynamical systems and difference equations. The material may be downloaded from

    $$http://www.math.miami.edu/\sim hk/elaydi/.$$

    2. Maple and Mathematica: Alternates to PHASER are the computer algebra systems Maple and Mathematica. Programs written in both may be downloaded from the Taylor and Francis website under "Electronic Products" then "Downloads and Updates" found in the right pane of the website at *http://www.crcpress.com/*. These programs may be used to solve problems in the text as well as for research in discrete dynamical systems and difference equations. Moreover, it does not take much time to be able to use these programs and no previous experience is necessary.

    Questions regarding technology and the book may be addressed to the author at selaydi@trinity.edu.

Preface to the First Edition

This book grew out of a course on discrete dynamical systems/difference equations that I taught at Trinity University since 1992. My students were juniors and seniors majoring in mathematics, computer science, physics, and engineering. The students have impacted not only the writing style and presentation of the material, but also the selection of exercises and material; I am greatly indebted to each and every one of them.

The only prerequisites for the course were calculus and linear algebra. The first three chapters deal with the dynamics of one-dimensional maps. Hence, a solid background in calculus suffices for understanding the material in these chapters. Chapters 4, 5, and 6 require a good knowledge of linear algebra (as provided by a standard course in linear algebra). The last chapter on complex dynamics requires some rudiments of complex variables.

The last three decades witnessed a surge of research activities in chaos theory and fractals. These two subjects have captured the imagination of both scientists and the population at large all over the world. The book *Chaos: Making a New Science* by James Gleick, as well as the movie, "Jurassic Park," have helped in popularizing chaos theory among millions of people around the globe.

Numerous books on chaos theory and fractals have appeared. They range from the very elementary to the most advanced and from the very mathematical to the most application-oriented. So, the question is, why do we need one more book on this subject when we already have many excellent books? There are basically four main reasons or justifications for writing this book. The most important is that I wanted to give a thorough exposition of stability theory in one and two dimensions including the famous method of Liapunov. Such an exposition is lacking in many current books on discrete dynamical systems. Stability theory is an important area of research in its own right, but it is also important in the context of chaos theory since the route to chaos is through stability. The second reason is that I wanted to present a readable and accessible account of fractals and the mathematics behind them. The third justification is that I wanted to include applications from real-world phenomena that would be beneficial to a wider readership. The last reason for writing this book is to show that the division between discrete dynamical systems and difference equations is an artificial one. In my view, the main difference for the most part, is the notation used. This book integrates both notations, which makes it easier for students and researchers alike to read literature and books with either title.

This book treats the modern theory of difference equations and may be regarded as a completion of my book, *An Introduction to Difference Equations*, which presents the classical aspects of difference equations.

Other important features of the book are as follows:

1. The book contains a very extensive and carefully selected set of exercises at the end of each section. The exercises form an integral part of the book; they range from routine problems designed to build basic skills to more challenging problems that produce deep understanding and build techniques. The problems denoted by an asterisk are the most challenging and they may be suitable for term projects or research problems for independent study courses or solving-problem groups.

2. Great efforts were made to present even the most difficult material in a

format that makes it accessible to students and scientists with varying backgrounds and interests.

3. The book encourages readers to make mathematical discoveries through computer experimentation. All of the graphs in the book were generated by Maple programs. You may download these programs from *http://www.crcpress.com/*. Readers are encouraged to improve these programs and to develop their own.

In Chapter 1, we present an extensive exposition of the dynamics of one-dimensional maps. It provides criteria for the stability of hyperbolic and nonhyperbolic fixed and periodic points. Period-doubling route to chaos is also introduced. As an application, a genotype selection model is presented in Section 1.9.

Chapter 2 deals with Sharkovsky's theorem and includes its proof in an appendix. Its converse is also provided. Rudiments of bifurcation theory is presented in Section 2.5. This chapter also includes the Lorenz map. In

In Chapter 3, we give a brief introduction to metric spaces, followed by a thorough presentation of chaos theory. This is facilitated by the introduction of symbolic dynamics and conjugacy.

Chapter 4 provides an extensive treatment of stability of two-dimensional maps. This includes an investigation of linear maps as well as second-order difference equations. Stability of nonlinear maps is studied by the method of Liapunov and by linearization. A brief introduction to the Hartman-Grobman theorem and the stable manifold theorem are included. At the end of the chapter, we give three applications: the kicked rotator and the Hénon map, a discrete epidemic model for gonorrhea, and perennial grass.

In Chapter 5, we study chaos of two-dimensional maps. This includes the study of toral automorphisms, symbolic dynamics, subshifts of finite type, and the horseshoe of Smale. The chapter concludes with a section on center manifolds and another on bifurcation. Of interest is the Neimark-Sacker (Hopf) bifurcation—a phenomenon that is not present in one-dimensional settings.

Chapter 6 gives a readable and extensive account on fractals, generated by affine transformations, and the underlying theory of iterated function system. A section on image compression is added to arouse the reader's interest in this important application of fractals.

Finally, Chapter 7 investigates the dynamics of one-dimensional maps in the complex plane. This chapter may be considered as an extension of Chapter 1 to the set of complex numbers. The Julia set and its topological properties are presented in Sections 7.3 and 7.4. In Section 7.5 Newton's method in the complex domain and the associated basin of attraction are studied. Then, in Section 7.6, the famous Mandelbrot set is examined and its connection with the bifurcation diagram of the real-valued quadratic map explored.

The following diagram shows the interdependence of the chapters:

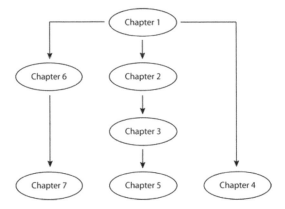

Suggestions for a One-Semester Course

There is enough material for a two-semester course. The instructor has a lot of flexibility in choosing material for a one-semester course.

Here are a few suggestions:

1. A course on chaos theory may consist of only Chapters 1, 2, 3, and 5.

2. A course on one-dimensional chaos and stability theory may consist of Chapters 1, 2, 3, and 4.

3. A course on chaos and fractals may consist of Chapters 1, 3, 6, and 7; or Chapters 1, 3, 4, and 7; or Chapters 1, 2, 3, and 7.

Acknowledgment

I am indebted to Ronald Mickens, who read the manuscript thoroughly and made numerous suggestions that improved the final draft of the book. I am grateful for his encouragement and support during the entire project. My sincere appreciation goes to Michael J. Field, who read the entire manuscript with tremendous care and insight. He caught many errors and typos in the earlier drafts. His recommendations have significantly improved this book. My sincere thanks go to Abdul-Aziz Yakubu, who made many helpful comments and caught numerous errors and misprints. I am grateful to Hassan Sedaghat, who made many helpful suggestions that improved portions of the book.

While writing this book, I benefited from discussions with Roberto Hasfura, Mario Martelli, Allan Peterson, Jim Cushing, Gerry Ladas, Timothy Sauer, Linda Allen, Marwan Awartani, Fathi Allan, Klara Janglajew, and Jia Li. I am grateful for their insight and interest.

I would like to thank the REU students Aaron Heap, Phillip Lynch, John May, Bryant Mathews, and Nick Neumann for their critical reading of the manuscript and for many helpful discussions and ideas. I am thankful to my students Brooke Liddle, Joshua Buckner, Cristina Cashman, Rachel Shepherd, Wendy Hendricks, and Ashley Moore, who caught many errors and misprints and made numerous suggestions to improve the style and substance in some earlier drafts of the manuscript.

Ellen Phifer and Stephanie Moore did a wonderful job in writing Maple programs to generate the graphs in the book; I am grateful for their help. My thanks are due to Ming Chiu and Scott Schaefer for generating some of the graphs and for many helpful discussions about various topics in the book.

I am thankful to Aletheia Barber, who meticulously typed the manuscript in LaTex in its many revised versions. Finally, I thank the Department of Mathematics at Trinity University for their unwavering support for the duration of this project.

Saber Elaydi
San Antonio, Texas, 1999

Author Biography

Saber Elaydi is the co-founder and co-editor-in-chief of the *Journal of Difference Equations and Applications* (JDEA) and the *Journal of Biological Dynamics* (JBD). He is also the author of the best seller, *An Introduction of Difference Equations* (2005). He has written over 100 publications in dynamical systems, differential equations, difference equations, and mathematical biology. Elaydi's current research deals with the stability and bifurcation of nonautonomous discrete dynamical systems and their application to populations with fluctuating habitats. He is a professor of mathematics at Trinity University and served as chair from 1991 to 1999. He has been serving as president of the International Society of Difference Equations (ISDE) since 2005.

LIST OF SYMBOLS

Foreword by Lord Robert M. May

When I was in graduate school, in the Physics Department at Sydney University roughly half a century ago, the first big computers were just being built. Given that we already knew the basic equations – the Navier-Stokes equations – governing atmospheric flow, it seemed only a matter of time before even more accurate forecasting of local weather would be available. This was a time when the Newtonian dream, although rarely explicitly articulated, still prevailed: the world is governed by rules; if the rules are sufficiently simple, the outcomes are predictable; more complicated situations – the spin of a ball in a roulette wheel, for instance – may seem unpredictable, but with sufficiently accurate observation of initial conditions and ever increasing computational power, such apparent randomness would eventually give way to predictability. Work by Poincare in the late 1800s, and by Cartwright and Littlewood in the 1930s, had cast a shadow over this dream, but only a handful of mathematicians were aware of this.

Today that Newtonian dream of an entirely predictable world is dead. We now realise that there are very simple systems – a forced pendulum, literally Newtonian clockwork, can provide an example – where we can know the rules exactly and where there are no random elements whatsoever, yet where the "chaotic" dynamics are so sensitive to initial conditions that long-term prediction is impossible. This awareness spread rapidly among the various branches of science in the 1970s. Its origins were two-fold. One was Lorenz's meteorological metaphor for thermal convection, in the form of a simple but nonlinear 3-dimensional system of differential equations (that is, with time a continuous variable). The other came from studies of even simpler 1-dimensional nonlinear difference equations, proposed as metaphors for the dynamics of fish or insect populations with discrete generations (that is, where time is a discrete variable). Lorenz's 3-dimensional system of differential equations, although relatively simple as things go, is still fairly complicated. The 1-dimensional difference equations, however, reveal the chaotic nature of things relatively simply: you know the equations, but an error in the tenth decimal place of the initial condition soon drives dynamical trajectories widely apart.

Subsequent work, beginning in the late 1980s and early 1990s, has explored what might be called the flip-side of the chaos coin. Given that we now know that relatively simple rules might give you dynamical trajectories that look like random noise, what are we to make of movements on the currency exchange markets, or in fluctuating fish populations? Are we to treat them as purely random, or might they be, at least in part, produced by simple –

but nonlinear – rules, giving rise to low-dimensional chaos? And if the latter, to what extent might it open new doors to predictability? To give just one explicit example, the algorithm used to generate random numbers for the first big computer, called Maniac, which was built at the Institute for Advanced Study in Princeton by Von Neumann and Ulam in 1948, used a 1-dimensional difference equation in a regime which we would now call chaotic. If you look at the numbers thus generated, and test them by conventional statistical methods (Kolmogorov-Smirnov test, for example) you will be reassured that you are indeed looking at random numbers. But newly developed methods enable us to see that we are dealing with a 1-dimensional chaotic system, and thus enable us to predict the next few "random numbers" in this sequence with high confidence, although such predictability fades as we move outside the "Liapunov horizon" and chaos really asserts itself (Sugihara and May, 1990).

The present textbook gives an excellent introduction to this new, and potentially revolutionary, territory. It will take the reader, with clarity and precision, from simple beginnings with 1-dimensional difference equations (and their cascades of period doubling en route to chaos), on to 2- and 3-dimensional systems, and beyond this to fractals and relationships between geometry and dynamics. The final chapter deals with the Julia and Mandelbrot sets, where in my opinion mathematical elegance and pure aesthetic beauty begin to merge.

Throughout, this text is further enhanced by an exceptionally well-selected set of problems that illustrate the principles and challenge the reader.

I ended my 1976 review article by saying "not only in research, but also in the everyday world of politics and economics, we would all be better off if more people realised that simple nonlinear systems do not necessarily possess simple dynamical properties." As this text shows, things have come a long way since then. Users of this text will enjoy the journey.

Robert M May
Zoology Department
Oxford University
Oxford, UK
June 2007

1

The Stability of One-Dimensional Maps

First-order difference equations arise in many contexts in the biological, economic, and social sciences. Such equations, even though simple and deterministic, can exhibit a surprising array of dynamical behavior, from stable points, to a bifurcation hierarchy of stable cycles, to apparently random fluctuations.

Robert M. May

Robert M. May (1936-)

Lord Robert McCredie May is credited with creating the new field of "chaotic dynamics in biology." In 1976, he published his most popular article "Simple mathematical models with very complicated dynamics" in the journal *Nature*, where he showed that first-order nonlinear difference equations can exhibit an astonishing array of dynamical behavior, ranging from stable fixed points to chaotic regions.

May's current research deals with factors influencing the diversity and abundance of plant and animal species and with the rates, causes and consequences of extinction. In 2000, he teamed up with Martin Nowak to write a seminal book on Virus dynamics titled *The mathematical foundation of immunology and virology.*

Trained as a physicist at Sydney University, where he received his PhD in 1959, he moved to biology for good in 1973, and later he became a professor of zoology at Princeton University and later at Oxford University.

1.1 Introduction

Difference equations have been increasingly used as mathematical models in many disciplines including genetics, eipdemiology, ecology, physiology, neural networks, psychology, engineering, physics, chemistry and social sciences. Their amenability to computerization and their mathematical simplicity have attracted researchers from a wide range of disciplines. As we will see in Section 1.2, difference equations are generated by maps (functions). Section 1.3 illustrates how discretizing a differential equation would yeild a difference equation. Discretization algorithms are part of a discipline called numerical analysis which belong to both mathematics and computer science. As most differential equations are unsolvable, one needs to resort to computers for help. However, computers understand only recursions or difference equations; thus the need to discretize differential equations.

1.2 Maps vs. Difference Equations

Consider a map $f : \mathbb{R} \to \mathbb{R}$ where $\mathbb{R}$ is the set of real numbers. Then the (positive) **orbit** $O(x_0)$ of a point $x_0 \in \mathbb{R}$ is defined to be the set of points

$$O(x_0) = \{x_0, f(x_0), f^2(x_0), f^3(x_0), \ldots\}$$

where $f^2 = f \circ f, f^3 = f \circ f \circ f$, etc.

Since most maps that we deal with are noninvertible, positive orbits will be called orbits, unless otherwise stated.

If we let $x(n) := f^n(x_0)$, then we obtain the first-order difference equation

$$\boxed{x(n+1) = f(x(n))} \tag{1.1}$$

with $x(0) = x_0$.

In population biology, $x(n)$ may represent a population size in generation n. Equation (1.1) models a simple population system with seasonal breeding whose generations do not overlap (e.g., orchard pests and temperate zone insects). It simply states that the size $x(n+1)$ of a population in generation $n+1$ is related to the size $x(n)$ of the population in the preceding generation n by the function f.

In epidemiology, $x(n)$ represents the fraction of the population infected at time n. In economics, $x(n)$ may be the price per unit in time n for a certain commodity. In the social sciences, $x(n)$ may be the number of bits of information that can be remembered after a period n.

Example 1.1

(The Logistic Map). Let $x(n)$ be the size of a population of a certain species at time n. Let μ be the rate of growth of the population from one generation to another. Then a mathematical model that describes the size of the population take the form

$$x(n+1) = \mu x(n), \ \mu > 0. \tag{1.2}$$

If the initial population $x(0) = x_0$, then by a simple iteration we find that

$$x(n) = \mu^n x_0. \tag{1.3}$$

is the solution of Equation (1.2).

If $\mu > 1$, then the population $x(n)$ increases without any bound to infinity. If $\mu = 1$, $x(n) = x_0$ and the population stays constant forever. Finally, for $\mu < 1$, $\lim\limits_{n \to \infty} x(n) = 0$, and the population eventually becomes extinct.

We observe that for most species none of the above scenarios are valid; the population increases until it reaches a certain maximum value. Then limited resources would force members of the species to fight and compete for those limited resources. This competition is proportional to the number of squabbles among them, given by $x^2(n)$. Consequently, a more reasonable model is given by

$$x(n+1) = \mu x(n) - bx^2(n) \tag{1.4}$$

where $b > 0$ is the proportionality constant of interaction among members of the species.

To simplify Equation (1.4), we let $y(n) = \frac{b}{\mu} x(n)$. Hence,

$$y(n+1) = \mu y(n)(1 - y(n)). \tag{1.5}$$

Equation (1.5) is called the logistic equation and the map $f(y) = \mu y(1-y)$ is called the logistic map. It is a reasonably good model for seasonably breeding populations in which generations do not overlap.

This equation/map will be the focus of our study throughout Chapter 1. By varying the value of μ, this innocent-looking equation/map exhibits complicated dynamics.

Surprisingly, a closed form solution of Equation (1.5) is not possible, except for $\mu = 2, 4$. ▯

A map f is called **linear** if it is of the form $f(x) = ax$ for some constant a. In this case, Equation (1.1) is called a **first-order linear** difference equation. Otherwise, f [or Equation (1.1)] is called **nonlinear** (or **density-dependent** in biology).

One of the main objectives in dynamical systems theory is the study of the behavior of the orbits of a given map or a class of maps. In the language of difference equations, we are interested in investigating the behavior of solutions of Equation (1.1). By a **solution** of Equation (1.1), we mean a sequence $\{\varphi(n)\}, n = 0, 1, 2, \ldots$, with $\varphi(n+1) = f(\varphi(n))$ and $\varphi(0) = x_0$, i.e., a sequence that satisfies the equation.

1.3 Maps vs. Differential Equations

1.3.1 Euler's Method

Consider the differential equation

$$x'(t) = g(x(t)), x(0) = x_0 \qquad (1.6)$$

where $x'(t) = \frac{dx}{dt}$.

For many differential equations such as Equation (1.6), it may not be possible to find a "closed form" solution. In this case, we resort to numerical methods to approximate the solution of Equation (1.6). In the Euler algorithm, for example, we start with a discrete set of points $t_0, t_1, \ldots, t_n, \ldots$, with $h = t_{n+1} - t_n$ as the **step size**. Then, for $t_n \leq t < t_{n+1}$, we approximate $x(t)$ by $x(t_n)$ and $x'(t)$ by $\dfrac{x(t_{n+1}) - x(t_n)}{h}$. Equation (1.6) now yields the difference equation

$$x(t_{n+1}) = x(t_n) + hg(x(t_n))$$

which may be written in the simpler form

$$x(n+1) = x(n) + hg(x(n)) \qquad (1.7)$$

where $x(n) = x(t_n)$.

Note that Equation (1.7) is of the form of Equation (1.1) with

$$f(x) = f(x, h) = x + hg(x).$$

Now given the initial data $x(0) = x_0$, we may use Equation (1.7) to generate the values $x(1), x(2), x(3), \ldots$ These values approximate the solution of the differential Equation (1.6) at the "grid" points $t_1, t_2, t_3, \ldots$, provided that h is sufficiently small.

Example 1.2

Let us now apply Euler's method to the differential equation:

$$x'(t) = 0.7x^2(t) + 0.7, \quad x(0) = 1, \quad t \in [0, 1]. \quad (DE)^1$$

---

[1] $DE \equiv$ differential equation.

TABLE 1.1

n	t	(ΔE) Euler $(h = 0.2)$ $x(n)$	(ΔE) Euler $(h = 0.1)$ $x(n)$	Exact (DE) $x(t)$
0	0	1	1	1
1	0.1		1.14	1.150
2	0.2	1.28	1.301	1.328
3	0.3		1.489	1.542
4	0.4	1.649	1.715	1.807
5	0.5		1.991	2.150
6	0.6	2.170	2.338	2.614
7	0.7		2.791	3.286
8	0.8	2.969	3.406	4.361
9	0.9		4.288	6.383
10	1	4.343	5.645	11.681

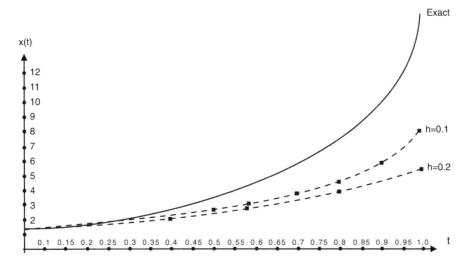

FIGURE 1.1
Comparison of exact and approximate numerical solutions for Example 1.2.

Using the separation of variable method, we obtain

$$\frac{1}{0.7} \int \frac{dx}{x^2 + 1} = \int dt.$$

Hence

$$\tan^{-1}(x(t)) = 0.7t + c.$$

Letting $x(0) = 1$, we get $c = \frac{\pi}{4}$. Thus, the exact solution of this equation is given by $x(t) = \tan\left(0.7t + \frac{\pi}{4}\right)$.

The corresponding difference equation using Euler's method is

$$x(n + 1) = x(n) + 0.7h(x^2(n) + 1), \quad x(0) = 1. \quad (\Delta E)^2$$

Table 1.1 shows the Euler approximations for $h = 0.2$ and 0.1, as well as the exact values. Figure 1.1 depicts the $n - x(n)$ diagram or the "time series." Notice that the smaller the step size we use, the better the approximation we have. □

Note that discretization schemes may be applied to nonlinear and higher order differential equations.

Example 1.3

(An Insect Population). Let us contemplate a population of aphids. These are plant lice, soft bodied, pear shaped insects which are commonly found on nearly all indoor and outdoor plants, as well as vegetables, field crops, and fruit trees.

Let

$$a(n) = \text{number of adult females in the } n\text{th generation,}$$
$$p(n) = \text{number of progeny (offspring) in the } n\text{th generation,}$$
$$m = \text{fractional mortality in the young aphids,}$$
$$q = \text{number of progeny per female aphid,}$$
$$r = \text{ratio of female aphids to total adult aphids.}$$

Since each female produces q progeny, it follows that

$$p(n + 1) = qa(n). \tag{1.8}$$

Now of these $p(n + 1)$ progeny, $rp(n + 1)$ are female young aphids of which $(1 - m)rp(n + 1)$ survives to adulthood. Thus

$$a(n + 1) = r(1 - m)p(n + 1). \tag{1.9}$$

---

[2]$\Delta E \equiv$ difference equation.

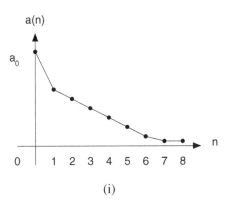

(i)

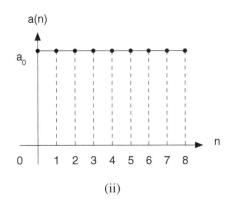

(ii)

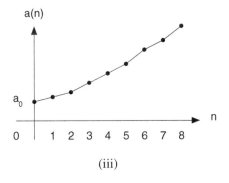

(iii)

(iv)

FIGURE 1.2

(i) $a(n)$ goes to extinction.

(ii) $a(n) = a_0$, constant population.

(iii) $a(n) \to \infty$ as $n \to \infty$, exponential growth.

(iv) Aphids.

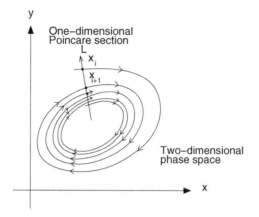

FIGURE 1.3
The Poincaré map is defined by $P(x_i) = x_{i+1}$.

Substituting from Equation 1.8 yields

$$a(n+1) = rq(1-m)a(n). \qquad (1.10)$$

Hence

$$a(n) = [rq(1-m)]^n a(0). \qquad (1.11)$$

There are three cases to consider.

(i) If $rq(1-m) < 1$, then $\lim\limits_{n\to\infty} a(n) = 0$ and the population of aphids goes to extinction.

(ii) If $rq(1-m) = 1$, then $a(n) = a_0$, and we have a constant population size.

(iii) If $rq(1-m) > 1$, then $\lim\limits_{n\to\infty} a(n) = \infty$, and the population grows exponentially to ∞.

☐

1.3.2   Poincaré Map

One of the most interesting ways on which a differential equation leads to a map, called a Poincaré map, is through the study of periodic solutions of a system of two differential equations

$$\frac{dx}{dt} = f(x, y)$$
$$\frac{dy}{dt} = g(x, y)$$

which has a periodic orbit (closed curve) in the plane. Now choose a line L that intersects this periodic orbit at a right angle. For any x_0 on the line L, $x_1 = P(x_0)$ is the point of intersection of the orbit starting at x_0 after it returns to the line L for the first time. Consequently, x_i is the intersection point of the orbit starting at x_0 after it returns to the line L for the ith time. This defines the Poincaré map associated with our differential equation (Figure 1.3). We will return to this method in Section 2.9.

1.4   Linear Maps/Difference Equations

The simplest maps to deal with are the linear maps and the simplest difference equations to solve are the linear ones. Consider the linear map

$$f(x) = ax,$$

then

$$f^n(x) = a^n x.$$

In other words, the solution of the difference equation

$$x(n+1) = ax(n), x(0) = x_0 \tag{1.12}$$

is given by

$$x(n) = a^n x_0. \tag{1.13}$$

We can make the following conclusions about the limiting behavior of the orbits of f or the solutions of Equation (1.12):

1. If $|a| < 1$, then $\lim_{n\to\infty} |f^n(x_0)| = 0$ $\left(\text{or } \lim_{n\to\infty} |x(n)| = 0\right)$ [see Fig. 1.4 (b) and (c)].

2. If $|a| > 1$, then $\lim_{n\to\infty} |f^n(x_0)| = \infty$ $\left(\text{or } \lim_{n\to\infty} |x(n)| = \infty\right)$ if $x_0 \neq 0$ [see Fig. 1.4 (a) and (d)].

3. (a) If $a = 1$, then f is the identity map where every point is a fixed point of f.

   (b) If $a = -1$, then $f^n(x_0) = \begin{cases} x_0 & \text{if } n \text{ is even} \\ -x_0 & \text{if } n \text{ is odd} \end{cases}$
   and the solution $x(n) = (-1)^n x_0$ of Equation (1.12) is said to be periodic of period 2.

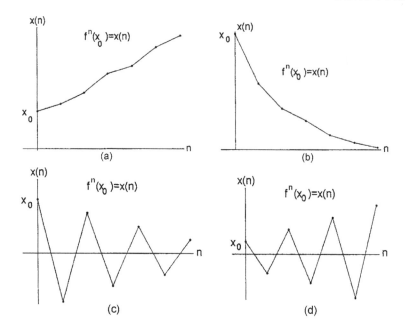

FIGURE 1.4
Time series $[n - x(n)]$ graphs (a) $a = 1.2$, (b) $a = 0.7$, (c) $a = -0.7$, (d) $a = -1.2$. Solutions of Eqs. (1.12) for different values of the parameter a.

Next, let us look at the **affine** map $f(x) = ax + b$. By successive iteration, we get

$$f^2(x) = a^2 x + ab + b$$
$$f^3(x) = a^3 x + a^2 b + ab + b$$
$$\vdots$$
$$f^n(x) = a^n x + \sum_{j=0}^{n-1} a^{n-j-1} b.$$

In other words, the solution of the difference equation

$$x(n+1) = ax(n) + b, x(0) = x_0 \tag{1.14}$$

is given by

$$x(n) = a^n x_0 + \sum_{j=0}^{n-1} a^{n-j-1} b$$

$$= a^n x_0 + b \left(\frac{a^n - 1}{a - 1} \right), \quad \text{if } a \neq 1 \tag{1.15}$$

$$x(n) = \left(x_0 + \frac{b}{a - 1} \right) a^n + \frac{b}{1 - a}, \quad \text{if } a \neq 1. \tag{1.16}$$

Using the formula of Equation (1.16), the following conclusions can be easily verified:

1. If $|a| < 1$, then $\lim_{n \to \infty} f^n(x_0) = \frac{b}{1 - a}$ $\left(\text{or } \lim_{n \to \infty} x(n) = \frac{b}{1 - a} \right)$.

2. If $|a| > 1$, then $\lim_{n \to \infty} f^n(x_0) = \pm \infty$, depending on whether $x_0 + \frac{b}{a - 1}$ is positive or negative, respectively.

3. (a) If $a = 1$, then $f^n(x_0) = x_0 + nb$, which tends to ∞ or $-\infty$ as $n \to \infty$ (or $x(n) = x_o + nb$).

   (b) If $a = -1$, then $f^n(x_0) = (-1)^n x_0 + \begin{cases} b & \text{if } n \text{ is odd} \\ 0 & \text{if } n \text{ is even} \end{cases}$

   $\left(\text{or } x(n) = (-1)^n x_0 + \begin{cases} b & \text{if } n \text{ is odd} \\ 0 & \text{if } n \text{ is even} \end{cases} \right)$.

Notice that the solution of the differential equation

$$\frac{dx}{dt} = ax(t), \quad x(0) = x_0$$

is given by

$$x(t) = e^{at} x_0. \tag{1.17}$$

Comparing (1.14) and (1.17) we see that the exponential e^{at} in the differential equation corresponds to a^n, the nth power of a, in the difference equation. The solution of the nonhomogeneous differential equation

$$\frac{dx}{dt} = ax(t) + b, \quad x(0) = x_0 \tag{1.18}$$

is given by

$$x(t) = e^{at}x_0 + \int_0^t e^{a(t-s)} b \, ds$$

$$= e^{at}x_0 + \frac{b}{a}(e^{at} - 1)$$

$$= \left(x_0 + \frac{b}{a}\right) e^{at} - \frac{b}{a}. \qquad (1.19)$$

In cases 1, 2, 3, the behavior of the difference equation (1.15) depends on whether a is inside the interval $(-1, 1)$, on its boundary, or outside it. However for differential equations, the behavior of the solution of Equation (1.18) depends on whether $a < 0$, $a = 0$, or $a > 0$, respectively. Consequently,

1. $a < 0$, $\lim\limits_{t \to \infty} x(t) = -\dfrac{b}{a}$ as $e^{at} \to 0$ as $t \to \infty$,

2. $a = 0$, $x(t) = x_0$ since $\dfrac{dx}{dt} = 0$,

3. $a > 0$, $\lim\limits_{t \to \infty} x(t) = \infty$ since $e^{at} \to \infty$ since $t \to \infty$.

Example 1.4

A drug is administered every six hours. Let $D(n)$ be the amount of the drug in the blood system at the nth interval. The body eliminates a certain fraction p of the drug during each time interval. If the amount administered is D_0, find $D(n)$ and $\lim\limits_{n \to \infty} D(n)$.   ▯

SOLUTION   The first step in solving this example is to write down a difference equation that relates the amount of drug in the patient's system $D(n+1)$ at the time interval $(n+1)$ with $D(n)$. Now, the amount of drug $D(n+1)$ is equal to the amount $D(n)$ minus the fraction p of $D(n)$ that has been eliminated from the body plus the new dose D_0. This yields

$$D(n+1) = (1-p)D(n) + D_0.$$

From Equations (1.14) and (1.15), we obtain

$$D(n) = (1-p)^n D_0 + D_0 \left(\frac{1 - (1-p)^n}{p}\right)$$

$$= \left(D_0 - \frac{D_o}{p}\right)(1-p)^n + \frac{D_o}{p}.$$

Thus,

$$\lim_{n \to \infty} D(n) = \frac{D_o}{p}. \qquad ∎$$

Exercises - (1.2–1.4)

1. Find the solution of the difference equation $x(n+1) - \frac{1}{2}x(n) = 2$, $x(0) = c$.

2. Find the solution of the equation $x(n+1) + 2x(n) = 3$, $x(0) = 1$.

3. (Pielou Logistic Equation). In population biology, the following equation, commonly called Pielou Logistic equation, is used to model populations with nonoverlapping generations

$$x(n+1) = \frac{\alpha x(n)}{1 + \beta x(n)}$$

   (a) Use the substitution $x(n) = \frac{1}{z(n)}$ to transform the equation into a linear equation.

   (b) Show that

$$\lim_{n \to \infty} x(n) = \begin{cases} (\alpha - 1)/\beta & \text{if } |\alpha| > 1, \\ 0 & \text{if } \alpha = 1 \text{ or } |\alpha| < 1, \\ \{x_0, -x_0/(1 + \beta x_0)\} & \text{if } \alpha = -1. \end{cases}$$

4. Find the exact solution of the logistic difference equation

$$x(n+1) = 2x(n)(1 - x(n)).$$

   (Hint: Let $x(n) = \frac{1}{2}(1 - y(n))$, then use iteration)

5. Find the exact solution of the logistic difference equation

$$x(n+1) = 4x(n)(1 - x(n)).$$

   (Hint: Let $x(n) = \sin^2 \theta(n)$)

6. The temperature of a body is measured as $100°F$. It is observed that the temperature change each period of 3 hours is -0.3 times the difference between the previous period's temperature and the room temperature, which is $65°F$.

   (a) Write a difference equation that describes the temperature $T(n)$ of the body at the end of n periods.

   (b) Find $T(n)$.

7. Consider the aphids population considered in Example 1.3 with $r = \frac{2}{3}$, $q = 4$, $m = \frac{1}{4}$.

    (a) Find a formula for $a(n)$.

    (b) If $a(0) = 10$, compute $a(1), a(2), \ldots, a(10)$.

    (c) Draw the time series $(n - a(n))$ graph.

8. Suppose that in each generation of female aphids, one-third of them is removed.

    (a) Write down the modified difference equation that models the female aphids.

    (b) Draw the time series $(n - a(n))$ graph for $r = \frac{2}{3}$, $q = 4$, $m = \frac{1}{4}$, $a(0) = 10$.

9. Suppose that in each generation of female aphids, nine are removed.

    (a) Write down the modified difference equation that models the female aphids.

    (b) Draw the time series $(n - a(n))$ graph for $r = \frac{2}{3}$, $q = 4$, $m = \frac{1}{4}$, $a(0) = 10$.

In Problems 10–12:

    (a) Find the associated difference equation by applying Euler's algorithm on the given differential equation.

    (b) Draw the graph of the solution of the difference equation in part (a).

    (c) Find the exact solution of the given differential equation and draw its graph on the same plot in part (b).[3]

10. $y' + 0.5y = 0$, $y(0) = 0.8$, $0 \le t \le 1$, $h = 0.2$

11. $y' = -y + 1$, $y(0) = 0$, $0 \le t \le 1$, $h = 0.25$

12. $y' + 2y = 0$, $y(0) = 0.5$, $0 \le t \le 1$, $h = 0.1$

1.5 Fixed (Equilibrium) Points

In Section 1.4, we were able to obtain closed form solutions of first-order linear difference equations. In other words, it was possible to write down an explicit formula for points $f^n(x_0)$ in the orbit of a point x_0 under the linear or

---
[3] Optional

affine map f. However, the situation changes drastically when the map f is nonlinear. For example, one cannot find a closed form solution for the simple difference equation $(\Delta E) : x(n+1) = \mu x(n)(1 - x(n))$, except when $\mu = 2$ or 4. For those of you who are familiar with first-order differential equations, this may be rather shocking. We may solve the corresponding differential equation $(DE^4 : x'(t) = \lambda x(t)(1 - x(t)))$ by simply separating the variables x and t and then integrating both sides of the equation. The solution of (DE) may be written in the form

$$x(t) = \frac{x_0 e^{\lambda t}}{1 + x_0(e^{\lambda t} - 1)}.$$

Note that the behavior of this solution is very simple: for $\lambda > 0$, $\lim_{t \to \infty} x(t) = 1$ and for $\lambda < 0$, $\lim_{t \to \infty} x(t) = 0$. Unlike those of (DE), the behavior of solutions of (ΔE) is extremely complicated and depends very much on the values of the parameter μ. Since we cannot, in general, solve (ΔE), it is important to develop qualitative or graphical methods to determine the behavior of their orbits. Of particular importance is finding orbits that consist of one point. Such points are called **fixed points**, or **equilibrium points (steady states)**.

Let us consider again the difference equation

$$x(n+1) = f(x(n)). \tag{1.20}$$

DEFINITION 1.1   *A point x^* is said to be a **fixed** point of the map f or an **equilibrium** point of Equation (1.20) if $f(x^*) = x^*$.*

Note that for an equilibrium point x^*, the orbit is a singleton and consists of only the point x^*. Moreover, to find all equilibrium points of Equation (1.20), we must solve the equation $f(x) = x$. Graphically speaking, a fixed point of a map f is a point where the curve $y = f(x)$ intersects the diagonal line $y = x$. For example, the fixed points of the cubic map $f(x) = x^3$ can be obtained by solving the equation $x^3 = x$ or $x^3 - x = 0$. Hence, there are three fixed points -1, 0, 1 for this map (see Fig. 1.5).

Closely related to fixed points are the **eventually fixed points**. These are the points that reach a fixed point after finitely many iterations. More explicitly, a point x is said to be an **eventually fixed point** of a map f if there exists a positive integer r and a fixed point x^* of f such that $f^r(x) = x^*$, but $f^{r-1}(x) \neq x^*$.

We denote the set of all fixed points by $Fix(f)$, the set of all eventually fixed points by $EFix(f)$, and the set of all eventually fixed points of the fixed points x^* by $EFix_{x^*}(f)$.

---

[4]From Equation (1.7), this DE leads to $y(n+1) = y(n) + h\lambda y(n)(1 - y(n))$ or $y(n+1) = (1 + h\lambda)y(n)[1 - \frac{h\lambda}{1+h\lambda} y(n)]$. Now, setting $x(n) = \frac{h\lambda}{1+h\lambda} y(n)$ leads to the above ΔE.

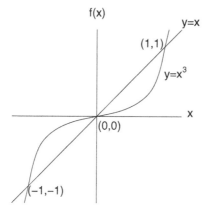

FIGURE 1.5
The fixed points of $f(x) = x^3$ are the intersection points with the diagonal line.

Given a fixed point x^* of a map f, then one can easily construct eventually fixed points by computing the ancestor set $f^{-1}(x^*) = \{x \neq x^* : f(x) = x^*\}$, $f^{-2}(x^*) = \{x : f^2(x) = x^*\}$, ..., $f^{-n}(x^*) = \{x : f^n(x) = x^*\}$, ...
Thus one may show that

$$\boxed{EFix_{x^*}(f) = \{x : f^n(x) = x^*, \quad n \in \mathbb{Z}^+\}.}$$ (1.21)

Note that the set $EFix(f) \setminus \{x^*\}$ may be empty, finite, or infinite as demonstrated by the following example.

Example 1.5

(i) Consider the logistic map $f(x) = 2x(1 - x)$. Then there are two fixed points $x^* = 0$ and $y^* = \frac{1}{2}$. A simple computation reveals that

$$f^{-1}(x) = \frac{1}{2}[1 \pm \sqrt{1 - 2x}].$$

Thus $f^{-1}\left(\frac{1}{2}\right) = \frac{1}{2}$ and $EFix_{y^*}(f) \setminus \{\frac{1}{2}\} = \emptyset$. Moreover, $f^{-1}(0) = \{0, 1\}$, and $EFix_{x^*}(f) = \{0, 1\}$. We conclude that we have only one "genuine" eventually fixed point, namely $x = 1$.

(ii) Let us now contemplate a more interesting example, $f(x) = 4x(1 - x)$. There are two fixed points, $x^* = 0$, and $y^* = \frac{3}{4}$. Clearly $EFix_{x^*}(f) = \{0, 1\}$. Notice that $f^{-1}(x) = \frac{1}{2}[1 \pm \sqrt{1 - x}]$. Hence

$$f^{-1}\left(\frac{3}{4}\right) = \frac{1}{2}\left[1 \pm \sqrt{1 - \frac{3}{4}}\right] = \frac{1}{2}\left[1 \pm \frac{1}{2}\right]$$

which equals either $\frac{3}{4}$ or $\frac{1}{4}$. Now $f^{-1}\left(\frac{1}{4}\right) = \frac{1}{2}\left[1 \pm \sqrt{1 - \frac{1}{4}}\right]$ which equals either $\frac{1}{2}\left[1 + \frac{\sqrt{3}}{2}\right]$ or $\frac{1}{2}\left[1 - \frac{\sqrt{3}}{2}\right]$. Repeating this process we may generate an infinitely many eventually fixed point, that is the set $EFix_{y^*}(f)$ is infinite. The following diagram shows some of the eventually fixed points.

$$1 \to 0$$

$$\frac{1}{4} \to \frac{3}{4}$$

$$\left(\frac{1}{2} - \frac{\sqrt{3}}{4}\right) \to \frac{1}{4} \to \frac{3}{4}$$

$$\left(\frac{1}{2} + \frac{\sqrt{3}}{4}\right) \to \frac{1}{4} \to \frac{3}{4}$$

$$\left[\frac{1}{2} - \frac{1}{2}\sqrt{\frac{1}{2} + \frac{\sqrt{3}}{2}}\right] \to \left[\frac{1}{2} - \frac{\sqrt{3}}{2}\right] \to \frac{1}{4} \to \frac{3}{4}$$

☐

It is interesting to note that the phenomenon of eventually fixed points does not have a counterpart in differential equations, since no solution can reach an equilibrium point in a finite time.

Next we introduce one of the most interesting examples in discrete dynamical systems: the tent map T.

Example 1.6
(The Tent Map). The tent map T is defined as

$$T(x) = \begin{cases} 2x, & \text{for } 0 \le x \le \frac{1}{2} \\ 2(1-x), & \text{for } \frac{1}{2} < x \le 1. \end{cases} \qquad ☐$$

This map may be written in the form

$$T(x) = 1 - 2\left|x - \frac{1}{2}\right|.$$

Note that the tent map is a piecewise linear map (see Fig. 1.6). The tent map possesses a rich dynamics and in Chapter 3 we show it is in fact "chaotic."

There are two equilibrium points $x_1^* = 0$ and $x_2^* = \frac{2}{3}$. Moreover, the point $\frac{1}{4}$ is an eventual equilibrium point since $T(\frac{1}{4}) = \frac{1}{2}, T^2(\frac{1}{4}) = T(\frac{1}{2}) = 1, T^3(\frac{1}{4}) = T(1) = 0$. It is left to you to show that if $x = \frac{k}{2^n}$, where k, and n are positive

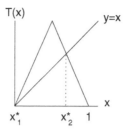

FIGURE 1.6
The tent map has two fixed points $x_1^* = 0$ and $x_2^* = \frac{2}{3}$.

integers with $0 < \frac{k}{2^n} \leq 1$, then x is an eventually fixed point (Problem 9). Numbers of this form are called **dyadic rationals**.

REMARK 1.1   Note that not every map has a fixed point. For example, the map $f(x) = x + 1$ has no fixed points since the equation $x + 1 = x$ has no solution.   ■

Now, our mathematical curiosity would lead to the following question: under what conditions does a map have a fixed point? Well, for continuous maps, there are two simple and interesting results that guarantee the presence of fixed points.

THEOREM 1.1

Let $f : I \to I$ be a continuous map, where $I = [a, b]$ is a closed interval in $\mathbb{R}$. Then, f has a fixed point.

PROOF   Define $g(x) = f(x) - x$. Then, $g(x)$ is also a continuous map. If $f(a) = a$ or $f(b) = b$, we are done. So assume that $f(a) \neq a$ and $f(b) \neq b$. Hence, $f(a) > a$ and $f(b) < b$. Consequently, $g(a) > 0$ and $g(b) < 0$. By the intermediate value theorem,[5] there exists a point $c \in (a, b)$ with $g(c) = 0$. This implies that $f(c) = c$ and c is thus a fixed point of f.

The above theorem says that for a continuous map f if $f(I) \subset I$, then f has a fixed point in I. The next theorem gives the same assertion if $f(I) \supset I$.
■

---

[5]**The intermediate value theorem**: Let $f : I \to I$ be a continuous map. Then, for any real number r between $f(a)$ and $f(b)$, there exists $c \in I$ such that $f(c) = r$.

THEOREM 1.2
Let $f : I = [a, b] \to \mathbb{R}$ be a continuous map such that $f(I) \supset I$. Then f has a fixed point in I.

PROOF   The proof is left to the reader as Problem 10.   ∎

Even if fixed points of a map do exist, it is sometimes not possible to compute them algebraically. For example, to find the fixed points of the map $f(x) = 2 \sin x$, one needs to solve the transcendental equation $2 \sin x - x = 0$.

Clearly $x = 0$ is a root of this equation and thus a fixed point of the map f. However, the other two fixed points may be found by graphical or numerical methods. They are approximately ± 1.944795452.

1.6   Graphical Iteration and Stability

One of the main objectives in the theory of dynamical systems is the study of the behavior of orbits near fixed points, i.e., the behavior of solutions of a difference equation near equilibrium points. Such a program of investigation is called **stability theory**, which henceforth will be our main focus. We begin our exposition by introducing the basic notions of stability. Let $\mathbb{Z}^+$ denote the set of nonnegative integers.

DEFINITION 1.2   *Let $f : I \to I$ be a map and x^* be a fixed point of f, where I is an interval in the set of real numbers $\mathbb{R}$. Then*

1. *x^* is said to be **stable** if for any $\varepsilon > 0$ there exists $\delta > 0$ such that for all $x_0 \in I$ with $|x_0 - x^*| < \delta$ we have $|f^n(x_0) - x^*| < \varepsilon$ for all $n \in \mathbb{Z}^+$. Otherwise, the fixed point x^* will be called **unstable** (see Figs. 1.7 and 1.8).*

2. *x^* is said to be **attracting** if there exists $\eta > 0$ such that $|x_0 - x^*| < \eta$ implies $\lim_{n \to \infty} f^n(x_0) = x^*$ (see Fig. 1.9).*

3. *x^* is **asymptotically stable**[6] if it is both stable and attracting (see Fig. 1.10). If in (2) $\eta = \infty$, then x^* is said to be **globally asymptotically stable**.*

Henceforth, unless otherwise stated, "stable" (asymptotically stable) always means "locally stable" (asymptotically stable).

---

[6] In the literature, x^* is sometimes called a sink.

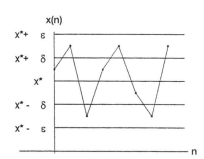

FIGURE 1.7
Stable fixed point x^*.

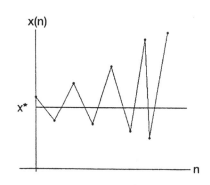

FIGURE 1.8
Unstable fixed point x^*.

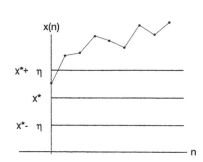

FIGURE 1.9
Unstable nonoscillating fixed point x^*.

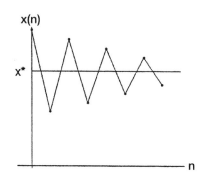

FIGURE 1.10
Asymptotically stable fixed point x^*.

The Cobweb Diagram:

One of the most effective graphical iteration methods to determine the stability of fixed points is the **cobweb diagram.**[7] On the $x-y$ plane, we draw the curve $y = f(x)$ and the diagonal line $y = x$ on the same plot (see Fig. 1.11).

We start at an initial point x_0. Then we move vertically until we hit the graph of f at the point $(x_0, f(x_0))$. We then travel horizontally to meet the line $y = x$ at the point $(f(x_0), f(x_0))$. This determines $f(x_0)$ on the x axis. To find $f^2(x_0)$, we move again vertically until we strike the graph of f at the point $(f(x_0), f^2(x_0))$; and then we move horizontally to meet the line $y = x$ at the point $(f^2(x_0), f^2(x_0))$. Continuing this process, we can evaluate all of the points in the orbit of x_0, namely, the set $\{x_0, f(x_0), f^2(x_0), \ldots, f^n(x_0), \ldots\}$ (see Fig. 1.11).

Example 1.7
Use the cobweb diagram to find the fixed points for the quadratic map $Q_c(x) = x^2 + c$ on the interval $[-2, 2]$, where $c \in [-2, 0]$. Then determine the stability of all fixed points.   ▯

SOLUTION To find the fixed point of Q_c, we solve the equation $x^2 + c = x$ or $x^2 - x + c = 0$. This yields the two fixed points $x_1^* = \frac{1}{2} - \frac{1}{2}\sqrt{1-4c}$ and $x_2^* = \frac{1}{2} + \frac{1}{2}\sqrt{1-4c}$. Since we have not developed enough machinery to treat the general case for arbitrary c, let us examine few values of c. We begin with $c = -0.5$ and an initial point $x_0 = 1.1$. It is clear from Fig. 1.12 that the fixed point $x_1^* = \frac{1}{2} - \frac{\sqrt{3}}{2} \approx -0.366$ is asymptotically stable, whereas the second fixed point $x_2^* = \frac{1}{2} + \frac{\sqrt{3}}{2} \approx 1.366$ is unstable.   ∎

Example 1.8
Consider again the tent map of Example 1.6. Find the fixed points and determine their stability.   ▯

SOLUTION The fixed points are obtained by putting $2x = x$ and $2(1 - x) = x$. From the first equation, we obtain the first fixed point $x_1^* = 0$; and from the second equation, we obtain the second fixed point $x_2^* = \frac{2}{3}$. Observe from the cobweb diagram (Fig. 1.13) that both fixed points are unstable.   ∎

REMARK 1.2 If one uses the language of difference equations, then in the Cobweb diagrams, the x-axis is labeled $x(n)$ and the y-axis is labeled $x(n+1)$.   ∎

---

[7]It is also called the stair-step diagram.

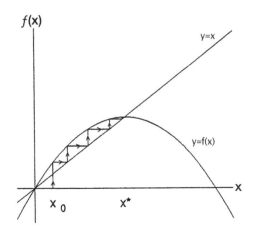

FIGURE 1.11

The Cobweb diagram: asymptotically stable fixed point x^*, $\lim\limits_{n\to\infty} f^n(x_0) = x^*$.

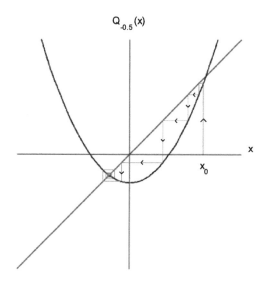

FIGURE 1.12

The Cobweb diagram of $Q_{-0.5}$: x_1^* is asymptotically stable but x_2^* is unstable.

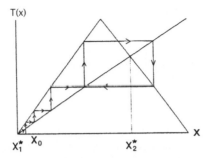

FIGURE 1.13

Both equilibrium points $x_1^* = 0$ and $x_2^* = \frac{2}{3}$ are unstable.

Exercises - (1.5 and 1.6)

Use Phaser, Mathematica, or Maple.

1. Find all fixed and eventually fixed points of the map $f(x) = |x - 1|$.

2. Consider the logistic map $F_\mu(x) = \mu x(1 - x)$.

   (a) Draw the cobweb diagram for $\mu = 2, 2.5, 3.2$.

   (b) Determine the stability of the equilibrium points for the values of μ in part (a).

3. (a) Find a function with four fixed points, all of which are unstable.

   (b) Find a function with no fixed points.

   (c) Find a function with a stable and an unstable fixed point.

4. Find the equilibrium points and determine their stability for the map $f(x) = 5 - \frac{6}{x}$.

5. **Pielou's logistic equation.** Pielou referred to the following equation as the discrete logistic equation:

$$x(n + 1) = \frac{\alpha x(n)}{1 + \beta x(n)}, \quad \alpha > 1, \ \beta > 0.$$

   (a) Find the positive equilibrium point.

   (b) Demonstrate, using the cobweb diagram, that the positive equilibrium point is asymptotically stable for $\alpha = 2$ and $\beta = 1$.

6. **Newton's method for computing the square root of a positive number.** The equation $x^2 = b$ can be written in the form $x = \frac{1}{2}(x + \frac{b}{x})$. This form leads to Newton's method:

$$x(n+1) = \frac{1}{2}\left(x(n) + \frac{b}{x(n)}\right).$$

(a) Show that this difference equation has two equilibrium points, $-\sqrt{b}$ and $\sqrt{b}$.

(b) Sketch cobweb diagrams for $b = 3$; $x_0 = 1$, $x_0 = -1$.

(c) What can you conclude from part (b)?

(d) Investigate the case when $b = -3$ and try to form an explanation of your results.

7. Consider the difference equation $x(n+1) = f(x(n))$, where $f(0) = 0$.

(a) Prove that $x(n) \equiv 0$ is a solution of the equation.

(b) Show that the function depicted in Fig. 1.14 cannot possibly be a solution of the equation.

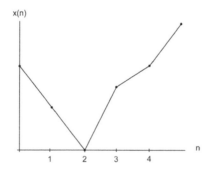

FIGURE 1.14
Problem 7(b)

8. Consider the family of quadratic maps $Q_c(x) = x^2 + c$, where c is a parameter.

(a) Draw the cobweb diagram for $c > \frac{1}{4}$, $c = \frac{1}{4}$, or $c < \frac{1}{4}$.

(b) Determine the stability of the fixed points for the values of c in part (a).

9. Show that if $x = \frac{k}{2^n}$, where k and n are positive integers with $0 < \frac{k}{2^n} \leq 1$, then x is an eventually fixed point of the tent map (see Example 1.6).

10. Prove Theorem 1.2.

In Problems 11–14, determine the stability of the fixed points of the maps using the Cobweb-diagram.

11. $f(x) = 0.5 \sin(\pi x)$

12. $f(x) = x + \frac{1}{\pi} \sin(2\pi x)$

13. $f(x) = 2xe^{-x}$

14. A population of birds is modeled by the difference equation

$$x(n+1) = \begin{cases} 3.2x(n) & \text{for } 0 \leq x(n) \leq 1, \\ 0.5x(n) + 2.7 & \text{for } x(n) > 1. \end{cases}$$

where $x(n)$ is the number of birds in year n. Find the equilibrium points and then determine their stability.

1.7 Criteria for Stability

In this section, we will establish some simple but powerful criteria for local stability of fixed points. Fixed (equilibrium) points may be divided into two types: **hyperbolic** and **nonhyperbolic**. A fixed point x^* of a map f is said to be **hyperbolic** if $|f'(x^*)| \neq 1$. Otherwise, it is nonhyperbolic. We will treat the stability of each type separately.

1.7.1 Hyperbolic Fixed Points

The following result is the main tool in detecting local stability.

THEOREM 1.3
Let x^* be a hyperbolic fixed point of a map f, where f is continuously differentiable at x^*. The following statements then hold true:

1. *If $|f'(x^*)| < 1$, then x^* is asymptotically stable.*

2. *If $|f'(x^*)| > 1$, then x^* is unstable.*

PROOF   1. Suppose that $|f'(x^*)| < M < 1$ for some $M > 0$. Then, there is an open interval $I = (x^* - \varepsilon, \ x^* + \varepsilon)$ such that $|f'(x)| \leq M < 1$ for all $x \in I$ (Why? Problem 10). By the mean value theorem,[8] for any $x_0 \in I$, there exists c between x_0 and x^* such that

$$|f(x_0) - x^*| = |f(x_0) - f(x^*)| = |f'(c)||x_0 - x^*| \le M|x_0 - x^*|. \qquad (1.22)$$

Since $M < 1$, inequality (1.22) shows that $f(x_0)$ is closer to x^* than x_0. Consequently, $f(x_0) \in I$. Repeating the above argument on $f(x_0)$ instead of x_0, we can show that

$$|f^2(x_0) - x^*| \le M|f(x_0) - x^*| \le M^2|x_0 - x^*|. \qquad (1.23)$$

By mathematical induction, we can show that for all $n \in \mathbb{Z}^+$,

$$|f^n(x_0) - x^*| \le M^n|x_0 - x^*|. \qquad (1.24)$$

To prove the stability of x^*, for any $\varepsilon > 0$, we let $\delta = \min(\varepsilon, \tilde{\varepsilon})$. Then, $|x_0 - x^*| < \delta$ implies that $|f^n(x_0) - x^*| \le M^n|x_0 - x^*| < \varepsilon$, which establishes stability. Furthermore, from Inequality (1.24) $\lim_{n\to\infty} |f^n(x_0) - x^*| = 0$ and thus $\lim_{n\to\infty} f^n(x_0) = x^*$, which yields asymptotic stability. The proof of part 2 is left to you as Problem 14. ∎

The following examples illustrate the applicability of the above theorem.

Example 1.9
Consider the map $G_\lambda(x) = 1 - \lambda x^2$ defined on the interval $[-1, 1]$, where $\lambda \in (0, 2]$. Find the fixed points of $G_\lambda(x)$ and determine their stability. ▢

SOLUTION   To find the fixed points of $G_\lambda(x)$ we solve the equation $1 - \lambda x^2 = x$ or $\lambda x^2 + x - 1 = 0$. There are two fixed points:

$$x_1^* = \frac{-1 - \sqrt{1 + 4\lambda}}{2\lambda} \quad \text{and} \quad x_2^* = \frac{-1 + \sqrt{1 + 4\lambda}}{2\lambda}.$$

Observe that $G_\lambda'(x) = -2\lambda x$. Thus, $|G_\lambda'(x_1^*)| = 1 + \sqrt{1 + 4\lambda} > 1$, and hence, x_1^* is unstable for all $\lambda \in (0, 2]$. Furthermore, $|G_\lambda'(x_2^*)| = \sqrt{1 + 4\lambda} - 1 < 1$ if and only if $\sqrt{1 + 4\lambda} < 2$. Solving the latter inequality for λ, we obtain $\lambda < \frac{3}{4}$. This implies by Theorem 1.3 that the fixed point x_2^* is asymptotically stable if $0 < \lambda < \frac{3}{4}$ and unstable if $\lambda > \frac{3}{4}$ (see Fig. 1.15). When $\lambda = \frac{3}{4}$, $G_\lambda'(x_2^*) = -1$. This case will be treated in Section 1.7.2. ∎

---

[8]**The mean value theorem.** If f is continuous on the closed interval $[a, b]$ and is differentiable on the open interval (a, b), then there is a number c in (a, b) such that $f'(c) = \frac{f(b) - f(a)}{b - a}$. This implies that $|f(b) - f(a)| = |f'(c)||b - a|$.

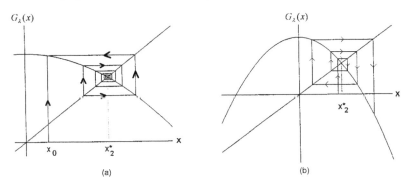

FIGURE 1.15
(a) $\lambda = \frac{1}{2}, x_2^*$ is asymptotically stable while (b) $\lambda = \frac{3}{2}, x_2^*$ is unstable.

Example 1.10
(Raphson-Newton's Method). Raphson-Newton's method is one of the simplest and oldest numerical methods for finding the roots of the equation $g(x) = 0$. The Newton algorithm for finding a zero r of $g(x)$ is given by the difference equation

$$x(n + 1) = x(n) - \frac{g(x(n))}{g'(x(n))}. \tag{1.25}$$

where $x(0) = x_0$ is our initial guess of the root r. Equation (1.25) is of the form of Equation (1.20) with

$$f_N(x) = x - \frac{g(x)}{g'(x)} \tag{1.26}$$

where f_N is called **Newton's function.**   ☐

THEOREM 1.4 (Taylor's Theorem)
Let f be differentiable of all orders at x_0. Then

$$f(x) = f(x_0) + (x - x_0)f'(x_0) + \frac{(x - x_0)^2}{2!}f''(x_0) + \dots$$

for all x in a small open interval containing x_0.

Formula (1.25) may be justified using Taylor's Theorem. A linear approximation of $f(x)$ is given by the equation of the tangent line to $f(x)$ at x_0:

$$f(x) = f(x_0) + (x - x_0)f'(x_0).$$

The intersection of this tangent line with the x-axis produces the next point x_1 in Newton's algorithm (Fig. 1.16). Letting $f(x) = 0$ and $x = x_1$ yields

$$x_1 = x_0 - \frac{f(x_0)}{f'x_0}.$$

By repeating the process, replacing x_0 by x_1, x_1 by x_2, ..., we obtain formula (1.25).

We observe first that if r is a root of $g(x)$, i.e., $g(r) = 0$, then from Equation (1.26) we have $f_N(r) = r$ and thus r is a fixed point of f_N (assuming that $g'(r) \neq 0$). On the other hand, if x^* is a fixed point of f_N, then from Equation (1.26) again we get $\frac{g(x^*)}{g'(x)} = 0$. This implies that $g(x^*) = 0$, i.e., x^* is a zero of $g(x)$. Now, starting with a point x_0 close to a root r of $g(x) = 0$, then Algorithm (1.25) gives the next approximation $x(1)$ of the root r. By applying the algorithm repeatedly, we obtain the sequence of approximations

$$x_0 = x(0), \; x(1), \; x(2), \ldots, \; x(n), \ldots$$

(see Fig. 1.16). The question is whether or not this sequence converges to the root r. In other words, we need to check the asymptotic stability of the fixed point $x^* = r$ of f_N. To do so, we evaluate $f'_N(r)$ and then use Theorem 1.3,

$$|f'_N(r)| = \left| 1 - \frac{[g'(r)]^2 - g(r)g''(r)}{[g'(r)]^2} \right| = 0, \; \text{ since } \; g(r) = 0.$$

Hence, by Theorem 1.3, $\lim_{n \to \infty} x(n) = r$, provided that x_0 is sufficiently close to r.

For $g(x) = x^2 - 1$, we have two zero's $-1, 1$. In this case, Newton's function is given by $f_N(x) = x - \frac{x^2-1}{2x} = \frac{x^2+1}{2x}$. The cobweb diagram of f_N shows that Newton's algorithm converges quickly to both roots (see Fig. 1.17).

1.7.2   Nonhyperbolic Fixed Points

The stability criteria for nonhyperbolic fixed points are more involved. They will be summarized in the next two results, the first of which treats the case when $f'(x^*) = 1$ and the second for $f'(x^*) = -1$.

THEOREM 1.5

Let x^* be a fixed point of a map f such that $f'(x^*) = 1$. If $f'(x)$, $f''(x)$, and $f'''(x)$ are continuous at x^*, then the following statements hold:

1. *If $f''(x^*) \neq 0$, then x^* is unstable (semistable).[9]*

---

[9]See the definition in Problem 17. The assumption that $f'''(x)$ is continuous at x^* is not needed in part 1.

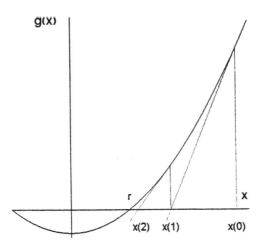

FIGURE 1.16
Newton's method for $g(x) = x^2 - 1$.

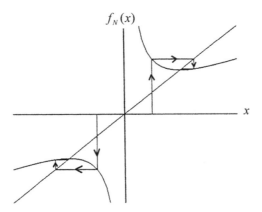

FIGURE 1.17
Cobweb diagram for Newton's function f_N when $g(x) = x^2 - 1$.

2. *If $f''(x^*) = 0$ and $f'''(x^*) > 0$, then x^* is unstable.*

3. *If $f''(x^*) = 0$ and $f'''(x^*) < 0$, then x^* is asymptotically stable.*

PROOF   1. Assume that $f'(x^*) = 1$ and $f''(x^*) \neq 0$. Then, the curve $y = f(x)$ is either concave upward ($f''(x^*) > 0$) or concave downward ($f''(x^*) < 0$), as shown in Fig. 1.18(a) and (b). Now, if $f''(x^*) > 0$, then $f'(x)$ is increasing in a small interval containing x^*. Hence, $f'(x) > 1$ for all $x \in (x^*, x^* + \delta)$, for some small $\delta > 0$ [see Fig. 1.18(a)]. Using the same proof as in Theorem 1.3, we conclude that x^* is unstable. Similarly, if $f''(x^*) < 0$ then $f'(x)$ is decreasing in a small neighborhood of x^*. Therefore, $f'(x) > 1$ for all $x \in (x^* - \delta, x^*)$, for some small $\delta > 0$, and again we conclude that x^* is unstable [see Fig. 1.18(b)]. Proofs of parts 2 and 3 are left to you as Problem 15.   ∎

Example 1.11
Let $f(x) = -x^3 + x$. Then $x^* = 0$ is the only fixed point of f. Note that $f'(0) = 1$, $f''(0) = 0$, $f'''(0) < 0$. Hence by Theorem 1.5, 0 is asymptotically stable.   □

The preceding theorem may be used to establish stability criteria for the case when $f'(x^*) = -1$. But before doing so, we need to introduce the notion of the **Schwarzian derivative**.

DEFINITION 1.3   *The* **Schwarzian derivative**, Sf, *of a function f is defined by*

$$Sf(x) = \frac{f'''(x)}{f'(x)} - \frac{3}{2}\left[\frac{f''(x)}{f'(x)}\right]^2. \tag{1.27}$$

And if $f'(x^*) = -1$, then

$$Sf(x^*) = -f'''(x^*) - \frac{3}{2}[f''(x^*)]^2. \tag{1.28}$$

THEOREM 1.6
Let x^* be a fixed point of a map f such that $f'(x^*) = -1$. If $f'(x)$, $f''(x)$, and $f'''(x)$ are continuous at x^*, then the following statements hold:

1. *If $Sf(x^*) < 0$, then x^* is asymptotically stable.*

2. *If $Sf(x^*) > 0$, then x^* is unstable.*

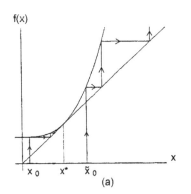

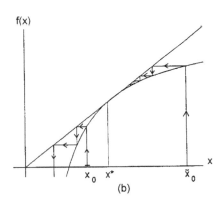

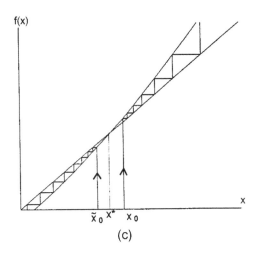

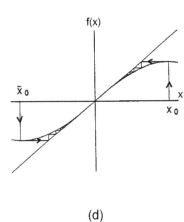

FIGURE 1.18

(a) $f'(x^*) = 1$, $f''(x^*) > 0$, unstable fixed point, semi-stable from the left.

(b) $f'(x^*) = 1$, $f''(x^*) < 0$, unstable fixed point, semi-stable from the right.

(c) $f'(x^*) = 1$, $f''(x^*) = 0$, $f'''(x^*) > 0$, unstable fixed point.

(d) $f'(x^*) = 1$, $f''(x^*) = 0$, $f'''(x^*) < 0$, asymptotically stable fixed point.

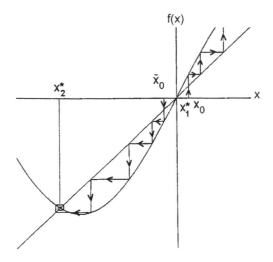

FIGURE 1.19

An asymptotically stable nonhyperbolic fixed point x_2^*.

PROOF   The main idea of the proof is to create an associated function g with the property that $g'(x^*) = 1$, so that we can use Theorem 1.5. This function is indeed $g = f \circ f = f^2$. Two important facts need to be observed here. First, if x^* is a fixed point of f, then it is also a fixed point of g. Second, if x^* is asymptotically stable (unstable) with respect to g, then it is also asymptotically stable (unstable) with respect to f (Why? Problem 16). By the chain rule:

$$g'(x) = \frac{d}{dx} f(f(x)) = f'(f(x))f'(x). \tag{1.29}$$

Hence,

$$g'(x^*) = [f'(x^*)]^2 = 1$$

and Theorem 1.5 now applies. For this reason we compute $g''(x^*)$. From Equation (1.29), we have

$$g''(x) = f'(f(x))f''(x) + f''(f(x))[f'(x)]^2 \tag{1.30}$$

$$g''(x^*) = f'(x^*)f''(x^*) + f''(x^*)[f'(x^*)]^2$$

$$= 0 \quad (\text{since } f'(x^*) = -1). \tag{1.31}$$

Computing $g'''(x)$ from Equation 1.31, we get

$$g'''(x^*) = -2f'''(x^*) - 3[f''(x^*)]^2. \tag{1.32}$$

It follows from Equation (1.29)

$$g'''(x^*) = 2Sf(x^*) \tag{1.33}$$

Statements 1 and 2 now follow immediately from Theorem 1.5. ∎

REMARK 1.3   Note that if $f'(x^*) = -1$ and $g = f \circ f$, then from (1.31) we have

$$\mathrm{S}f(x^*) = \frac{1}{2}g'''(x^*).$$ ∎   (1.34)

Furthermore,

$$g''(x^*) = 0.$$ (1.35)

We are now ready to give an example of a nonhyperbolic fixed point.

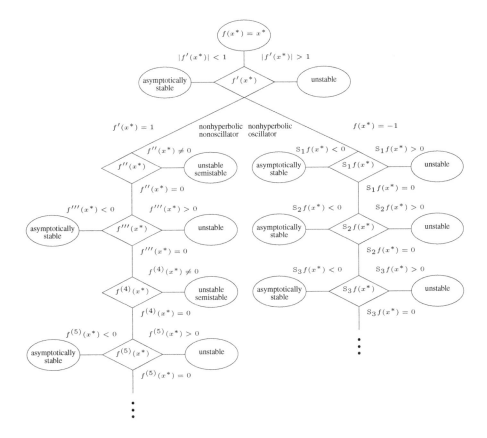

FIGURE 1.20
Classification of fixed points.

34   *Discrete Chaos*

Example 1.12

Consider the map $f(x) = x^2 + 3x$ on the interval $[-3, 3]$. Find the equilibrium points and then determine their stability.   ☐

SOLUTION   The fixed points of f are obtained by solving the equation $x^2 + 3x = x$. Thus, there are two fixed points: $x_1^* = 0$ and $x_2^* = -2$. So for x_1^*, $f'(0) = 3$, which implies by Theorem 1.3 that x_1^* is unstable. For x_2^*, we have $f'(-2) = -1$, which requires the employment of Theorem 1.6. We observe that

$$Sf(-2) = -f'''(-2) - \frac{3}{2}[f''(-2)]^2 = -6 < 0.$$

Hence, x_2^* is asymptotically stable (see Fig. 1.19).   ∎

Diagram 1.20 provides a complete classification of fixed points which goes beyond the material in this section. Detailed analysis of the contents in the diagram may be found in [22].

In [22] the cases when $Sf(x^*) = 0$ and $f'''(x^*) = 0$ were investigated. In the diagram, we have $S_1 f(x) = Sf(x)$, $S_2 f(x) = \frac{1}{2}g^{(5)}(x)$, where $g = f^2$, and more generally $S_k f(x) = \frac{1}{2}g(2k+1)(x)$.

Exercises - (1.7)

In Problems 1–8, find the fixed points and determine their stability.

1. $f(x) = x^2$

2. $f(x) = \frac{1}{2}x^3 + \frac{1}{2}x$

3. $f(x) = 3x(1 - x)$

4. $f(x) = \tan^{-1}(x)$

5. $f(x) = xe^{1.5}(1 - x)$

6. $f(x) = \begin{cases} 0.8x; & \text{if } x \leq \frac{1}{2} \\ 0.8(1 - x); & \text{if } x > \frac{1}{2} \end{cases}$

7. $f(x) = -x^3 - x$

8. $f(x) = \begin{cases} 2x; & \text{if } 0 \leq x \leq \frac{1}{2} \\ 2x - 1; & \text{if } \frac{1}{2} < x \leq 1 \end{cases}$

9. Find the equilibrium points of the equation

$$x(n+1) = \frac{\alpha x(n)}{1 + \beta x(n)}, \alpha > 1, \beta > 0.$$

Then determine the values of the parameters α and β for which a given equilibrium point is asymptotically stable or unstable.

10. Assume that f is continuously differentiable at x^*. Show that if $|f'(x^*)| < 1$, for a fixed point x^* of f, then there exists an interval $I = (x^*-\varepsilon, x^*+\varepsilon)$ such that $|f'(x)| \le M < 1$ for all $x \in I$ and for some constant M.

11. Let $f(x) = ax^2 + bx + c, a \neq 0$, and x^* be a fixed point of f. Prove the following statements:

    (a) If $f'(x^*) = 1$, then x^* is unstable.
    (b) If $f'(x^*) = -1$, then x^* is asymptotically stable.

12. Suppose that for a root x^* of a function g, we have $g(x^*) = g'(x^*) = 0$ where $g''(x^*) \neq 0$ and $g''(x)$ is continuous at x^*. Show that its Newton function f_N, defined by Equation (1.26), is defined on x^*. (*Hint: Use L'Hopital's rule.*)

13. Find the equilibrium points of the equation

$$x(n+1) = \alpha x(n)\left(\frac{1+\alpha}{\alpha} - x(n)\right).$$

Then determine the values of the parameter α for which a given equilibrium point is asymptotically stable or unstable.

14. Prove Theorem 1.3, part 2.

15. Prove Theorem 1.5, parts 2 and 3.

16. Let x^* be a fixed point of a continuous map f. Show that if x^* is asymptotically stable with respect to the map $g = f^2$, then it is asymptotically stable with respect to the map f.

17. **Semistability definition**: A fixed point x^* of a map f is semistable (from the right) if for any $\varepsilon > 0$ there exists $\delta > 0$ such that if $0 < x_0 - x^* < \delta$ then $|f^n(x_0) - x^*| < \varepsilon$ for all $n \in \mathbb{Z}^+$. If, in addition, $\lim_{n\to\infty} f^n(x_0) = x^*$ whenever $0 < x_0 - x^* < \eta$ for some $\eta > 0$, then x^* is said to be semiasymptotically stable (from the right). Semistability (semiasymptotic stability) from the left is defined analogously. Suppose that $f'(x^*) = 1$ and $f''(x^*) \neq 0$. Prove that x^* is

    (a) Semiasymptotically stable from the right if $f''(x^*) < 0$.

(b) Semiasymptotically stable from the left if $f''(x^*) > 0$.

In Problems 18 and 19, determine whether or not the fixed point $x^* = 0$ is semiasymptotically stable from the left or from the right.

18. $f(x) = x^3 + x^2 + x$

19. $f(x) = x^3 - x^2 + x$

1.8   Periodic Points and their Stability

The notion of periodicity is one of the most important notion in the field of dynamical systems. Its importance stems from the fact that many physical phenomena have certain patterns that repeat themselves. These patterns produce cycles (or periodic cycles), where a cycle is understood to be the orbit of a periodic point. In this section, we address the questions of existence and stability of periodic points.

DEFINITION 1.4   *Let $\bar{x}$ be in the domain of a map f. Then,*

1. *$\bar{x}$ is said to be a **periodic point** of f with period k if $f^k(\bar{x}) = \bar{x}$ for some positive integer k. In this case $\bar{x}$ may be called k-periodic. If in addition $f^r(\bar{x}) \neq \bar{x}$ for $0 < r < k$, then k is called the **minimal period** of $\bar{x}$. Note that $\bar{x}$ is k-periodic if it is a fixed point of the map f^k.*

2. *$\bar{x}$ is said to be an **eventually** periodic point of a period k and delay m if $f^{k+m}(\bar{x}) = f^m(\bar{x})$ for some positive integer k and $m \in \mathbb{Z}^+$ (see Fig. 1.21). Notice that if $k = 1$, then $f(f^m(\bar{x})) = f^m(\bar{x})$ and $\bar{x}$ is then an eventually fixed point, and if $m = 0$, then $\bar{x}$ is k-periodic. In other words, $\bar{x}$ is eventually periodic if $f^k(\bar{x})$ is periodic, for some positive integer k.*

The orbit of a k-periodic point is the set

$$O(\bar{x}) = \{\bar{x}, f(\bar{x}), f^2(\bar{x}), \dots, f^{k-1}(\bar{x})\}$$

and is often called a k-periodic cycle. Graphically, a k-**periodic point** is the x coordinate of a point at which the graph of the map f^k meets the diagonal line $y = x$.

Next we turn our attention to the question of stability of periodic points.

DEFINITION 1.5   *Let $\bar{x}$ be a periodic point of f with minimal period k. Then,*

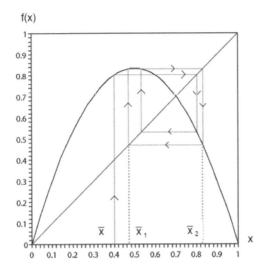

FIGURE 1.21
An eventually periodic point $\overline{x}$: The orbit of $\overline{x}$ goes into a 2-periodic cycle $\{\overline{x}_1, \overline{x}_2\}$.

1. $\overline{x}$ is **stable** *if it is a stable fixed point of* f^k.

2. $\overline{x}$ is **asymptotically stable** *if it is an asymptotically stable fixed point of* f^k.

3. $\overline{x}$ is **unstable** *if it is an unstable fixed point of* f^k.

Thus, the study of the stability of k-periodic solutions of the difference equation
$$x(n+1) = f(x(n)) \tag{1.36}$$
reduces to studying the stability of the equilibrium points of the associated difference equation
$$y(n+1) = g(y(n)) \tag{1.37}$$
where $g = f^k$.

The next theorem gives a practical criteria for the stability of periodic points based on Theorem 1.3 in the preceding section.

THEOREM 1.7
Let $O(\overline{x}) = \{\overline{x}, f(\overline{x}), \ldots, f^{k-1}(\overline{x})\}$ be the orbit of the k-periodic point $\overline{x}$, where f is a continuously differentiable function at $\overline{x}$. Then the following statements hold true:

1. $\overline{x}$ is asymptotically stable if

$$|f'(\overline{x}_1)f'(f(\overline{x}_2))\ldots f'(f^{k-1}(\overline{x}_k))| < 1. \qquad (1.38)$$

2. $\overline{x}$ is unstable if

$$|f'(\overline{x})f'(f(\overline{x}))\ldots f'(f^{k-1}(\overline{x}))| > 1. \qquad (1.39)$$

PROOF   By using the chain rule, we can show that

$$\frac{d}{dx}f^k(\overline{x}) = f'(\overline{x})f'(f(\overline{x}))\ldots f'(f^{k-1}(\overline{x})).$$

Conditions (1.38) and (1.39) now follow immediately by application of Theorem 1.3 to the composite map $g = f^k$.   ∎

Example 1.13

Consider the difference equation $x(n+1) = f(x(n))$ where $f(x) = 1 - x^2$ is defined on the interval $[-1, 1]$. Find all the 2-periodic cycles, 3-periodic cycles, and 4-periodic cycles of the difference equation and determine their stability.
∎

SOLUTION   First, let us calculate the fixed points of f out of the way. Solving the equation $x^2 + x - 1 = 0$, we find that the fixed points of f are $x_1^* = -\frac{1}{2} - \frac{\sqrt{5}}{2}$ and $x_2^* = -\frac{1}{2} + \frac{\sqrt{5}}{2}$. Only x_2^* is in the domain of f. The fixed point x_2^* is unstable. To find the two cycles, we find f^2 and put $f^2(x) = x$. Now, $f^2(x) = 1 - (1 - x^2)^2 = 2x^2 - x^4$ and $f^2(x) = x$ yields the equation

$$x(x^3 - 2x + 1) = x(x-1)(x^2 + x - 1) = 0.$$

Hence, we have the 2-periodic cycle $\{0, 1\}$; the other two roots are the fixed points of f. To check the stability of this cycle, we compute $|f'(0)f'(1)| = 0 < 1$. Hence, by Theorem 1.7, the cycle is asymptotically stable (Fig. 1.22).

Next we search for the 3-periodic cycles. This involves solving algebraically a sixth-degree equation, which is not possible in most cases. So, we resort to graphical (or numerical) analysis. Figure 1.23 shows that there are no 3-periodic cycles. Moreover, Fig. 1.24 shows that there are no 4-periodic cycles. Later, in Chapter 2, we will prove that this map has no periodic points other than the above 2-periodic cycle.

Since $f^{-1}(x) = \sqrt{1-x}$, it follows that the point $f^{-1}(x_2^*) = \sqrt{\frac{3-\sqrt{5}}{2}}$ is an eventually fixed point. Let $g = f^2$. Then $g^{-1}(x) = \sqrt{1 + \sqrt{1-x}}$. Now $g^{-1}(0) = \sqrt{2}$ which is outside the domain of f. Hence f has no eventually periodic points.   ∎

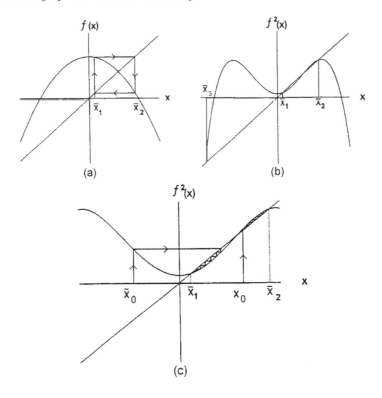

FIGURE 1.22

(a) A 2-periodic cycle $\{\overline{x}_1, \overline{x}_2\}$; (b) Periodic points of $f : \overline{x}_1$, and $\overline{x}_2$ are fixed points of f^2; (c) Periodic points of $f : \overline{x}_1$, and $\overline{x}_2$ are asymptotically stable fixed points of f^2.

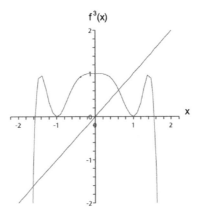

FIGURE 1.23

f^3 has no "genuine" fixed points, it has a fixed point x^* which is a fixed point of f, f has no points of period 3.

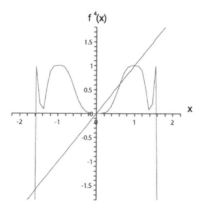

FIGURE 1.24
f^4 has no "genuine" fixed points, it has three fixed points, a fixed point x^* of f and two fixed points $\bar{x}_1$, $\bar{x}_2$ of f^2, f has no 4-periodic cycles.

Example 1.14
(The Tent Map Revisited). The tent map T is defined as

$$T(x) = \begin{cases} 2x; & 0 \leq x \leq \frac{1}{2} \\ 2(1-x); & \frac{1}{2} < x \leq 1. \end{cases}$$

It may be written in the compact form

$$T(x) = 1 - 2\left|x - \frac{1}{2}\right|.$$

Find all the 2-periodic cycles and the 3-periodic cycles of T and determine their stability. ⬚

SOLUTION   First, we observe that the fixed points of T are $x_1^* = 0$ and $x_2^* = \frac{2}{3}$; they are unstable since $|T'| = 2$. To find the 2-periodic cycles, we compute T^2. After some computation, we obtain

$$T^2(x) = \begin{cases} 4x; & 0 \leq x < \frac{1}{4} \\ 2(1-2x); & \frac{1}{4} \leq x < \frac{1}{2} \\ 4(x-\frac{1}{2}); & \frac{1}{2} \leq x < \frac{3}{4} \\ 4(1-x); & \frac{3}{4} \leq x \leq 1. \end{cases}$$

There are four fixed points of T^2 : $0, \frac{2}{5}, \frac{2}{3}, \frac{4}{5}$, two of which $(0, \frac{2}{3})$, are fixed points of T. Thus, $\{\frac{2}{5}, \frac{4}{5}\}$ is the only 2-periodic cycle [see Fig. 1.25(b)]. Since $|T'(\frac{2}{5})T'(\frac{4}{5})| = 4 > 1$, this 2-periodic cycle is unstable (Theorem 1.7). From Fig. 1.25(c), we observe that T^3 has eight fixed points, two of which are fixed points of T. Thus, there are two periodic cycles of period 3. It is easy to check that these cycles are $C_1 = \{\frac{2}{7}, \frac{4}{7}, \frac{6}{7}\}$ and $C_2 = \{\frac{2}{9}, \frac{4}{9}, \frac{8}{9}\}$, both of which are unstable.

Note that the point $\frac{3}{5}$ is an eventually 2-periodic point as $\frac{3}{5} \rightarrow \frac{2}{5} \rightarrow \frac{4}{5}$. Moreover, the point $\frac{3}{7}$ is an eventually 3-periodic point since $\frac{3}{7} \rightarrow \frac{6}{7} \rightarrow \frac{2}{7} \rightarrow \frac{4}{7}$. A general result characterizing periodic and eventually periodic points of the tent map will be given in Section 3.2. ∎

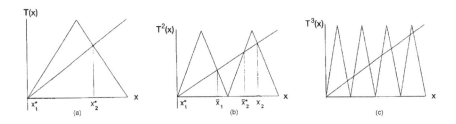

FIGURE 1.25
(a) The tent map T has two fixed points; (b) T^2 has 4 fixed points, 2 periodic points $\bar{x}_1$, $\bar{x}_2$, and 2 fixed points x_1^*, x_2^* of T; (c) T^3 has 8 fixed points, two cycles of period 3 and two fixed maps of T.

Exercises - (1.8)

In Problems 1–6, find the 2-periodic points and determine their stability.

1. $Q(x) = x^2 - 0.85$

2. $p(x) = \frac{1}{2}x^2 - x + \frac{1}{2}$

3. $f(x) = xe^{2(1-x)}$

4. $g(x) = 5 - \frac{6}{x}$

5. $h(x) = \frac{1-x}{3x+1}$

6. $f(x) = |x - 1|$

7. Let $Q(x) = ax^2 + bx + c$, where $a, b, c \in \mathbb{R}$ and $a \neq 0$.

   (a) If $\{x_0, x_1\}$ is a 2-periodic cycle such that $Q'(x_0)Q'(x_1) = -1$, prove that the 2-periodic cycle is asymptotically stable.

   (b) If $\{x_0, x_1\}$ is a 2-periodic cycle such that $Q'(x_0)Q'(x_1) = 1$, determine whether the cycle is stable or unstable.

8. Let $g(x) = ax^3 - bx + 1$, where $a, b \in \mathbb{R}$. Find the values of a and b for which the cycle $\{0, 1\}$ is asymptotically stable.

Problems 9 and 10 deal with Baker's map on the interval $[0, 1]$, which is defined as follows:

$$B(x) = \begin{cases} 2x; & 0 \le x \le \frac{1}{2} \\ 2x - 1; & \frac{1}{2} < x \le 1. \end{cases}$$

9. (a) Find the 2-periodic cycles of B and determine their stability.

   (b) Find the number of k-periodic points of B including those points that are not of prime period k.

10. Show that if m is an odd positive integer, then $\bar{x} = \frac{k}{m}$ is periodic for $k = 1, 2, \ldots, m - 1$.

In Problems 11–14, use Carvalho's lemma.

Carvalho's Lemma [19].

If k is a positive integer and $x(n)$ is a periodic sequence of period k, then the following hold true:

(i) If $k > 1$ is odd and $m = \frac{k-1}{2}$, then

$$x(n) = c_0 + \sum_{j=1}^{m} \left[c_j \cos\left(\frac{2jn\pi}{k}\right) + d_j \sin\left(\frac{2jn\pi}{k}\right) \right],$$

for all $n \ge 1$.

(ii) If k is even and $k = 2m$, then

$$x(n) = c_0 + (-1)^n c_m + \sum_{j=1}^{m-1} \left[c_j \cos\left(\frac{2jn\pi}{k}\right) + d_j \sin\left(\frac{2jn\pi}{k}\right) \right],$$

for all $n \ge 1$. For example, a 2-periodic cycle is of the form $x(n) = c_0 + (-1)^n c_1$; a 3-periodic cycle is of the form $x(n) = c_0 + c_1 \cos\left(\frac{2n\pi}{3}\right) + d_1 \sin\left(\frac{2n\pi}{3}\right)$.

11. Consider the logistic map $F_\mu(x) = \mu x(1-x)$. Use Carvalho's lemma to find the values of μ where the map has a 2-periodic cycle, $0 < \mu \leq 4$. Then find these 2-periodic cycles.

12.* (Term project). Find the values of μ where the equation in Problem 11 has a 3-periodic cycle.

13. Find the values of α and β for which the difference equation $x(n+1) = \alpha x(n)/(1 + \beta x(n))$, where $\alpha, \beta \in \mathbb{R}$ has a 2-periodic cycle.

14. The population of a certain species is modeled by the difference equation $x(n+1) = \mu x(n)e^{-x(n)}$, $x(n) \geq 0$, $\mu > 0$. For what values of μ does the equation have a 2-periodic cycle?

1.9   The Period-Doubling Route to Chaos

We end this chapter by studying in detail the **logistic map**:

$$F_\mu(x) = \mu x(1-x), \tag{1.40}$$

which gives rise to the logistic difference equation

$$x(n+1) = \mu x(n)(1 - x(n)), \tag{1.41}$$

where $x \in [0,1]$ and $\mu \in (0,4]$.

1.9.1   Fixed Points

Let us begin our exposition by examining the equilibrium points of Equation (1.41). There are two fixed points of F_μ: $x_1^* = 0$ and $x_2^* = \frac{\mu-1}{\mu}$. We now examine the stability of each fixed point separately.

1. The fixed point $x_1^* = 0$: observe that $F_\mu'(0) = \mu$. Therefore, from Theorem 1.3, we conclude

   (a) x_1^* is asymptotically stable if $0 < \mu < 1$ [see Fig. 1.26(a)].

   (b) x_1^* is unstable if $\mu > 1$ [see Fig. 1.26(c)].

   The case where $\mu = 1$ needs special attention, for we have $F_1'(0) = 1$ and $F_1''(0) = -2 \neq 0$. By applying Theorem 1.5, we may conclude that 0 is unstable. This is certainly true if we consider negative as well as positive initial points in the neighborhood of 0. Since negative initial points are not in the domain of F_μ, we may discard them and consider only initial points in neighborhoods of 0 of the form $(0, \delta)$. Now, Problem 17a in

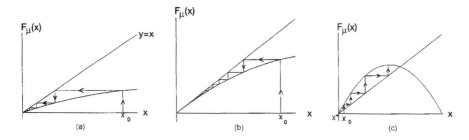

FIGURE 1.26

(a) $0 < \mu < 1$: 0 is asymptotically stable; (b) $\mu = 1$: 0 is asymptotically stable; (c) $\mu > 1$: 0 is unstable.

Exercises 1.7 tells us that the fixed point is semiasymptotically stable from the right. In other words, $x_1^* = 0$ is asymptotically stable in the domain $[0, 1]$ [see Fig. 1.26(b)].

2. The fixed point $x_2^* = \frac{\mu-1}{\mu}$: Clearly x_2^* will be in the interval $(0, 1]$ if $\mu > 1$. Moreover, $F_\mu'(\frac{\mu-1}{\mu}) = \mu - 2\mu(\frac{\mu-1}{\mu}) = 2 - \mu$. Thus, by Theorem 1.3, x_2^* is asymptotically stable if $|2 - \mu| < 1$. Solving this inequality for μ, we obtain $1 < \mu < 3$ as the values of μ where x_2^* is asymptotically stable [see Fig. 1.27(a)]. When $\mu = 3$, we have $F_3'(x_2^*) = F_3'(\frac{2}{3}) = -1$, and x_2^* is therefore nonhyperbolic. In this case, we need to compute the Schwarzian derivative: $SF_3'(x_2^*) = -\frac{3}{2}(36) < 0$. Hence, by Theorem 1.6, the equilibrium point $x_2^* = \frac{2}{3}$ is asymptotically stable under F_3 [see Fig. 1.27(a)]. Furthermore, by Theorem 1.3, the fixed point x_2^* is unstable for $\mu > 3$. We now summarize our findings.

   (a) x_2^* is asymptotically stable for $1 < \mu \leq 3$.[10]

   (b) x_2^* is unstable for $\mu > 3$ [see Fig. 1.27(b)].

Looking at Fig. 1.27(b), we observe that the orbit of x_0 flips around x_2^* and then settles bouncing between two points, which indicates the appearance of a 2-periodic cycle.

1.9.2   2-Periodic Cycles

To find the 2-periodic cycles we solve the equation $F_\mu^2(x) = x$, or

$$\mu^2 x(1 - x)[1 - \mu x(1 - x)] - x = 0. \qquad (1.42)$$

---

[10]In Section 2.4, we will show that in fact all points in $(0, 1)$ are attracted to x_2^*, that is x_2^* is globally asymptotically stable.

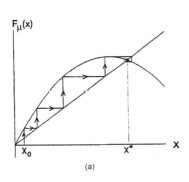
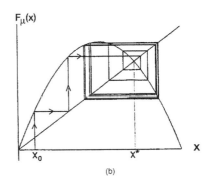

(a) (b)

FIGURE 1.27
(a) $1 < \mu \le 3$, x_2^* is asymptotically stable, and (b) $\mu > 3$ x_2^* is unstable.

Discarding the equilibrium points 0 and $\frac{\mu-1}{\mu}$ by dividing the left side of Equation (1.42) by $x(x - \frac{\mu-1}{\mu})$, we obtain

$$\mu^2 x^2 - \mu(\mu+1)x + (\mu+1) = 0.$$

Solving this equation,

$$\overline{x}_1 = \frac{(1+\mu) - \sqrt{(\mu-3)(\mu+1)}}{2\mu} \quad \text{and}$$

$$\overline{x}_2 = \frac{(1+\mu) + \sqrt{(\mu-3)(\mu+1)}}{2\mu}. \tag{1.43}$$

Clearly $\overline{x}_1$ and $\overline{x}_2$ are defined only if $\mu > 3$. Next, we investigate the stability of this 2-periodic cycle. By Theorem 1.7, this 2-periodic cycle is asymptotically stable if

$$|F_\mu'(\overline{x}_1)\, F_\mu'(\overline{x}_2)| < 1$$

or

$$-1 < \mu^2(1 - 2\overline{x}_1)(1 - 2\overline{x}_2) < 1$$
$$-1 < \mu^2 \left(1 - \frac{(1+\mu) - \sqrt{(\mu^2 - 2\mu - 3)}}{\mu}\right) \left(1 - \frac{(1+\mu) + \sqrt{(\mu^2 - 2\mu - 3)}}{\mu}\right) < 1$$
$$-1 < -\mu^2 + 2\mu + 4 < 1.$$

Solving the last two inequalities yields the range: $3 < \mu < 1 + \sqrt{6}$ for asymptotic stability. Now, for $\mu = 1 + \sqrt{6}$,

$$F_\mu'(\overline{x}_1)\, F_\mu'(\overline{x}_2) = -1.$$

In this case, we need to apply Theorem 1.6 on F_μ^2 to determine the stability of the periodic points $\overline{x}_1$ and $\overline{x}_2$ of F_μ. After some computation, we conclude that $SF_\mu^2(\overline{x}_1) < 0$ and $SF_\mu^2(\overline{x}_2) < 0$, which implies that the cycle $\{\overline{x}_1, \overline{x}_2\}$ is asymptotically stable (Problem 1). Moreover, the periodic cycle $\{\overline{x}_1, \overline{x}_2\}$ is unstable for $\mu > 1 + \sqrt{6}$.

In summary:

1. $3 < \mu \leq 1 + \sqrt{6}$: The 2-periodic cycle $\{\overline{x}_1, \overline{x}_2\}$ is asymptotically stable.

2. $\mu > 1 + \sqrt{6}$: The 2-periodic cycle $\{\overline{x}_1, \overline{x}_2\}$ is unstable.

Thus, the positive equilibrium point is asymptotically stable for $1 < \mu \leq 3$, where it loses its stability after $\mu_1 = 3$. For $\mu > \mu_1$, an asymptotically stable 2-periodic cycle appears where it loses its stability after a second magic number $\mu_2 = 1 + \sqrt{6} \approx 3.44949\ldots$, etc.

1.9.3 2^2-Periodic Cycles

The search for 4-periodic cycles can be successful if one is able to solve the equation $F_\mu^4(x) = x$. This involves solving a twelfth-degree equation, which is not possible in general. So we turn to graphical or numerical analysis to help us find the 4-periodic cycles (see Fig. 1.28). It turns out that there is one 2^2

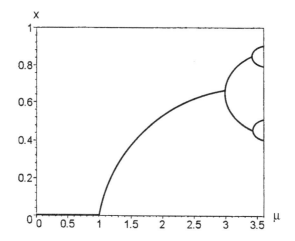

FIGURE 1.28
The appearance of a 4-periodic cycle. An exchange of stability occurs at $\mu = 1$ between $x_1^* = 0$ and $x_2^* = (\mu - 1)/\mu$.

cycle when $\mu > 1 + \sqrt{6}$ which is asymptotically stable for $1 + \sqrt{6} < \mu \leq 3.54409$. This 2^2 cycle loses its stability when $\mu > 3.54409$. Again, the story repeats itself, when $\mu > \mu_3$, the 2^2 cycle bifurcates into an asymptotically stable 2^3

cycle. This process of double bifurcation continues indefinitely and produces a sequence $\{\mu_n\}_{n=1}^{\infty}$. Table 1.2 sheds some light on some remarkable patterns:

TABLE 1.2

n	μ_n	$\mu_n - \mu_{n-1}$	$\frac{\mu_n - \mu_{n-1}}{\mu_{n+1} - \mu_n}$
1	3	–	–
2	3.449489 ...	0.449489 ...	-
3	3.544090 ...	0.094601 ...	4.751419 ...
4	3.564407 ...	0.020317 ...	4.656248 ...
5	3.568759 ...	0.0043521 ...	4.668321 ...
6	3.569692 ...	0.00093219 ...	4.668683 ...
7	3.569891 ...	0.00019964 ...	4.669354 ...

From Table 1.2, we make the following observations (which can be proved, at least numerically):

1. The sequence $\{\mu_n\}$ seems to tend to a specific number, $\mu_\infty \approx 3.570$.

2. The window size ($\mu_n - \mu_{n-1}$) between successive μ_i values gets narrower and narrower, eventually approaching zero.

3. The ratio $\dfrac{\mu_n - \mu_{n-1}}{\mu_{n+1} - \mu_n}$ approaches a constant called **Feigenbaum number** δ named after its discoverer, Mitchell Feigenbaum [39]. In fact,

$$\delta = \lim_{n \to \infty} \frac{\mu_n - \mu_{n-1}}{\mu_{n+1} - \mu_n} \approx 4.669201609\ldots \tag{1.44}$$

Feigenbaum discovered that the number δ is universal and does not depend on the family of maps under discussion; it is the same for a large class of maps, called **unimodal** maps.[11]

Formula (1.44) may be used to generate the sequence $\{\mu_n\}$ with good accuracy. We let $\delta = \dfrac{\mu_n - \mu_{n-1}}{\mu_{n+1} - \mu_n}$ and solve for μ_{n+1}. Then, we obtain

$$\mu_{n+1} = \mu_n + \frac{\mu_n - \mu_{n-1}}{\delta}. \tag{1.45}$$

For example, given $\mu_1 = 3$ and $\mu_2 = 1 + \sqrt{6}$ (in Table 1.2), then from Formula (1.45) we get $\mu_3 = (1 + \sqrt{6}) + \frac{(1+\sqrt{6})-3}{4.6692} \approx 3.54575671$, which is a good approximation of the actual μ_3 in Table 1.2.

---

[11]A map f on the interval $[0, 1]$ is said to be unimodal if $f(0) = f(1) = 0$ and f has a unique critical point between 0 and 1.

Mitchell J. Feigenbaum (1944-)

Mitchell J. Feigenbaum is famed for his discovery in 1975 of the period-doubling route to chaos, commonly called the Feigenbaum scenario, and the universality of the Feigenbaum number δ in nonlinear systems. Everything from purely mechanical systems, fluid dynamics and the weather to the patterns of biological growth in nature and the dynamics of heart, hormone and brain rhythms have been found to exhibit aspects of the Feigenbaum scenario.

Mitchell Feigenbaum received his PhD in Physics in 1970 from the Massachusetts Institute of Technology. While working at the Los Alamos National Laboratory in New Mexico, he was inspired by a lecture given by the mathematician Steven Smale on nonlinear dynamics. Using a hand-held calculator, Feigenbaum studied the logistic map and other one-hump maps. Observing the changes in the behavior of the orbits caused by changing the values of the parameter led him to the discovery of the universality property. Dr. Mitchell Feigenbaum is currently the Chairman of the Physics Department and the Center for the Study of Physics and Biology at The Rockefeller University in New York.

The best way to illustrate the above discussion is to draw the so-called **bifurcation diagram**.

Bifurcation Diagram

We let the horizontal axis represent the parameter μ and the vertical axis representing higher iterates $F_\mu^n(x_0)$ of a specific initial point x_0, so the diagram will show the limiting behavior of the orbit of x_0. The computer-generated bifurcation diagram (see Fig. 1.30) is obtained by the following procedure:

1. Choose an initial value x_0[12] from the interval $[0, 1]$ and iterate, say, 500 times to find

$$x_0, F_\mu(x_0), F_\mu^2(x_0), \ldots, F_\mu^{400}(x_0), F_\mu^{401}(x_0), \ldots, F_\mu^{500}(x_0).$$

---

[12]We select the critical value $x_0 = 0.5$ since by Singer's Theorem (Chapter 2), the orbit of the critical point must approach the attracting fixed point.

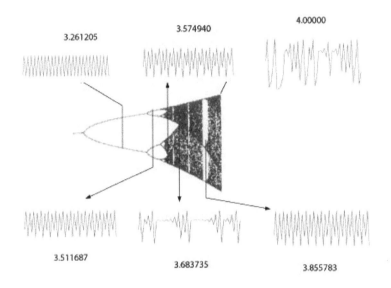

FIGURE 1.29
Montage of time series at different iterates of the logistic map.

2. Drop the first 400 iterations $x_0, F_\mu(x_0), \ldots, F_\mu^{400}(x_0)$ and plot the rest of the iterations $F_\mu^{401}(x_0), \ldots, F_\mu^{500}(x_0)$ in the bifurcation diagram.

3. The procedure is done repeatedly for values of μ between 0 and 4 taking increments of $\frac{1}{100}$.

4. (a) Note that for $0 < \mu \leq 1$, only the value $x = 0$ shows up in the diagram. This is because the orbit of $x_0 = 0.5$ converges to the fixed point $x_1^* = 0$. Thus, high iterates beyond the 50th iterates will not be distinguishable from 0. Once $x_1^* = 0$ loses its stability beyond $\mu = 1$, it will disappear from the graph.

   (b) For $1 < \mu \leq 3$, the orbit of $x_0 = 0.5$ converges to the positive fixed point $x_2^* = \frac{\mu-1}{\mu}$. For example, for $\mu = \frac{3}{2}$, the point $(\frac{3}{2}, \frac{1}{3})$ appears in the diagram; for $\mu = 2$, there corresponds the point $(2, \frac{1}{2})$ and at $\mu_1 = 3$ we can see the point $(3, \frac{2}{3})$. Beyond $\mu_1 = 3$, x_2^* loses its stability and makes a disappearing act.

   (c) For $\mu_1 < \mu \leq \mu_2$, the orbit of 0.5 converges to a 2-periodic cycle. Hence, to each μ in this range there corresponds two points in the diagram. Beyond μ_2, the 2-periodic cycle loses its stability, disappears from the diagram, and then gives birth to a 4-periodic cycle. This double bifurcation continues until $\mu = \mu_\infty$ (see Fig. 1.30).

x

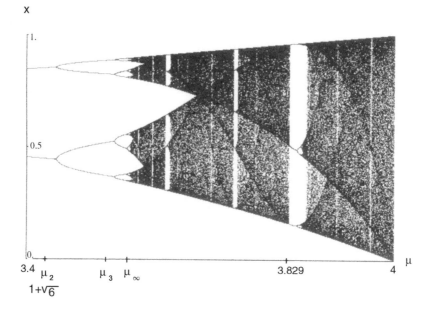

FIGURE 1.30

Bifurcation diagram $3.4 < \mu \leq \mu_\infty$.

1.9.4 Beyond μ_∞

Let us now turn our attention to the parameter values $\mu > \mu_\infty$. The situation
here is much more complicated than the period-doubling region $1 < \mu \leq \mu_\infty$
(where only stable cycles appear in the bifurcation diagram). The best way
to explain the dynamics of the orbit of x_0 is to start from $\mu = 4$ and march
backward to μ_∞. At $\mu = 4$, we see only one band covering the whole interval
$[0, 1]$ (see Fig. 1.31). This band slowly narrows as μ decreases but then
bifurcates into two parts at $\mu = \lambda_1$. Then, it bifurcates again into four parts
at $\mu = \lambda_2$. The splitting continues indefinitely, where at λ_k we will have 2^k
bands. This decreasing sequence $\{\lambda_n\}$ converges to $\lambda_\infty = \mu_\infty$. Furthermore,
the quotient $\dfrac{\lambda_n - \lambda_{n-1}}{\lambda_{n+1} - \lambda_n}$ tends to the Feigenbaum's number $\delta \approx 4.6692$.

Periodic Windows

The biggest window in the bifurcation diagram occurs for values of μ between
3.828 and 3.857. This is called **period 3-window**. An asymptotically stable
3-periodic cycle appears first at $\mu = 1 + \sqrt{8} \approx 3.828$, after which the period-
doubling phenomenon takes over. This 3-periodic cycle then loses its stability
and gives birth to an asymptotically stable 6-periodic cycle. The period dou-
bling continues until $\mu \approx 3.8415\ldots$ (corresponds to μ_∞ in the first part of the

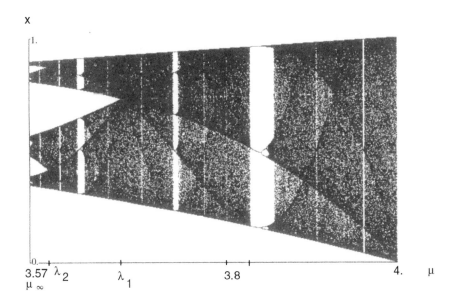

FIGURE 1.31
Appearance of the sequence $\lambda_1, \lambda_2, \ldots$ from right to left.

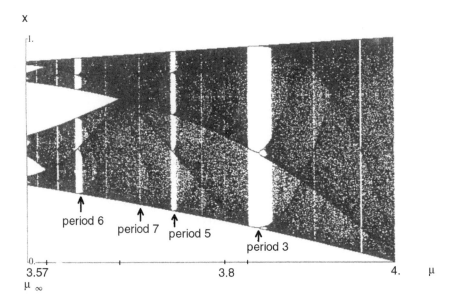

FIGURE 1.32
The appearance of odd periods.

diagram) after which we get into a complicated dynamics region. Windows of all odd periods appear to the left of the period 3-window (see Fig. 1.32). We will present more details about this in the next chapter.

Period Doubling Near the Feigenbaum Limit

> Fourteen lines accomodate
> The points I've picked to illustrate.
>
> In some systems you will find
>
> Orbits moving toward a station,
> Then show themselves to have a mind
>   To move no more on iteration.
>
> But tweak an additive parameter
>   And where before you was the stall,
>   Now they, surprisingly, begin
>   A two-step foxtrot on the floor,
>
> Like iams in a strict tetrameter,
>   Or like an active ping pong ball.
>   More tweaks, and doubling comes again;
>   Yet more, redoubling as before.

—J.D. MEMORY
NORTH CAROLINA STATE UNIVERSITY
RALEIGH, NC 27695-8021

Exercises - (1.9)

In problems 1 and 2, we consider the logistic map $F_\mu(x) = \mu x(1 - x)$.

1. Show that for $\mu = 1 + \sqrt{6}$, the 2-periodic cycle $\{\bar{x}_1, \bar{x}_2\}$ defined by Equation (1.43) is asymptotically stable under F_μ.

2. Find the number of k-periodic cycles (of prime period k) of F_μ for $k = 2, 3, 4, 5, 6$ if they exist. Do you detect any pattern here?

In Problems 3–7:

(a) Find the fixed points and determine their stability.

(b) Find the 2-periodic cycles and determine their stability.

(c) Find the sequence $\mu_1, \mu_2, \mu_3, \mu_4$, where μ_i is the first value of μ at which the 2^i−cycle appears (use Formula (1.45)).

(d) Draw the bifurcation diagram using Phaser, Maple, or Mathematica.

3. $Q_c(x) = x^2 + c$, where $c \in [-2, 0]$ and $x \in [-2, 2]$.

4. $P_\mu(x) = 1 - \mu x^2$ on the interval $[-1, 1]$ and $\mu \in (0, 2]$.

5. $G_\mu(x) = \mu \sin \pi x$, $0 < \mu \leq 1$ and $-1 \leq x \leq 1$.

6. $H_\mu(x) = \mu \arctan x$ for all x with $\mu > 0$.

7.

$$T_\mu(x) = \begin{cases} \mu x, & \text{for } x \leq \frac{1}{2} \\ \mu(1 - x), & \text{for } x \geq \frac{1}{2} \end{cases}$$

on the interval $[0, 1]$ for $\mu \in (0, 2]$.

8.* Use Carvalho's lemma (see Exercises 1.8, Problem 11) to find the values of μ for which F_μ has a 3-periodic cycle. Find the 3-periodic cycle and determine its stability.

9.* Use Carvalho's lemma (see Exercises 1.8, Problem 11) to find 4-periodic cycles of F_μ.

10.* Use Carvalho's lemma to find the values of c for which the quadratic map

$$Q_c(x) = x^2 + c, \ c \in [-2, 0]$$

has a 3-periodic cycle and then determine its stability.

Superattracting Fixed Points and Periodic Points

Let $\bar{x}$ be a k-periodic point of the logistic map $F_\mu(x) = \mu x(1 - x)$, $k = 1, 2, 3, \ldots$ Then, $\bar{x}$ is said to be **superattracting** if $(F_\mu^k)'(\bar{x}) = 0$. In other words, $\bar{x}$ is superattracting if it coincides with a critical point of F_μ^k.

11. (Term Project).

(a) Find μ at which the fixed point $x^* = \frac{\mu-1}{\mu}$ is superattractive and call this μ "s_1."

(b) Explain why the fixed point in part (a) is called superattractive.

(c) Find μ at which the 2-periodic cycle of F_μ is superattractive. Put this $\mu = s_2$.

(d) Use Newton's method to find $S_{2^n}, n = 2, 3, 4, \ldots$

(e) If we let $\alpha = \lim\limits_{n \to \infty} \dfrac{S_{2^n} - S_{2^{n-1}}}{S_{2^{n+1}} - S_{2^n}}$, what do you think the relationship is between α and the Feigenbaum number δ?

12. Repeat Problem 13 for the family of quadratic maps $Q_c = x^2 + c$, $c \in [-2, 0]$, $x \in [-2, 2]$.

q-Curves

In Problems 13 and 14, define the kth polynomial $q_k(\mu) = F_\mu^k\left(\frac{1}{2}\right)$ as the kth iterate of F_μ starting at the critical value 1 where $\left(F_\mu^k\right)'\left(\frac{1}{2}\right) = 0$. For example, $q_1(\mu) = F_\mu\left(\frac{1}{2}\right) = \frac{1}{4}\mu$, $q_2(\mu) = \frac{1}{4}\mu^2\left(1 - \frac{1}{4}\mu\right), \ldots, q_{k+1}(\mu) = \mu q_k(\mu)(1 - q_k(\mu))$. Hence, each $q_k(\mu)$ is a 2^{k-1} degree polynomial. The graph of any q_k as a function of μ is called a **q curve** [71].

13. (a) Use any graphing device to draw the graphs of $q_1, q_2, q_3, q_4, q_5, q_6$.

　　(b) Draw the curve q_i, $1 \le i \le 6$ and the bifurcation diagram of F_μ on the same plot.

　　(c) Show graphically that the curve q_i crosses the line $x = 0.5$ in the window of period i in the bifurcation diagram.

14.* **Superattracting root theorem** [71]. For any $k \in \mathbb{Z}^+$ prove that $q_k(\mu) = 0$ and $q_j(\mu) \ne 0$ for $0 < j < k$ if and only if F_μ has a superattracting k-periodic point.

1.10   Applications

1.10.1   Fish Population Modeling

The growth of most biological populations is limited by factors including environmental variation, changes in rates of survival or reproduction, disease, competitive interactions, and predator-prey relationships. Those factors that can influence populations in relation to its size are referred to as "density dependent" factors. We have density dependent whenever our underlying difference equation is nonlinear. The population size to which a population will tend to return to in response to density dependent factors is known as the "equilibrium" population or the carrying capacity, normally denoted by K. In the sequel, we will examine two population models that are widely used in biology.

(I) Beverton-Holt Model (Beverton and Holt, 1957)

The Beverton-Holt model depicts density dependent recruitment of a population with limited resources in which resources are not shared equally. It assumes that the per capita number of offspring is inversely proportional to a linearly increasing function of the number of adults.

Let $p(n)$ be the size of a population in generation n, and $p(n+1)$ be the size of their offspring (generation $n + 1$). Suppose that μ is the net reproductive rate, that is the number of offspring that each individual leaves before dying if there are no limitation in resources. Then the Beverton-Holt model is given by

$$\frac{p(n+1)}{p(n)} = \frac{\mu}{1 + [(\mu - 1)/K]p(n)}, \quad \mu > 0, \quad K > 0,$$

or

$$p(n+1) = \frac{\mu K p(n)}{K + (\mu - 1)p(n)}. \tag{1.46}$$

This equation represent the map (Figure 1.33)

$$G(p) = \frac{\mu K p}{K + (\mu - 1)p}, \quad p \in [0, \infty). \tag{1.47}$$

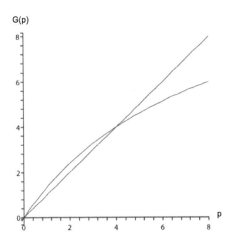

FIGURE 1.33

The Beverton-Holt map with $\mu = 1.5$, $K = 4$.

The reader may now compare Figure 1.33 with Figure 1.35 obtained from real data. The horizontal axis represents the "spawning stock biomass (SSB)" which is the total weight in metric tons (MT) of all sexually mature winter flounder in the gulf of Main for 1982–2001 year classes. The vertical axis represents the number in millions of age-1 year fish (Santa Ana watershed advisory). We now present two methods to study the dynamics of the Beverton-Holt map.

<u>Method 1</u>. Let $x(n) = \frac{1}{p(n)}$. Then we obtain the linear difference equation

$$x(n+1) = \frac{1}{\mu}x(n) + \frac{(\mu - 1)}{\mu K}. \tag{1.48}$$

Using formula (1.20) we have for $x(0) = x_0$,

$$x(n) = \left(x_0 - \frac{1}{K} \right) \mu^{-n} + \frac{1}{K}.$$

Hence

$$\lim_{n \to \infty} x(n) = \begin{cases} \infty & \text{if } \mu < 1, \\ \frac{1}{K} & \text{if } \mu > 1, \\ x_0 & \text{if } \mu = 1. \end{cases}$$

Since $p(n) = \frac{1}{x(n)}$, it follows that

$$\lim_{n \to \infty} p(n) = \begin{cases} 0 & \text{if } \mu < 1 \text{ (Extinction)}, \\ K & \text{if } \mu > 1 \text{ (Stability)}, \\ p(0) & \text{if } \mu = 1 \text{ (Constant)}. \end{cases}$$

Method 2. Notice that $p_1^* = 0$ and $p_2^* = K$ are the only fixed points of our map. Now

$$G'(p) = \frac{\mu K^2}{[K + (\mu - 1)x]^2}.$$

Since $G'(0) = \mu$, it follows by Theorem 1.5 that the fixed point $p_1^* = 0$ is asymptotically stable if $\mu < 1$, and unstable if $\mu > 1$. However, $G'(K) = \frac{1}{\mu}$ implies that $x_2^* = K$ is unstable if $\mu < 1$ and asymptotically stable if $\mu > 1$. The case $\mu = 1$ can be studied by inspecting the map and noticing that every point is a fixed point. Moreover, it may be shown using the monotonicity of the map that for every $\mu < 1$, $p_1^* = 0$ is in fact globally asymptotically stable, and if $\mu > 1$, $p_2^* = K$ is globally asymptotically stable on $(0, \infty)$. We now provide the main ingredients to prove the latter statement. The three main facts that are needed are

(i) if $0 < x < K$, then $f(x) > x$,

(ii) if $K < x < \infty$, then $f(x) < x$,

(iii) $x < y$ if and only if then $G(x) < G(y)$.

(II) Ricker Model (Ricker, 1975)

In contrast to the Beverton-Holt model, the Ricker model predicts declining recruitment (offspring) $p(n + 1)$ at high stock levels (adults) $p(n)$ according to the equation

$$p(n + 1) = p(n)e^{r[1 - (p(n)/K)]}.$$

Letting $u(n) = \frac{r}{K}p(n)$ yields

$$u(n + 1) = u(n)e^{r - u(n)} \tag{1.49}$$

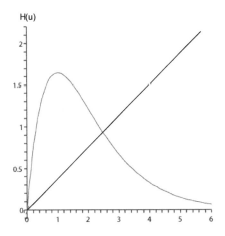

FIGURE 1.34
Ricker map with $r = 1.5$.

which represents the map (see Figure 1.34).

$$H(u) = ue^{r-u}, \quad r > 0. \tag{1.50}$$

This map has two fixed points $u_1^* = 0$, $u_2^* = r$.

Now $H'(0) = e^r > 1$ for $r > 0$ implies that $p_1^* = 0$ is unstable. Moreover, $H'(r) = 1 - r$. Hence $p_2^* = r$ is asymptotically stable if $0 < r < 2$. It will be left as a term project to show that in fact p_2^* is globally asymptotically stable on $(0, \infty)$ if $0 < r < 2$. For $r > 2$, the map goes through period-doubling in its route to chaos.

The question that now arises: which one of these two models would fit actual data. Figure 1.35 for the Coho Salmon and Anchoveta shows uncertainty and inconclusiveness in the answer. For example, although there is a reasonably good fit of the Beverton-Holt and Ricker curves to data for Coho Salmon, population data for Anchoveta show considerable variation about the hypothetical stock-recruitment curves.

Fisheries Management

Notice that in either model the derivative of the map at 0 indicates how steep the curve is. In the Holt-Beverton map, $G'(0) = \mu$ and for the Ricker map, $H'(0) = e^r$. Hence the greater μ or r, the greater the expected "compensatory" response of the population to density changes and the larger the harvestable portion of the stock.

Figure 1.36 depcits the graph of four populations. Population A has the strongest "compensatory" response, while population D has the weakest and

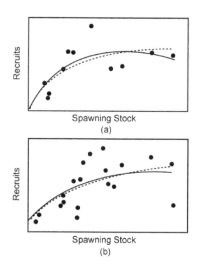

FIGURE 1.35
The Ricker curve (solid line) and Beverton-Holt curve (dotted line) fitted to data for (a) Coho Salmon and (b) Anchoveta.

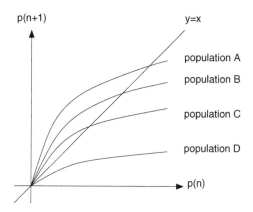

FIGURE 1.36
Hypothetical curves for four populations A, B, C, D.

indicates the tendency of the population to go extinct. Population D requires immediate interference to prevent extinction of the given fish.

Exercises - (1.10)

Term Project 1. Consider the Beverton-Holt model. Show that

(a) For $0 < \mu < 1$, the fixed point $p_1^* = 0$ is globally asmpotitically stable.

(b) For $\mu > 1$, the fixed point $p_2^* = K$ is globally asymptotically stable on $(0, \infty)$.

Term Project 2. Consider the Ricker model. Show that

(a) For $0 < r < 2$, show that $u^* = r$ is globally asmpotitically stable.

(b) For $r > 2$, develop the bifurcation diagram and analyze the dynamics of the map.

2

Attraction and Bifurcation

> Period three implies chaos.
>
> *Li and Yorke*

James A. Yorke (1941 -)

James Yorke teamed up with T. Y. Li to write in 1975 the best known paper to the general public "Period three implies Chaos." In this paper the word "Chaos" was coined and henceforth became one of the most celebrated subjects in mathematics and a wide range of disciplines. Yorke was awarded the 2003 Japan Prize in recognition of his contributions as a founder and leader of Chaos Theory. (He shared the prize with Benoit Mandelbrot). His recent research interests range from chaos theory and weather prediction to genome research to the population dynamics of the HIV/AIDS epidemics. Yorke received his PhD. in Mathematics from the University of Maryland in 1966. He is a distinguished professor of mathematics and Physics at the University of Maryland and serves as a consulting editor of the *Journal of Difference Equations and Applications*.

2.1 Introduction

In Chapter 1 we have encountered asymptotically stable fixed points and periodic cycles. The former and the latter sets are commonly called stable attractors (or just attractors). Here we broaden and deepen our analysis firstly by considering attractors with infinitely many points, and secondly by investigating the nature of the "basin of attraction," that is, the maximal set that is attracted to our attractor. Then Singer's Theorem is invoked to find the maximum number of stable attractors.

We will delve deeply into stable attractors and global stable attractors and include the latest results including Elaydi-Yakubu Theorem. The rest of the Chapter will continue the study of bifurcation that started in Chapter 1.

2.2   Basin of Attraction of Fixed Points

It is customary to call an asymptotically stable fixed point or a cycle an attractor. This name makes sense since in this case all nearby points tend to the attractor. The maximal set that is attracted to an attractor M is called the *basin of attraction* of M. Our analysis applies to cycles of any period.

We start our exposition with fixed points.

DEFINITION 2.1   *Let x^* be a fixed point of map f. Then the basin of attraction (or the stable set) $W^s(x^*)$ of x^* is defined as*

$$W^s(x^*) = \{x : \lim_{n \to \infty} f^n(x) = x^*\}.$$

In other words, $W^s(x^*)$ consists of all points that are forward asymptotic to x^*.

Observe that if x^* is an asymptotically stable fixed point, $W^s(x^*)$ contains an open interval around x^*. The maximal interval in $W^s(x^*)$ that contains x^* is called the *immediate basin of attraction* and is denoted by $\mathcal{B}(x^*)$.

Example 2.1
The map $f(x) = x^2$ has one attracting fixed point $x^* = 0$. Its basin of attraction $W^s(0) = (-1, 1)$. Note that 1 is an unstable fixed point and –1 is an eventually fixed point that goes to 1 after one iteration.   ⬜

Example 2.2
Let us now modify the map f. Consider the map $g : [-2, 4] \to [-2, 4]$ defined as

$$g(x) = \begin{cases} x^2 & \text{if } -2 \leq x \leq 1, \\ 3\sqrt{x} - 2 & \text{if } 1 < x \leq 4. \end{cases}$$

The map g has three fixed points $x_1^* = 0$, $x_2^* = 1$, $x_3^* = 4$. The basin of attraction of $x_1^* = 0$, $W^s(0) = (-1, 1)$, while the basin of attraction of $x_3^* = 4$, $W^s(4) = [-2, -1) \cup (1, 4]$. Moreover, the immediate basin of attractions of $x_1^* = 0$ is $\mathcal{B}(0) = W^s(0) = (-1, 1)$, while $\mathcal{B}(4) = (1, 4]$.   ⬜

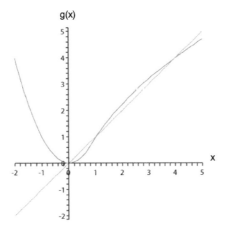

FIGURE 2.1
The basin of attraction $W^s(0) = (-1, 1)$ and $W^s(4) = [-2, -1) \cup (1, 4]$. The immediate basin of attraction $\mathcal{B}(4) = (1, 4]$.

REMARK 2.1 Observe that in the preceding example, the basins of attraction of the two fixed points $x_1^* = 0$ and $x_3^* = 4$ are disjoint. This is no accident and is, in fact, generally true. This is due to the uniqueness of the limit of a sequence. In other words, if $\lim_{n \to \infty} f^n(x) = L_1$ and $\lim_{n \to \infty} f^n(x) = L_2$, then certainly $L_1 = L_2$. ∎

It is worth noting here that finding the basin of attraction of a fixed point is in general a difficult task. The most efficient method to determining the basin of attraction is the method of **Liapunov functions**, which will be developed later in Chapter 4. In this section, we will develop some of the basic topological properties of the basin of attraction. Henceforth, all our maps are assumed to be *continuous*. We begin our exposition by defining the important notion of invariance.

DEFINITION 2.2 *A set M is* positively invariant *under a map f if $f(M) \subseteq M$. In other words, for every $x \in M$, the orbit $\mathcal{O}(x) \subseteq M$. Since we are only considering forward iterations of f, the prefix "positively" will, henceforth, be dropped.*

Clearly an orbit of a point is invariant.

Next we show that the basin of attraction of an attracting fixed point is invariant and open.

THEOREM 2.1

Let $f : I \to \mathbb{R}$ be a continuous map from an interval I to $\mathbb{R}$ and let x^* be an asymptotically stable fixed point of f. Then $W^s(x^*)$ is a nonempty open and (positively) invariant set.

PROOF   Since $B(x^*) \subset W^s(x^*)$, it follows that $W^s(x^*) \neq \emptyset$. To prove invariance, let $y \in W^s(x^*)$. Then $\lim_{n\to\infty} f^n(y) = x^*$. Since f is continuous,

$$f\left(\lim_{n\to\infty} f^n(y)\right) = \lim_{n\to\infty} f^{n+1}(y) = \lim_{n\to\infty} f^n(f(y)) = f(x^*) = x^*.$$

Hence $f(y) \in W^s(x^*)$ which proves the invariance of $W^s(x^*)$. It remains to show that $W^s(x^*)$ is open. This means that $W^s(x^*)$ is the union of open intervals of the form (c, d), and if our domain I has an end point, $I = [a, \infty)$ or $[a, b]$, then $[a, c)$ and $(c, b]$ are also allowed. Assume that $W^s(x^*)$ is not open. Then there exists $a \in W^s(x^*)$ such that a is not an end point of I and a is not an interior point of $W^s(x^*)$. This means that any open interval containing a must contain points not in $W^s(x^*)$. Hence there exists a sequence of points $\{x_n\}$ such that $x_n \notin W^s(x^*)$ for all $n \geq 1$ and $\lim_{n\to\infty} x_n = a$. Now since $a \in W^s(x^*)$, $\lim_{n\to\infty} f^n(a) = x^* \in B(x^*)$. Hence for some positive integer $f^m(a) \in B(x^*)$. Moreover, there exists a small interval $J = (f^m(a) - \delta, f^m(a) + \delta)$ containing $f^m(a)$ such that $J \subset B(x^*)$. Now $x_n \to a$ implies by the continuity of f that $f^m(x_n) \to f^m(a)$. Hence for some positive integer N, $f^m(x_N) \in J \subset B(x^*)$. Hence $x_N \in W^s(x^*)$, a contradiction. This completes the proof of the theorem.   ∎

REMARK 2.2   Theorem 2.1 showed that the basin of attraction of an asymptotically stable fixed point is the union (finite or infinite) of intervals of the form $[a, c), (c, d), (d, b]$ if a and b are end points. Moreover, this allows intervals of the form $(-\infty, d)$, and (c, ∞).   ∎

There are several (popular) maps such as the logistic map and Ricker map in which the basin of attraction, for the attracting fixed point, is the entire space with the exception of one or two points (fixed or eventually fixed). For the logistic map $F_\mu(x) = \mu x(1 - x)$ and $1 < \mu < 3$, the basin of attraction $W^s(x^*) = (0, 1)$ for the fixed point $x^* = \frac{\mu-1}{\mu}$. And for the Ricker map $R_p(x) = xe^{p-x}$, $0 < p < 2$, the basin of attraction $W^s(x^*) = (0, \infty)$, for $x^* = p$. Here we will consider only the logistic map and leave it to the reader to prove the statement concerning the Ricker map.

Notice that $|F'_\mu(x)| = |\mu - 2\mu x| < 1$ if and only if $-1 < \mu - 2\mu x < 1$. This implies that $\frac{\mu-1}{2\mu} < x < \frac{\mu+1}{2\mu}$. Hence $|F'_\mu(x)| < 1$ for all $x \in \left(\frac{\mu-1}{2\mu}, \frac{\mu+1}{2\mu}\right)$. Observe that $x^* = \frac{\mu-1}{\mu} \in \left(\frac{\mu-1}{2\mu}, \frac{\mu+1}{2\mu}\right)$ if and only if $1 < \mu < 3$. Now

$F_\mu\left(\frac{\mu+1}{2\mu}\right) = F_\mu\left(\frac{\mu-1}{2\mu}\right) = \frac{1}{2}\left[\frac{(\mu-1)(\mu+1)}{2\mu}\right]$. Since $1 < \mu < 3$, it follows that $\frac{\mu-1}{2\mu} < \frac{1}{2} \cdot \frac{(\mu-1)(\mu+1)}{2\mu} < \frac{\mu+1}{2\mu}$. Hence $\left[\frac{\mu-1}{2\mu}, \frac{\mu+1}{2\mu}\right] \subset W^s(x^*)$.

If $z \in \left(0, \frac{\mu-1}{2\mu}\right)$, then $F'_\mu(z) > 1$. By the Mean Value Theorem, $\frac{F_\mu(z) - F_\mu(0)}{z-0} = F'_\mu(\gamma)$, for some γ with $0 < \gamma < z$. Thus $F_\mu(z) = F'_\mu(\gamma)z$. This implies that $F_\mu(z) \geq \beta z$, for some $\beta > 1$. Then for some $r \in \mathbb{Z}^+$, $F^r_\mu(z) \geq \beta^r z > \frac{\mu-1}{2\mu}$ and $F^{r-1}_\mu(z) < \frac{\mu-1}{2\mu}$. Moreover, since F is increasing on $\left[0, \frac{\mu-1}{2\mu}\right]$, $F^r_\mu(z) < F_\mu\left(\frac{\mu-1}{2\mu}\right) = \mu\left(\frac{\mu-1}{2\mu}\right)\left(1 - \frac{\mu-1}{2\mu}\right) = \frac{\mu-1}{\mu}\left(\frac{\mu+1}{4}\right) \leq x^*$. Thus $z \in W^s(x^*)$. On the other hand, $F_\mu\left(\frac{\mu+1}{2\mu}, 1\right) \subset (0, x^*)$ and hence $\left(\frac{\mu+1}{2\mu}, 1\right) \subset W^s(x^*)$. This shows that $W^s(x^*) = (0, 1)$.

In Summary:

LEMMA 2.1
For the logistic map $F_\mu(x) = \mu x(1 - x)$, $1 < \mu < 3$, $W^s(x^*) = (0, 1)$ *for* $x^* = \frac{\mu-1}{\mu}$.

REMARK 2.3 Not all locally asymptotically stable fixed points are globally asymptotically stable as it may be illustrated by the following example. ∎

Example 2.3
Consider the map

$$f(x) = x\exp[-1.9(x-1) + (7.6 - 8\ln 3)(x-1)^3]$$

defined on $[0, \infty)$. Notice that the fixed point $x^* = 1$ is locally asymptotically stable since $f'(1) = -0.9$. However, the map has a 2-periodic cycle $\{\frac{1}{2}, \frac{3}{2}\}$. The immediate basin of attraction of $x^* = 1$ is $\left(\frac{1}{2}, \frac{3}{2}\right)$ (see Fig. 2.2). Thus $x^* = 1$ is locally but not globally asymptotically stable which is due to the presence of a periodic 2-cycle in the interior of the domain of f. ▯

The above example leads to an important and difficult question: under what conditions do local asymptotic stability imply global asymptotic stability? Delving deeply into this subject is beyond the scope of this book. But the interested reader may consult with references [21], [85]. Nevertheless, we are going to state without a proof one of the fundamental results addressing this question. We now define a class of maps, called population models which are amenable to the promised theorem.

DEFINITION 2.3 *A continuous map* $f : [0, \infty) \to [0, \infty)$ *is said to be a population model if*

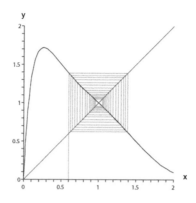

FIGURE 2.2
Locally but not globally asymptotical stable fixed point.

(i) $f(0) = 0$ and f has a unique positive fixed point x^*.

(ii) $f(x) > x$ if $0 < x < x^*$ and $f(x) < x$ if $x > x^*$.

(iii) If $f'(x_m) = 0$ and $x_m \leq x^*$, then $f'(x) > 0$ for $0 \leq x < x_m$ and $f'(x) < 0$ for $x > x_m$ and $f(x) > 0$.

THEOREM 2.2
The positive fixed point of a population model is globally asymptotically stable on $(0, \infty)$ if and only if it has no periodic 2-cycle.

2.3   Basin of Attraction of Periodic Orbits

We now extend the study in Section 2.2 to periodic orbits. Let $c_r = \{\overline{x}_1, \overline{x}_2, \ldots, \overline{x}_r\}$ be a periodic r-cycle. Then each point $\overline{x}_i \in c_r$ is a fixed point under f^r. Let $g = f^r$. Then the (immediate) basin of attraction $\{B(\overline{x}_i)\}W^s(\overline{x}_i)$ is defined, respectively, as in Definition 2.3 under the map g. This leads to the following definition.

DEFINITION 2.4   *The basin of attraction $W^s(c_r)$ is defined as*

$$W^s(c_r) = \bigcup_{i=1}^{r} W^s(\overline{x}_i)$$

and the immediate basin of attraction $B(c_r)$ is defined as

$$B(c_r) = \bigcup_{i=1}^{r} B(\overline{x}_i).$$

LEMMA 2.2
Let $c_r = \{\overline{x}_1, \overline{x}_2, \ldots, \overline{x}_r\}$ be a periodic r-cycle. Then for $i \neq j$, $W^s(\overline{x}_i) \cap W^s(\overline{x}_j) = \emptyset$.

PROOF   Let $y \in W^s(\overline{x}_i) \cap W^s(\overline{x}_j)$. Then

$$\lim_{n \to \infty} f^{nr}(y) = \overline{x}_i \quad \text{and} \quad \lim_{n \to \infty} f^{nr}(y) = \overline{x}_j.$$

By the uniqueness of the limit of a sequence, we conclude that $\overline{x}_i = \overline{x}_j$, a contradiction.   ∎

In the previous section we have seen examples of continuous maps whose fixed points are globally asymptotically stable. Population models constitute a class of maps where such phenomenon occur.

The burning question now is whether or not we can have a globally asymptotically stable periodic cycles (not fixed points). This question has been settled recently[1] by Elaydi and Yakubu [34].

THEOREM 2.3 (Elaydi-Yakubu)
Let $f : I \to \mathbb{R}$ be a continuous map on an interval I. Then f has no globally asymptotically stable periodic cycles. Assume that $c_r = \{\overline{x}_1, \overline{x}_2, \ldots, \overline{x}_r\}$ is a globally asymptitically stable periodic cycle.

PROOF   Let $g = f^r$. Then under g, $W^s(\overline{x}_i)$ is an open subset of I, for $1 \leq i \leq r$. Moreover, by Lemma 2.2, $W^s(\overline{x}_i) \cap W^s(\overline{x}_j) = \varnothing$ as in Lemma 2.2, $i \neq j$. If $W^s(c_r) = \bigcup_{i=1}^{r} W^s(\overline{x}_i)$, then I is the union of disjoint open sets, which violates the assumption that I is an interval.   ∎

REMARK 2.4   Theorem 2.3 says that only a fixed point of a continuous map can be globally asymptotically stable.   ∎

---

[1]The result was proved for general connected metric spaces. The proof presented here can be easily modified to obtain the general result. Notice that the interval I may be replaced by a connected metric space.

Exercises - (2.2 and 2.3)

1. Investigate the basin of attraction of the fixed points of the map

$$f(x) = \begin{cases} x^2 & \text{if } -3 \le x \le 1, \\ 4\sqrt{x} - 3 & \text{if } 1 < x \le 9. \end{cases}$$

2. Let $f(x) = |x - 1|$. Find $W^s(\frac{1}{2})$.

3. Suppose that $f : I \to I$ is a continuous and onto map on an interval I. Let $\bar{x}$ be an asymptotically stable periodic point of period $k \ge 2$. Show that $W^s(f(\bar{x})) = f(W^s(\bar{x}))$.

4. Describe the basin of attraction of all fixed and periodic points of the maps:

   (a) $f(x) = x^2$,

   (b) $g(x) = x^3$,

   (c) $h(x) = 2xe^{-x}$,

   (d) $q(x) = -\frac{4}{\pi} \arctan x$.

5. Investigate the basin of attraction of the fixed points for the map

$$f(x) = \begin{cases} \frac{x}{2} & \text{if } 0 \le x \le 0.2, \\ 3x - \frac{1}{2} & \text{if } 0.2 < x \le \frac{1}{2}, \\ 2 - 2x & \text{if } \frac{1}{2} < x \le 1. \end{cases}$$

6. Let f be a continuous map that has two periodic points x and y, $x \ne y$, with periods r and t, $r \ne t$, respectively. Prove that $W^s(x) \cap W^s(y) = \emptyset$.

7. Suppose that a set M is invariant under a one-to-one continuous map f. A point $x \in M$ is said to be an interior point if $(x - \delta, x + \delta) \subset M$ for some $\delta > 0$. Prove that the set of all interior points of M, denoted by $\text{int}(M)$, is invariant.

8. Let x^* be an attracting fixed point under a continuous map f. If the immediate basin of attraction $B(x^*) = (a, b)$, show that the set $\{a, b\}$ is invariant. Then conclude that there are only three scenarios in this case: (1) both a and b are fixed points, or (2) a or b is fixed and the other is an eventually fixed point, or (3) $\{a, b\}$ is a 2-cycle.

9.* Show that for Ricker map

$$R(x) = xe^{p-x}, \quad 0 < p < 2,$$
$$W^s(x^*) = (0, \infty), \quad \text{where } x^* = p.$$

10. Consider the Beverton-Holt map

$$f(p) = \frac{\mu K p}{K + (\mu - 1)p}, \quad K > 0.$$

(a) If $0 < \mu < 1$, find the basin of attraction of the fixed point $p_1^* - 0$.

(b) If $\mu > 1$, find the basin of attraction of the fixed point $p_2^* = K$.

11. Consider the map

$$g(x) = x\left(x - \frac{3}{2}\right)[-2 - (x - 1) - 6(x - 1)^2].$$

Show that the fixed point $x^* = 1$ is locally but not globally asymptotically stable.

12. Let f be a continuous function on $\mathbb{R}$ and let x^* be a fixed point of f. Let (y_1, y_2) be an interval containing x^* such that

$$x < f(x) < x^* \quad \text{for} \quad y_1 < x < x^*, \tag{2.1}$$
$$x^* < f(x) < x \quad \text{for} \quad x^* < x < y_2 \tag{2.2}$$

and f is increasing on the interval (y_1, y_2). Prove that $(y_1, y_2) \subset W^s(x^*)$.

13. (Term project). Consider the logistic map $F(x) = \mu x(1 - x)$ defined on $[0, 1]$, where $3 < \mu < 1 + \sqrt{6}$. Show that the basin attraction $W^s(c_2)$ of the periodic 2-cycle $W^S(c_2) = (0, 1)\backslash Efix_{x^*}(f)$, where $x^* = \frac{\mu - 1}{\mu}$.

14. (Term project). Consider the Ricker map $R(x) = xe^{p-x}$, defined on $[0, \infty)$, where $2 < p < 2.5264$. Show that the basin of attraction of the periodic 2-cycle c_2 is given by $W^S(c_2) = (0, \infty)\backslash EFix_{x^*}(f)$, where $x^* = p$.

15.* (Term project). Let c_r be the locally asymptotically stable periodic orbit of the logistic map with $r = 2^n$. What is $W^s(c_r)$? Prove your statement.

16.* Repeat Problem 2.3 for the Ricker map.

17.* Another proof of Elaydi-Yakubu Theorem. Use the following steps to prove Elaydi-Yakubu Theorem.

(i) **LEMMA 2.3**

Let $f : I \subset \mathbb{R} \to \mathbb{R}$ be a continuous map on an interval I. If there exists $x_0 \in I$ such that $\overline{O(x_0)}$ (closure of the orbit[2]) is bounded,

---

[2] $\overline{O(x_0)}$ is the union of the orbit $O(x_0)$ and the limit points of $O(x_0)$.

then f has a fixed point that is either a limit point of $O(x_0)$ or is located between points in $O(x_0)$.

(ii) Let $\{\overline{x}_1, \overline{x}_2\}$ be a globally asymptotically stable periodic 2-cycle. Then $O(\overline{x}_1) = \{\overline{x}_1, \overline{x}_2\}$ is compact and bounded. By the Lemma in (i), f must have a fixed point x^*, which is either one of the points $\overline{x}_1$, $\overline{x}_2$ or lies between them. This violates the assumption.

2.4 Singer's Theorem

We are still plagued with many unresolved issues concerning periodic attractors. The main question that we are going to address in this: How many periodic attractors can a continuous map possess? In 1978, David Singer [97] gave a satisfactory answer to this question for maps with negative Schwarzian derivations. Recall from Section 1.7 that the Schwarzian derivative Sf of a map f is defined by the formula

$$Sf(x) = \frac{f'''(x)}{f'(x)} - \frac{3}{2}\left(\frac{f''(x)}{f'(x)}\right)^2 \tag{2.3}$$

It turns out that many polynomials have negative Schwarzian derivatives as demonstrated by the following result.

LEMMA 2.4
Let $p(x)$ be a polynomial of degree n, $p(x) = a_0 x^n + a_1 x^{n-1} + \cdots + a_n$, such that all the roots of its derivative $p'(x)$ are distinct and real. Then $Sp(x) < 0$ ($-\infty$ is allowed).

PROOF Let $r_1, r_2, \ldots, r_{n-1}$ be the real and distinct roots of $p'(x)$. Then $p'(x) = a(x - r_1)(x - r_2)\ldots(x - r_{n-1})$, where $a = na_0$. Taking the natural logarithm of both sides yields

$$\ln p'(x) = \ln a + \sum_{i=1}^{n-1} \ln(x - r_i).$$

Differentiating with respect to x we obtain

$$\frac{p''(x)}{p'(x)} = \sum_{i=1}^{n-1} \frac{1}{(x - r_i)} \tag{2.4}$$

Differentiating one more time we get

$$\frac{p'(x)p'''(x) - (p''(x))^2}{(p'(x))^2} = \frac{p'''(x)}{p'(x)} - \left(\frac{p''(x)}{p'(x)}\right)^2$$

$$= -\sum_{i=1}^{n-1} \frac{1}{(x-r_i)^2} \tag{2.5}$$

Now using (2.4) and (2.5) yields

$$Sp(x) = \frac{p'''(x)}{p'(x)} - \left(\frac{p''(x)}{p'(x)}\right)^2 - \frac{1}{2}\left(\frac{p''(x)}{p'(x)}\right)^2$$

$$= -\sum_{i=1}^{n-1} \frac{1}{(x-r_i)^2} - \frac{1}{2}\left(\sum_{i=1}^{n-1} \frac{1}{(x-r_i)}\right)^2 < 0.$$

∎

The following lemma provides us with necessary conditions to identify maps that possess negative Schwarzian derivatives.

LEMMA 2.5
Assume that f is a C^3 map on $\mathbb{R}$ and $Sf(x) < 0$. Then the following statements hold true.

(i) *If u is a point at which $f'(x)$ has a local minimum, then $f'(u) \leq 0$, and if v is a point at which $f'(x)$ has a local maximum, then $f'(v) \geq 0$. In other words, $f'(x)$ cannot have a positive local minimum or a negative local maximum.*

(ii) *Let a_1, a_2, a_3 be three fixed points of f such that $a_1 < a_2 < a_3$. If $f'(a_2) \leq 1$, then f has a critical point in the interval (a_1, a_3), i.e., there exists a point $d \in (a_1, a_3)$ with $f'(d) = 0$.*

PROOF

(i) The proof of (i) is left to the reader as Problem 14.

(ii) Since $f(a_1) = a_1$ and $f(a_2) = a_2$, it follows by the Mean Value Theorem that for some $b_1 \in (a_1, a_2)$, $f'(b_1) = 1$. Similarly, there exists $b_2 \in (a_2, a_3)$ such that $f'(b_2) = 1$ (see Fig. 2.3). Furthermore, since f' is continuous on $[b_1, b_2]$ it must attain its minimum value at a point $c \in [b_1, b_2]$. If c is either b_1 or b_2, then $f'(a_2)$ is a local minimum of f'. Thus, without loss of generality, we may assume that $c \in (b_1, b_2)$, and $f'(c)$ is a local minimum of f'. Thus $f''(c) = 0$ and $f'''(c) > 0$. Now if $f'(c) = 0$, we are done as we have a critical point in the interval (a_1, a_3).

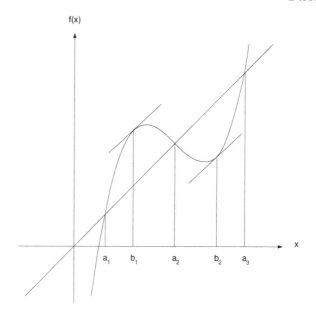

FIGURE 2.3
There exists a number $b_1 \in (a_1, a_2)$ with $f'(b_1) = 1$ and a number $b_2 \in (a_2, a_3)$ with $f'(b_2) = 1$. This implies the existence of a critical point d with $f'(d) = 0$.

On the other hand, if $f'(c) \neq 0$ we then have $0 > Sf(c) = \frac{f'''(c)}{f'(c)}$. Since $f'''(c) > 0$, we must have $f'(c) < 0$. And since $f'(b_1) = 1 > 0$, it follows by the Intermediate Value Theorem that $f'(d) = 0$ for some d between b_1 and c and the proof is now complete.

∎

The above lemma provides us with ways to eliminate maps that do not process negative Schwarzian derivative.

Example 2.4

(i) Consider the map $f(x) = \frac{1}{3}x^3 - \frac{1}{2}x^2 + x + \frac{1}{2}$. Then $f'(x) = x^2 - x + 1$ has no real roots. This is an indication that this polynomial may not have negative Schwarzian derivative in light of Lemma 2.4. But to be definite about it we invoke Lemma 2.5 (i). Notice that $f''(x) = 2x - 1 = 0$ implies that $x = \frac{1}{2}$ is a critical point of f'. And since $f'''(x) = 2 > 0$, f' has a local minimum at $x = \frac{1}{2}$ with the value $f'\left(\frac{1}{2}\right) = \frac{3}{4} > 0$. By Lemma 2.5 (i) we can affirm that the map f does not have a negative Schwarzian derivative (Figure 2.4)

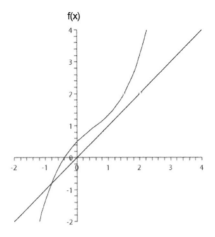

FIGURE 2.4
f cannot have a negative Schwarzian derivative.

(ii) The map $f(x) = \frac{3}{4}x + x^3$ cannot possess a negative Schwarzian derivative, it has three fixed points $-\frac{1}{2}, 0, \frac{1}{2}$ but f has no critical points in the
interval $\left(-\frac{1}{2}, \frac{1}{2}\right)$ (see Fig. 2.5).

$\square$

We need one more technical result before stating Singer's Theorem, the
product rule for the Schwarzian derivative.
Product Rule:

$$\boxed{\mathrm{S}(f \circ g)(x) = \mathrm{S}f(g(x)) \cdot (g'(x))^2 + \mathrm{S}g(x)} \tag{2.6}$$

The proof is left for the reader as Problem 7. As a consequence of the
product rule, we obtain the following result.

COROLLARY 2.1

(i) *If* $\mathrm{S}f < 0$ *and* $\mathrm{S}g < 0$, *then* $\mathrm{S}(f \circ g) < 0$.

(ii) *If* $\mathrm{S}f < 0$, *then* $\mathrm{S}f^k < 0$ *for all* $k \in \mathbb{Z}^+$.

PROOF

(i) Use the Product Rule (2.6); see Problem 13.

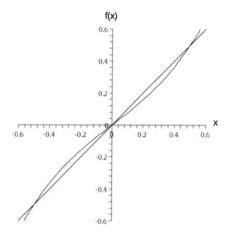

FIGURE 2.5

f cannot have a negative Schwarzian derivative.

(ii) Use mathematical induction (Problem 13)

THEOREM 2.4

Suppose that f is a C^3 map defined on the closed interval $I = [a, b]$, $a = -\infty$, $b = \infty$ are allowed, such that $Sf(x) < 0$ for all $x \in I$ ($Sf(x) = -\infty$ is allowed). If p is an attracting k-periodic point, then either

(i) *the immediate basin of attraction $W^s(p)$ extends to a or to b ($-\infty$ or $+\infty$), or*

(ii) *the immediate basin of attraction $W^s(p)$ contains a critical point x_c ($f'(x_c) = 0$).*

As an immediate consequence of this we get Singer's Theorem.

COROLLARY 2.2 (Singer, 1978)

Suppose that f is a C^3 map on a closed interval I such that $Sf(x) < 0$, for all $x \in I$. If f has n critical points in I, then for every $k \in \mathbb{Z}^+$, f has at most $(n + 2)$ attracting k-periodic orbits.

PROOF If there are n critical points, then by Theorem 2.4 (ii), there are n attracting k-periodic cycles for any given k. We may have two more additional attracting k-periodic cycle by Theorem 2.4 (i). Thus the maximum number of attracting periodic cycles is $n + 2$. ∎

PROOF OF THEOREM 2.4

Let $I = [a, b]$, where a may take the value of $-\infty$ and b may be ∞, and let p be an attracting k-periodic point of f. Then p is an attracting fixed point of the map $g = f^k$. Let $\mathcal{B}(p)$ be the immediate basin of attraction of p under g. If $p \in (a, b)$, then $\mathcal{B}(p)$ is of the form (c, d), $[a, c)$, or $(d, b]$. By Theorem 2.1, $g(\mathcal{B}(p)) \subset \mathcal{B}(p)$. Assume that $\mathcal{B}(p) = (c, d)$. Then g must map (c, d) into itself. However, g will not map the points c or d into the interior of interval (c, d) since neither c nor d is in $\mathcal{B}(p)$. Hence we have three cases to consider.

1. $g(c) = c$, and $g(d) = d$ (both fixed points of g)

2. $g(c) = d$, and $g(d) = c$ ($\{c, d\}$ is a 2-periodic orbit of g)

3. $g(c) = g(d) = c$ or d (one point is fixed and the other is eventually fixed points of g)

Case 1: $g(c) = c$, and $g(d) = d$.
Since p is an attraction fixed point of g, it follows that $-1 \le g'(p) \le 1$. By Corollary 2.1, $Sg < 0$. Hence by Lemma 2.5, for the fixed points p, $c < p < d$, g must have a critical point $\tilde{x}$ in the interval, (c, d), $g'(\tilde{x}) = 0$. But $0 = g'(\tilde{x}) = f'(f^{k-1}(\tilde{x})) \ldots f'(\tilde{x})$. Thus $f'(f^i(\tilde{x})) = 0$ for some i, $0 \le i \le k-1$. Since $f^i(\tilde{x}) \in (c, d)$, $\mathcal{B}(p)$ contains the critical point $f^i(\tilde{x})$ of f.
Case 2: $g(c) = d$, and $g(d) = c$.
This case reduces to Case 1 if one considers the map $h = g^2$. For then $h(c) = c$ and $h(d) = d$.
Case 3: $g(c) = g(d) = c$ or d.
By the Mean Value Theorem $g'(\tilde{x}) = 0$ for some $\tilde{x} \in (c, d)$. Then as in Case 1, one may show that this implies that f has a critical point in (c, d). ∎

Remarks about Singer's Theorem

1. The above proof shows that if the map has a negative Schwarzian derivative and the immediate basin of attraction $J = (c, d)$ of a periodic point p is bounded, then p must attract a critical point. This remark leads to the solution of the following two examples.

Example 2.5

The map $f(x) = 1 - 2x^2$ on $[-1, 1]$ has no attracting periodic points. To show this, note that 0 is the only critical point. But the orbit of 0 is $0, 1, -1, -1, -1, \ldots..$ Thus, -1 is a possible attracting fixed point of f. But, this is impossible because $f'(-1) = 4$ (Theorem 1.3). ∎

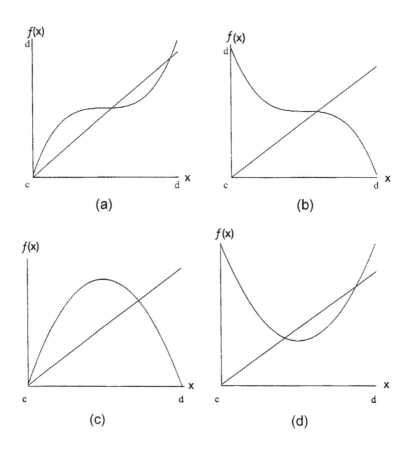

FIGURE 2.6
(a) $g(c) = c$, $g(d) = d$, (b) $g(c) = d$, $g(d) = c$, (c) $g(c) = g(d) = c$,
(d) $g(c) = g(d) = d$

Example 2.6

The logistic map $F_\mu(x) = \mu x(1-x), 0 < \mu \leq 4, x \in [0,1]$, has at most one attracting periodic cycle. As we saw in Chapter 1, for $0 < \mu \leq 1$, 0 is the only attracting fixed point where the region of attraction is [0,1]. For $1 < \mu < 4, F_\mu$ has only one critical point $\frac{1}{2}$. By Corollary 2.2, there are at most three attracting periodic cycles associated with intervals of the form $[0, c), (c, d)$, and $(d, 1]$ with $0 < c < d < 1$. Since $F'_\mu(0) = \mu > 1$, the fixed point 0 is unstable (Theorem 1.3); therefore, $[0, c)$ cannot be a basin of attraction. Furthermore, $F_\mu(1) = 0$ and hence $(d, 1]$ is not a basin of attraction either. We conclude that there is at most one attracting periodic cycle in (c, d) for a given $\mu \in (1, 4]$. Figure 2.7 is part of the bifurcation diagram of F_μ which shows a window with six horizontal curves. We may wonder whether this represents two attracting 3-periodic cycles, one attracting 6-periodic cycle, or another combination. The above example gives us the definite answer, namely, that this is an attracting 6-periodic cycle. Moreover, since there is only one attracting 3-periodic cycle, only one 3-periodic cycle will appear in the bifurcation diagram precisely at window 3 when $\mu = 1 + \sqrt{8}$ (see Fig. 2.8). ▯

2. If the map has a negative Schwarzian derivative and the basin of attraction for an attracting periodic cycle is unbounded or contains an end point, then this periodic cycle may not attract a critical point, as the following example shows.

Example 2.7

Let $G(x) = \mu\tan^{-1}(x), \mu \neq 0$. Then, $G'(x) = \frac{\mu}{1+x^2}$. Clearly $G(x)$ has no critical points. Now, if $|\mu| < 1$, then $x^* = 0$ is an asymptotically stable fixed point where the basin of attraction is $(-\infty, \infty)$ (Figure 2.10). While if $\mu > 1$, then G has two attracting fixed points x_1^* and x_2^* with basins of attraction of the form $(-\infty, 0)$ and $(0, \infty)$, respectively (see Fig. 2.9). Finally, if $\mu < -1$, then G has an attracting 2-cycle $\{\bar{x}_1, \bar{x}_2\}$ with a basin of attraction of the form $(-\infty, 0) \cup (0, \infty)$ (see Fig. 2.11). ▯

Exercise - (2.4)

1. Give an example of a polynomial that does not have a negative Schwarzian derivative.

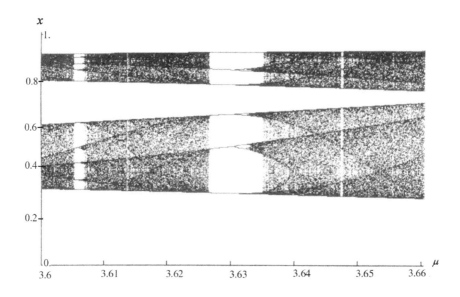

FIGURE 2.7
The appearance of a 6-cycle.

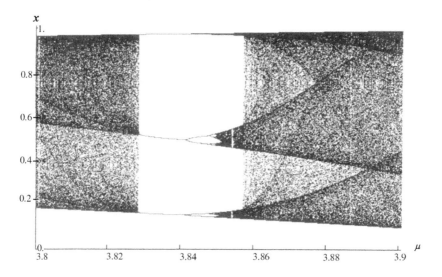

FIGURE 2.8
The appearance of a 3-periodic cycle.

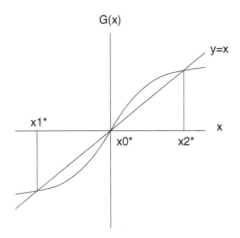

FIGURE 2.9
The map $G(x) = \mu \tan^{-1}(x)$ with $\mu > 1$, $W^s(x_1^*) = (-\infty, 0)$, $W^s(x_2^*) = (0, \infty)$.

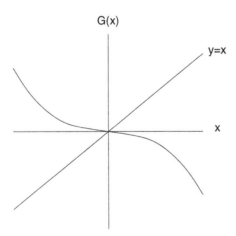

FIGURE 2.10
The map $G(x) = -\mu \tan^{-1}(x)$ with $\mu = \frac{1}{2}$. $x^* = 0$ is globally asymptotically stable.

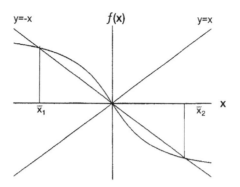

FIGURE 2.11

An attracting 2-cycle $\{\bar{x}_1, \bar{x}_2\}$ of the map G with basin of attraction $(-\infty, 0) \cup (0, \infty)$.

2. Give an example of a continuous function f that has an attracting fixed point but $Sf > 0$.

3. Sketch a graph of a continuous function that has two asymptotically stable fixed points.

4. Let f be a continuous map on $\mathbb{R}$. Prove that

    (a) if $f(b) = b$ and $x < f(x) < b$ for all $x \in [a, b)$, then $a \in W(b)$.

    (b) if $f(b) = b$ and $b < f(x) < x$ for all $x \in (b, c]$, then $c \in W(b)$.

5. Show that $Sf < 0$ for $f(x) = x \exp(r(1 - \frac{x}{k}))$, $r, k \in \mathbb{R}$.

6. Let $g(x) = \frac{ax+b}{cx+d}$ be a linear "fractional transformation." Show that $Sg(x) = 0$.

7. Prove Formula (2.6).

8. Consider the linear fractional transformation $g(x)$ defined in Problem 6 (a) shows that for any map $f \in c^3$, $S(g \circ f) = S(f)$.

9. Show that the logistic map F_μ has no attracting periodic points if $\mu > 2 + \sqrt{5}$.

10. Show that the logistic map $F_4(x) = 4x(1 - x)$ has no attracting periodic points.

11. Let $h(x) = \mu \sin x$ for $0 \le x \le \pi$. Determine the maximum possible number of attracting cycles of h for $0 < \mu < \pi$.

12. Let f be a C^3 map. Show that $Sf = 0$ if and only if $f(x) = (ax + b)/(cx + d)$ for some real numbers a, b, c, d.

13. Prove Corollary 2.1.

14. In the proof of Theorem 2.4, give a detailed proof of the existence of a critical point of f in Case 2 where $g(c) = d$ and $g(d) = c$.

2.5 Bifurcation

In this section, we resume our investigation of the bifurcation phenomena discussed in Sec. 1.9. But before embarking on such a task, we need to explain what bifurcation really means. Roughly speaking, the term bifurcation refers to the phenomenon of a system exhibiting new dynamical behavior as the parameter is varied. As we have seen in Chapter 1 (see Fig. 1.28), the logistic map $F_\mu(x) = \mu x(1 - x)$ undergoes a period doubling at an infinite sequence of values of the parameter $\mu : \mu_1, \mu_2, \mu_3, \ldots$, where $\mu_1 = 3, \mu_2 = 1 + \sqrt{6}, \ldots$. Note that for the fixed point $x^* = \frac{\mu-1}{\mu}, F'_{\mu_1} = -1$. Similarly, if $\{\overline{x}_1, \overline{x}_2\}$ is the 2-cycle of F_{μ_2} (or the fixed points of $F^2_{\mu_2}$), then $[F^2_{\mu_2}(\overline{x}_i)]' = -1$. This is a trademark of a period-doubling bifurcation that is always associated with the appearance of a slope of -1.

The logistic map F_μ undergoes another important type of bifurcation, commonly called **saddle node** or **tangent** bifurcation. This bifurcation is associated with the appearance of a slope of 1. Now, it can be shown that period 3 appears at $\tilde{\mu} = 1 + \sqrt{8} \approx 3.8284$ (see Saha and Strogatz [90] for an elementary derivation). Figure 2.12 depicts the graph of F^3_μ for $\mu < \tilde{\mu}(\mu = 3.75)$. Here, F^3_μ has only two fixed points that are fake 3-periodic cycles; they are fixed points of F_μ. Figure 2.13 shows how period 3 appears when $\mu = \tilde{\mu}$ at which $F^3_{\tilde{\mu}}$ touches the diagonal line at the 3-periodic cycle $\{\overline{x}_1, \overline{x}_2, \overline{x}_3\}$. Since $[F^3_{\tilde{\mu}}(\overline{x}_i)]' = 1$ and $[F^3_{\tilde{\mu}}(\overline{x}_i)]'' \neq 0$, it follows from Theorem 1.5 that this 3-periodic cycle is unstable (actually it is semiasymptotically stable from the left). Finally, Fig. 2.14 depicts the graph of F^3_μ for $\mu > \tilde{\mu}(\mu = 3.95)$ where each point in the 3-periodic cycle $\{\overline{x}_1, \overline{x}_2, \overline{x}_3\}$ gives rise to two new fixed points of F^3_μ, i.e., two new 3-periodic cycles. On one cycle (indicated by black dots in Fig. 2.14) the slope of F^3_μ is less than 1 in magnitude. Consequently, this cycle is attracting. However, the other cycle (indicated by open dots) is unstable since the slope of F^3_μ exceeds 1. By increasing μ on the attracting cycle, $(F^3_\mu)'$ decreases to -1, at which the 3-periodic cycle undergoes a double-period bifurcation and the appearance of $2^k \times 3$ cycles, $k = 1, 2, 3$, etc. (see Fig. 2.7). On the other hand, if we decrease μ (starting from $\tilde{\mu} = 1 + \sqrt{8}$), the 3-periodic cycle disappears (Fig. 2.12) and the system exhibits an interesting phenomenon known as **intermittency** (Pomeau and Mennaville [81]). Figure 2.15 shows the behavior of an orbit for F_μ with $\mu = 3.828$ where part of the orbit looks like a 3-periodic cycle. Then the orbit goes into an erratic

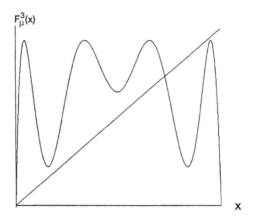

FIGURE 2.12

The disappearance of 3-periodic cycles when $\mu < \tilde{\mu} = 1 + \sqrt{8}$.

behavior to return again to the "ghostly" 3-periodic cycle. Such behavior will be repeated indefinitely and is always associated with the saddle node bifurcation.

Harrison and Biswas [46] gave an experimental example of intermittency in a laser. In Fig. 2.16, we plot the intensity of the emitted laser light as a function of time. The tilt of the mirror in the laser cavity (the parameter of the function) is then varied. As you see in the figure, the lowest panel demonstrates that the laser is pulsating periodically. Moving up in Fig. 2.16, we observe the appearance of ghostly 3-cycles with intermittent bouts of chaos.

The next example gives a detailed account of both types of bifurcation.

Example 2.8

Consider the family of maps $G_c(x) = c - x^2$, where c is scalar (Fig. 2.17). Then, for $c < -\frac{1}{4}$, there are no fixed points. At $c = -\frac{1}{4}$ we witness the appearance of the fixed point $x^* = -\frac{1}{2}$. Here we have a tangent bifurcation since $G_c'(-\frac{1}{2}) = 1$. To draw the bifurcation diagram $(c$-$x)$, we find the equation of the fixed points of G_c. This is obtained by putting $G_c(x) = x$. The equation is given by $c = x^2 + x$ $\left(\text{or } x = -\frac{1}{2} \pm \frac{1}{2}\sqrt{1 + 4c}\right)$. The upper branch in Fig. 2.18 is given by $x = -\frac{1}{2} + \frac{1}{2}\sqrt{1 + 4c}$ and the lower branch is given by $x = -\frac{1}{2} - \frac{1}{2}\sqrt{1 + 4c}$. At $c^* = -\frac{1}{4}$ and $x^* = -\frac{1}{2}$, $G_{c^*}'(x^*) = 1$, which indicates the appearance of a tangent bifurcation. The fixed point $x^* = -\frac{1}{2}$ is unstable (semistable from the right). Note that for (c, x) on the upper branch, $|G_c'(x)| < 1$ which implies that every fixed point on this branch is attracting. This state of affairs is indicated by a solid curve. On the other hand, the lower branch is a dashed curve since $|G_c'(x)| > 1$ and thus the fixed points are unstable.

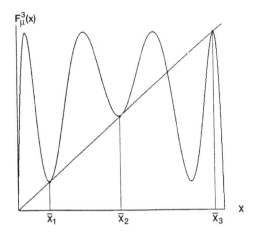

FIGURE 2.13

The appearance of a 3-periodic cycle when $\mu = \tilde{\mu} = 1 + \sqrt{8}$.

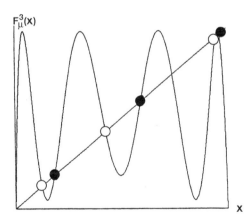

FIGURE 2.14

For $\mu = 3.9$, H_μ has two 3-periodic cycles. The black dots indicate the attracting 3-periodic cycle and the open dots indicate the unstable 3-perioidic cycle.

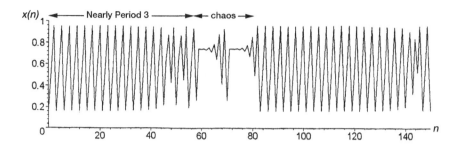

FIGURE 2.15

The orbit is nearly a 3-periodic cycle (ghostly 3-periodic cycle) for H_μ when $\mu = 3.828$.

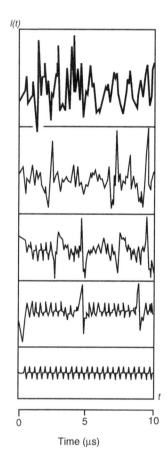

FIGURE 2.16

The intensity of the emitted laser light $I(t)$ as a function of time t.

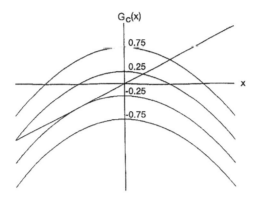

FIGURE 2.17
Graphs of parabolas $G_c(x) = c - x^2$ for various values of c.

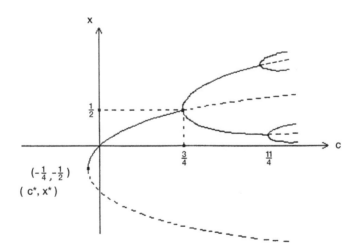

FIGURE 2.18
Bifurcation diagram for $G_c(x) = c - x^2$.

Things change dramatically at $\tilde{c} = \frac{3}{4}$. Here at the fixed point $\tilde{x} = \frac{1}{2}$, $G'_{\tilde{c}}(\tilde{x}) = -1$. This fixed point is attracting by Theorem 1.6. A 2-periodic cycle is born for $c > \frac{3}{4}$. To find the bifurcation equation of this 2-periodic cycle, we solve the equation $G_c^2(x) = x$ or $x^4 - 2cx^2 + x + c^2 - c = 0$. We factor out the bifurcation equation of the fixed points $x^2 + x - c = 0$ to obtain the bifurcation equation of the genuine 2-cycle: $c = x^2 - x + 1$ which has two branches $x_{1,2} = \frac{1}{2} \pm \sqrt{c - \frac{3}{4}}$. The 2-cycles $\{x_1, x_2\}$ are attracting for $\frac{3}{4} < c \leq \frac{11}{4}$.   ☐

We are now ready to formalize the above discussion. For notational convenience, a one-parameter family $H_\mu(x)$ may be written as a map $H(\mu, x)$ of two variables, i.e., $H(\mu, x) : \mathbb{R} \times \mathbb{R} \longrightarrow \mathbb{R}$. This is particularly convenient when we take partial derivatives with respect to the parameter μ.

THEOREM 2.5 (The Saddle-node Bifurcation)
Suppose that $H_\mu(x) \equiv H(\mu, x)$ is a C^2 one-parameter family of one-dimensional maps (i.e., both $\frac{\partial^2 H}{\partial x^2}$ and $\frac{\partial^2 H}{\partial \mu^2}$ exist and are continuous), and x^* is a fixed point of H_{μ^*}, with $H'_{\mu^*}(x^*) = 1$. Assume further that

$$A = \frac{\partial H}{\partial \mu}(\mu^*, x^*) \neq 0 \quad and \quad B = \frac{\partial^2 H}{\partial x^2}(\mu^*, x^*) \neq 0.$$

Then there exists an interval I around x^* and a C^2 map $\mu = p(x)$, where $p : I \longrightarrow \mathbb{R}$ such that $p(x^*) = \mu^*$, and $H_{p(x)}(x) = x$. Moreover, if $AB < 0$, the fixed points exist for $\mu > \mu^*$, and, if $AB > 0$, the fixed points exist for $\mu < \mu^*$.

REMARK 2.5   The conclusion of the preceding theorem states that two curves of fixed points given by $\mu = p(x)$ emanate from the fixed point (μ^*, x^*). Furthermore, the sign of AB determines the direction of the bifurcation; if $AB < 0$, then we have Fig. 2.19 and for $AB > 0$ we have Fig. 2.20.   ∎

To prove the above theorem, we need a version of the implicit function theorem (Protter and Morrey [82]) which we state here without proof.

THEOREM 2.6 (The Implicit Function Theorem)
Suppose that $G : \mathbb{R} \times \mathbb{R} \longrightarrow \mathbb{R}$ is a C^1 map in both variables such that for some $(\mu^*, x^*) \in \mathbb{R} \times \mathbb{R}, G(\mu^*, x^*) = 0$ and $\frac{\partial G}{\partial \mu}(\mu^*, x^*) \neq 0$. Then, there exists an open interval J around μ^*, an open interval I around x^*, and a C^1 map $\mu = p(x)$, where $p : I \longrightarrow J$ such that

1. $p(x^*) = \mu^*$.

2. $G(p(x), x) = 0$, for all $x \in I$.

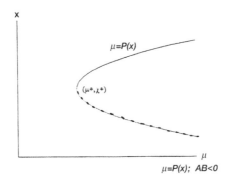

FIGURE 2.19
Saddle-node bifurcation for $F_\mu : \mu = p(x)$ when $AB < 0$.

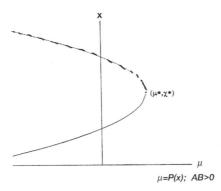

FIGURE 2.20
Saddle-node bifurcation for $F_\mu : \mu = p(x)$ when $AB > 0$.

PROOF OF THEOREM 2.5

Let $G(\mu, x) = H(\mu, x) - x$. Then it is easy to see that the map G satisfies the hypothesis of the implicit function theorem stated above. Hence, there exists a C^1 map $\mu = p(x)$ with $\mu^* = p(x^*)$, and $G(p(x), x) = 0$. Thus, for all x in some interval I,

$$H(p(x), x) = x. \tag{2.7}$$

Differentiating both sides of Equation (2.7) with respect to x and using the chain rule, we obtain

$$\frac{\partial H}{\partial \mu}(\mu^*, x^*)p'(x^*) + \frac{\partial H}{\partial x}(\mu^*, x^*) = 1. \tag{2.8}$$

Since $\frac{\partial H}{\partial x}(\mu^*, x^*) = 1$, $p'(x^*) = 0$. Thus, x^* is a critical point of the curve $\mu = p(x)$ as may be seen in Fig. 2.19.

Differentiating (2.8) one more time with respect to x yields

$$\frac{\partial^2 H}{\partial \mu^2}(\mu^*, x^*)[p'(x^*)]^2 + \frac{\partial H}{\partial \mu}(\mu^*, x^*)p''(x^*) + \frac{\partial^2 H}{\partial x^2}(\mu^*, x^*) = 0$$

Since $p'(x^*) = 0$, we have

$$p''(x^*) = -\frac{\partial^2 H(\mu^*, x^*)/\partial x^2}{\frac{\partial H}{\partial \mu}(\mu^*, x^*)} = -\frac{B}{A}. \qquad (2.9)$$

It follows from Formula (2.9) that if $AB < 0$, then $p''(x^*) > 0$ and the curve $p(x)$ is concave upward at $x = x^*$. Hence, the curve $p(x)$ opens to the right. The direction is reversed if $AB > 0$ (see Fig. 2.20). ∎

REMARK 2.6   Two types of bifurcation appear when $\frac{\partial H}{\partial x}(\mu^*, x^*) = 1$, but $\frac{\partial H}{\partial \mu}(\mu^*, x^*) = 0$. The first type is called **transcritical bifurcation**, which appears at $\mu^* = 1, x^* = 0$ for the logistic map $F_\mu(x) = \mu x(1 - x)$. In this bifurcation, two branches of fixed points, one attracting ($x = 0$) and another unstable ($x = \frac{\mu-1}{\mu}$), meet when $\mu = 1$. Beyond $\mu = 1$, the first branch $x = 0$ becomes unstable and the other branch becomes an attractor. In other words, exchange of stability occurs at $\mu = 1$ (see Fig. 2.21). Note that $\frac{\partial^2 F}{\partial x^2}(1, 0) = -2 \neq 0$. ∎

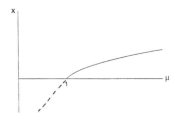

FIGURE 2.21
Exchange of stability at $\mu = 1$.

The map $H_\mu(x) = \mu x - x^3$ exhibits another type of bifurcation called **pitchfork bifurcation** at $x^* = 0$ and $\mu^* = 1$. This bifurcation, however, is different from the transcritical bifurcation because $\frac{\partial^2 H}{\partial x^2}(1, 0) = 0$. As shown in Fig. 2.22, for $0 < \mu \leq 1$ we have only one branch of attracting fixed points: $x = 0$. Beyond $\mu = 1$, this fixed point loses its stability and two new branches of attracting fixed points $x = \pm\sqrt{\mu - 1}$ appear.

As was pointed out in Alligood, et al. [2], these two bifurcations are not essential (generic) in the sense that any translation of the maps gives rise to the

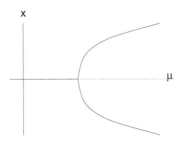

FIGURE 2.22
Pitchfork Bifurcation.

saddle node bifurcation. In other words, the maps $F_\mu(x) = \mu x(1-x) + \alpha, \alpha \neq 0$ and $H_\mu(x) = \mu x - x^3 + \alpha, \alpha \neq 0$ exhibit a saddle node bifurcation (see Problems 6 and 7).

Next we introduce to you the mathematical foundation of the familiar period-doubling bifurcation.

THEOREM 2.7 (Period-Doubling Bifurcation)
Suppose that

1. $H_\mu(x^*) = x^*$ *for all* μ *in an interval around* μ^*.

2. $H'_{\mu^*}(x^*) = -1$.

3. $\frac{\partial^2 H^2}{\partial \mu \, \partial x}(\mu^*, x^*) \neq 0$.

Then, there is an interval I *about* x^* *and a function* $p : I \longrightarrow \mathbb{R}$ *such that* $H_{p(x)}(x) \neq x$ *but* $H^2_{p(x)}(x) = x$.

PROOF Let $G(\mu, x) = H^2_\mu(x) - x$. Then, from Assumption (1),

$$\frac{\partial G}{\partial \mu}(\mu^*, x^*) = \left[\frac{\partial H}{\partial \mu}(\mu^*, x^*) \right]^2.$$

If $\frac{\partial G}{\partial \mu}(\mu^*, x^*) \neq 0$ and we can apply the implicit function theorem (Theorem 2.6) to obtain the desired result. On the other hand, if $\frac{\partial G}{\partial \mu}(\mu^*, x^*) = 0$ then it is not possible to apply the implicit function theorem. To rectify the situation, we introduce another function $B(\mu, x)$ defined as follows:

$$B(\mu, x) = \begin{cases} G(\mu, x)/(x - x^*) & \text{if } x \neq x^*, \\ \frac{\partial G}{\partial x}(\mu^*, x^*) & \text{if } x = x^*. \end{cases} \quad (2.10)$$

TABLE 2.1
Types of bifurcation of fixed points.

Bifurcation	Canonical Example	Noncanonical Example	$\frac{\partial H}{\partial x}$ (μ^*, x^*)	$\frac{\partial H}{\partial \mu}$ (μ^*, x^*)	$\frac{\partial^2 H}{\partial x^2}$ (μ^*, x^*)
Saddle-node (fold, tangent)	$H_\mu(x) = x + \mu \mp x^2$ $\mu^* = 0,\ x^* = 0$	$H_\mu(x) = \mu - x^2$ $\mu^* = -\frac{1}{4},\ x^* = -\frac{1}{2}$	1	$\neq 0$	$\neq 0$
Pitchfork	$H_\mu(x) = x + \mu \mp x^3$ $\mu^* = 0,\ x^* = 0$	$H_\mu(x) = \mu x - x^3$ $\mu^* = 1,\ x^* = 0$	1	0	0
Transcritical	$H_\mu(x) = x + \mu x \mp x^2$ $\mu^* = 0,\ x^* = 0$	$H_\mu(x) = \mu x(1 - x)$ $\mu^* = 1,\ x^* = 0$	1	0	$\neq 0$
Period-doubling (flip)	$H_\mu(x) = -x - \mu + x^3$ $\mu^* = 0,\ x^* = 0$	$H_\mu(x) = \mu - x^2$ $\mu^* = \frac{3}{4},\ x^* = \frac{1}{2}$ $H_\mu(x) = \mu x(1 - x)$ $\mu^* = 3,\ x^* = \frac{2}{3}$	-1	$\neq 0$	$\neq 0$

We first observe that

$$B(\mu^*, x^*) = \frac{\partial G}{\partial x}(\mu^*, x^*)$$
$$= [H'_{\mu^*}(x^*)]^2 - 1$$
$$= 0.$$

Moreover,

$$\frac{\partial B}{\partial \mu}(\mu^*, x^*) = \frac{\partial}{\partial \mu}\left(\frac{\partial G}{\partial x}(\mu^*, x^*)\right)$$
$$= \frac{\partial}{\partial \mu}\left(\frac{\partial}{\partial x}H^2(\mu^*, x^*) - 1\right)$$
$$= \frac{\partial^2}{\partial \mu \partial x}H^2(\mu^*, x^*)$$
$$\neq 0. \quad \blacksquare$$

Hence, by the implicit function theorem there is a C^1 map $p(x)$ defined on an interval around x^* such that $p(x^*) = \mu^*$ and $B(p(x), x) = 0$. Hence, $G(p(x), x)/(x - x^*) = 0, x \neq x^*$. Consequently, $H^2_{p(x)}(x) = x$ and thus x is of period 2 for $\mu = p(x)$. It is left to you in Problem 8 to verify that

$$p'(x^*) = 0. \tag{2.11}$$

In conclusion, we summarize our findings as in Table 2.1.

Example 2.9 [86]
  Consider $f_\mu(x) = -\mu x + ax^2 + bx^3,\ a, b > 0$. The map has the fixed point 0 since $f_\mu(0) = 0$. Notice that $f'_\mu(x) = \frac{\partial f_\mu(x)}{\partial x} = -\mu + 2ax + 3bx^2$ and $f'_\mu(0) = -\mu$. Since $f'_1(0) = -1$, the period-doubling bifurcation is present. To find the 2-periodic orbit, we need to solve the equation $f^2_\mu(x) = x$. Now

$$f^2_\mu(x) = \mu^2 x + (-a\mu + a\mu^2)x^2 + (-b\mu - 2a^2\mu - b\mu^3)x^3 + O(x^4)$$

where $O(x^4)$ consists of terms of degree 4 or higher. Since $f_\mu^2(x) - x = 0$ has 0 as a root, we need to factor out x from this equation to obtain

$$Q(\mu, x) = \frac{f_\mu^2(x) - x}{x}$$
$$= \mu^2 - 1 + (-a\mu + a\mu^2)x + (-b\mu - 2a^2\mu - b\mu^3)x^2 + O(x^3).$$

Notice that

$$Q(0, 1) = 0$$
$$\frac{\partial Q}{\partial x}(0, 1) = 0$$
$$\frac{\partial Q}{\partial \mu}(0, 1) = 2 \neq 0$$
$$\frac{\partial^2 Q}{\partial x^2}(0, 1) = -4(b + a^2).$$

Let $\mu = p(x)$ as in Theorem 2.1 and consider

$$Q(p(x), x) = \mu^2 - 1 + (-ap(x) + ap^2(x))x$$
$$+ (-bp(x) - 2a^2p(x) - bp^3(x))x^2 + O(x^3) = 0.$$

Differentiating gives[3]

$$\frac{dQ}{dx} = \frac{\partial Q}{\partial x}(p(x), x) + \frac{\partial Q}{\partial \mu}(p(x), x)p'(x) = 0$$
$$\left.\frac{dQ}{dx}\right|_{(1,0)} = 0 = 0 + 2p'(0).$$

Hence $p'(0) = 0$.

Differentiating again we get

$$0 = \frac{d^2Q}{dx^2}(p(x), x) + \frac{\partial^2 Q}{\partial x \partial \mu}(p(x), x)p'(x) + \frac{\partial^2 Q}{\partial x \partial \mu}(p(x), x)p'(x)$$
$$+ \frac{\partial^2 Q}{\partial \mu^2}(p(x), x)[p'(x)]^2 + \frac{\partial Q}{\partial \mu}(p(x), x)p''(x)$$

and at $x = 0$ and $\mu = 1$, we get

$$0 = -4(b + a^2) + 2p''(0)$$
$$p''(0) = -2(b + a^2).$$

---

[3] $\frac{dQ}{dx} = \frac{\partial Q}{\partial x} + \frac{\partial Q}{\partial \mu}\frac{\partial \mu}{\partial x}$

Expanding $\mu = p(x)$ around $x = 0$ using Taylor's Theorem, yields

$$\mu = p(x) = p(0) + p'(0)x + \frac{p''(0)x^2}{2} + O(x^3)$$

$$= 1 - \frac{(2(b + a^2)x^2}{2} + O(x^3).$$

If $b + a^2 \neq 0$, the 2-periodic orbit appears either for $\mu < 1$ or $\mu > 1$ as the quadratic term dominates $O(x^3)$ near $x = 0$. To find the stability of the periodic orbit we need the find the derivative of f^2 at $(0, 1)$.

By Taylor's Theorem, for multivariables, we get

$$\frac{\partial(f^2)}{\partial x}(p(x), x) = \frac{\partial(f^2)}{\partial x}(1, 0) + \frac{\partial^2(f^2)}{\partial x^2}(1, 0)x$$

$$+ \frac{\partial^2(f^2)}{\partial \mu \partial x}(0, 1)(p(x) - 1) + \frac{1}{2}\frac{\partial^3(f^2)}{\partial x^3}(1, 0)x^2 + \dots$$

$$= 1 + O(x) + 2(b + a^2)x^2 - 6(b + a^2)x^2 + O(x^3)$$

$$= 1 - 4(b + a^2)x^2 + O(x^3). \tag{2.12}$$

Now, if $b + a^2 > 0$, the 2-periodic orbit is asymptotically stable, and if $b + a^2 < 0$, the 2-periodic orbit is unstable. ▯

The reader is asked to verify the computation above of the four terms in the expression of $\frac{\partial(f^2)}{\partial x}(p(x), x)$ as Problem 14.

Exercises - (2.5)

1. Show that the map $H_\mu(x) = -x - \mu x + x^3$, undergoes a period-doubling bifurcation at $\mu^* = 0$. Draw the bifurcation diagram for the map.

2. (Ricker map). Consider the population model $x(n+1) = R_\mu(x(n))$ where $R_\mu(x) = \mu x e^{-x}$, where $x(n)$ is the population density in year n and $\mu > 0$ is the growth rate. Show that this map undergoes a period-doubling bifurcation at $\mu^* = e^2$. Draw the bifurcation diagram for this map and interpret your results.

3. Show that the map $H_\mu(x) = \mu + x + x^2$ exhibits a Saddle-node bifurcation at $\mu^* = 0$ and draw its bifurcation diagram.

4. Show that the map $G_\mu(x) = -(1 + \mu)x + x^3$ undergoes a period-doubling bifurcation at $\mu^* = 0$. Draw its bifurcation diagram.

5. Show that the map $H_\mu(x) = \mu x(1 - x) + \alpha, \alpha \neq 0$ exhibits a Saddle-node bifurcation and sketch its bifurcation diagram.

6. Show that the map $H_\mu(x) = \mu x - x^3 + \alpha$, $\alpha \neq 0$ undergoes a Saddle-node bifurcation and sketch its bifurcation diagram.

7. Prove Formula (2.11).

8. Determine the saddle node and period-doubling bifurcation for the map $H_\mu(x) = \mu \sin x$, for $|x| < \pi$ and $0 < \mu \leq \frac{\pi}{2}$.

9. Consider the map B defined by (2.10). Show that

   (a) $\frac{\partial B}{\partial x}(\mu^*, x^*) = \frac{1}{2} \frac{\partial^2 G}{\partial x^2}(\mu^*, x^*)$.

   (b) $\frac{\partial^2 B}{\partial x^2}(\mu^*, x^*) = \frac{1}{3} \frac{\partial^3 G}{\partial x^3}(\mu^*, x^*)$.

   (Hint: Use L'Hopital's Rule.)

10. Suppose that in addition to the hypothesis of Theorem 2.7 that $SH_{\mu^*}(x^*) \neq 0$, where S denotes the Schwarzian derivative. Show that the curve $\mu = p(x)$ satisfies $p''(x^*) \neq 0$.

    [Hint: Show that $p''(x) = -\frac{2}{3}SH_{\mu^*}(x^*)$.]

11.* Show that the period 3 window for the logistic map $F_\mu(x) = \mu x(1-x)$ starts at $\mu = 1 + \sqrt{8}$.

12. Show that Assumption (3) in Theorem 2.7 can be replaced by the assumption $\frac{\partial^2 H}{\partial \mu \partial x}(\mu^*, x^*) \neq 0$.

13. Verify the computations in (2.12).

14.* (Term project) Let $f_\mu(x) = \mu x - x^3$.

    (a) Carry out the bifurcation analysis as in Example 2.9.

    (b) Draw the bifurcation diagram.

    (c) Write an essay about the dynamics of the map and discuss the bifurcation diagram.

15.* (Term project: Reverse period-doubling) Do all one dimensional maps with a single maximum lead to a period-doubling route to chaos? The following map, called the Gaussian map, provides a negative answer. Let $G(x) = e^{-bx^2} + c$.

    (a) Let $b = 7.5$. Plot the graph of G for $c = -0.9$, and -0.3. Find the fixed points.

    (b) Plot the bifurcation diagram with $b = 7.5$ and c ranges from -1 to 1, with $x_0 = 0$. Carry out a qualitative analysis of this diagram and summarize your findings in an essay.

    (c) Repeat (b) but for $x_0 = 0.7$.

    (d) Let $b = 4$, $x_0 = 0$. Plot the bifurcation diagram of G for c between -1, 1. Write a quantitative analysis of the diagram.

2.6 Sharkovsky's Theorem

Aleksandr Nikolayevich Sharkovsky (1936 -)

In 1964, A. N. Sharkovsky discovered one of the most remarkable theorems (named after him) for continuous maps on the real line. Published in Russian, his theorem remained totally unknown in the West until the publication of the famous Li and York "Period three implies Chaos" ten years later. And in 1996, the Converse of Sharkovsky's theorem was publicized widely in the American Mathematical Monthly by the author of this book.

Sharkovsky's Theorem created a new field of Mathematics called Iteration Theory. His recent research is focused on turbulence, and boundary value problems. He is currently at the Institute of Mathematics, National Academy of Ukraine and a consulting editor of the Journal of Difference Equations and Applications.

2.6.1 Li-Yorke Theorem

In 1975, Li and Yorke [62] published the article, "Period three implies chaos" in the *American Mathematical Monthly*. In this paper, they proved that if a continuous map f has a point of period 3, then it must have points of any period k. Soon afterward, it was found that Li-Yorke's theorem is only a special case of a remarkable theorem published in 1964 by the Ukranian mathematician Alexander Nikolaevich Sharkovsky [93]. Sharkovsky introduced a new ordering $\triangleright$ of the positive integers in which 3 appears first. He proved that if $k \triangleright r$ and f has a k-periodic point, then it must have an r-periodic point. This clearly implies Li-Yorke's theorem. However, to their credit, Li and Yorke were the first to coin the word "chaos" and introduce it to mathematics.

It is worth mentioning that neither Li-Yorke's theorem nor Sharkovsky's theorem is intuitive. To illustrate this point, recall from Example 1.14 that the tent map $T(x) = 1 - 2|x - \frac{1}{2}|$ has two cycles of period 3 : $\{\frac{2}{7}, \frac{4}{7}, \frac{6}{7}\}$ and $\{\frac{2}{9}, \frac{4}{9}, \frac{8}{9}\}$.

Is it intuitively clear that the tent map has cycles of all periods? I do not think so.

Let us now turn our attention to Sharkovsky's ordering of the positive integers. This ordering is defined as follows:

$$3 \rhd 5 \rhd 7 \rhd \ldots \qquad\qquad 2 \times 3 \rhd 2 \times 5 \rhd 2 \times 7 \rhd \ldots$$
$$\text{odd integers} \qquad\qquad 2\times \text{ odd integers}$$

$$2^2 \times 3 \rhd 2^2 \times 5 \rhd 2^2 \times 7 \rhd \ldots \quad 2^n \times 3 \rhd 2^n \times 5 \rhd 2^n \times 7 \rhd \ldots$$
$$2^2 \times \text{ odd integers} \qquad\qquad 2^n \times \text{ odd integers}$$

$$\ldots \quad \ldots \quad \ldots \qquad\qquad \ldots \rhd 2^n \rhd \ldots \rhd 2^3 \rhd 2^2 \rhd 2 \rhd 1$$
$$\text{powers of 2.}$$

We first list all the odd integers except 1, then 2 times the odd integers, 2^2 times the odd integers, and, in general, 2^n times the odd integers for all $n \in \mathbb{Z}^+$. This is followed by powers of 2 in a descending order. It is easy to see that this ordering exhausts all of the positive integers. Notice that $m \rhd n$ signifies that m appears before n in the Sharkovsky's ordering.

THEOREM 2.8 (Sharkovsky's Theorem)

Let $f : I \to I$ be a continuous map on the interval I, where I may be finite, infinite, or the whole real line.

If f has a periodic point of period k, then it has a periodic point of period r for all r with $k \rhd r$.

PROOF　　See the Appendix at the end of this chapter. Proof of the theorem may also be found in Block and Coppel [12].

We will now make a few comments about the theorem and then give a proof of a consequence of it: the Li-Yorke theorem.

1. The only way that a continuous map f has finitely many periodic points is if f has only periods that are powers of 2. Otherwise, it has infinitely many periodic points. For example, if f has a periodic point of period $2^{10} \times 5$, then it has infinitely many periodic points of periods

$$2^{10} \times 5, 2^{10} \times 7, 2^{10} \times 9, \ldots 2^{11} \times 3, 2^{11} \times 5, 2^{11} \times 7, \ldots$$
$$\ldots 2^n, 2^{n-1}, \ldots, 2^2, 2, 1.$$

2. If $m \rhd n$, then there are continuous maps with periodic points of period n but not of period m (see the proof of Theorem 2.10).

3. Sharkovsky's theorem does not extend to two or higher dimensional Euclidean spaces. It is not even true for the unit circle S^1. For example, the map $f : S^1 \to S^1$ defined by $f(e^{i\theta}) = e^{i(\theta + \frac{2\pi}{3})}$ is of period 3 at all points in S^1, but f has no other periods. ∎

Now we go back and prove the Li-Yorke theorem.

THEOREM 2.9 (Li and Yorke)

Let $f : I \to I$ be a continuous map on an interval I. If there is a periodic point in I of period 3, then for every $k = 1, 2, \ldots$ there is a periodic point in I having period k.

To prove this theorem, we need some preliminary results.

LEMMA 2.6

Let $f : I \to R$ be continuous, where I is an interval. For any closed interval $J \subset f(I)$, there is a closed interval $Q \subset I$ such that $f(Q) = J$.

PROOF   Let $J = [f(p), f(q)]$, where $p, q \in I$. If $p < q$, let r be the largest number in $[p, q]$ with $f(r) = f(p)$ and let s be the smallest number in $[p, q]$ such that $f(s) = f(q)$ and $s > r$. We claim that $f([r, s]) = J$. We observe that by the intermediate value theorem, we have $f([r, s]) \supset J$. Assume that there exists t with $r < t < s$ such that $f(t) \notin J$. Without loss of generality, suppose that $f(t) > f(q)$. Applying the intermediate value theorem again yields $f([r, t]) \supset J$. Hence, there is $x \in [r, t)$ such that $f(x) = f(q)$, which contradicts our assumption that s is the smallest number in $[p, q]$ with $f(p) = f(q)$. The case where $p > q$ is similar. The proof is now complete. ∎

LEMMA 2.7

Let $f : I \to I$ be continuous and let $\{I_n\}_{n=0}^{\infty}$ be a sequence of closed and bounded intervals with $I_n \subset I$ and $I_{n+1} \subset f(I_n)$ for all $n \in \mathbb{Z}^+$. Then, there is a sequence of closed and bounded intervals Q_n such that $Q_{n+1} \subset Q_n \subset I_0$ and $f^n(Q_n) = I_n$ for $n \in \mathbb{Z}^+$.

PROOF   Define $Q_0 = I_0$. Then, $f^0(Q_0) = I_0$. If Q_{n-1} has been defined so that $f^{n-1}(Q_{n-1}) = I_{n-1}$, then $I_n \subset f(I_{n-1}) = f^n(Q_{n-1})$. By applying Lemma 2.6 on f^n, there is a closed bounded interval $Q_n \subset Q_{n-1}$ such that $f^n(Q_n) = I_n$. ∎

We are now well prepared to give the proof of Theorem 2.9.

PROOF OF THEOREM 2.9

Suppose that f has a 3-cycle $\{x, f(x), f^2(x)\}$. Then one may rename the elements of the cycle so that it will become $\{a, b = f(a), c = f(b)\}$ with either $a < b < c$ or $a > b > c$. For example, if $x < f^2(x) < f(x)$, we let $a = f(x), b = f(a), c = f^2(a)$ and thus we have $a > b > c$. Let us assume that $a < b < c$. Write $J = [a, b], L = [b, c]$. For any positive integer $k > 1$, let $\{I_n\}$ be a sequence of intervals with $I_n = L$ for $n = 0, 1, \ldots, k - 2$ and $I_{k-1} = J$, and define I_n to be periodic inductively, $I_{n+k} = I_n$ for $n \in \mathbb{Z}^+$. The sequence $\{I_n\}$ looks like

$$L, L, \ldots, L, J, \ L, L, \ldots, L, J, \ L, L, \ldots, L, J, \ldots$$
$$(k-1) \text{ times} \quad (k-1) \text{ times} \quad (k-1) \text{ times.}$$

If $k = 1$, let $I_n = L$ for all $n \in \mathbb{Z}^+$. Since $f(a) = b$, $f(b) = c$, and $f(c) = a$, it follows by the intermediate value theorem that $L, J \subset f(L)$ and $L \subset f(J)$ (see Fig. 2.23).

Hence, one may apply Lemma 2.7 to produce a sequence $\{Q_n\}$ of closed, bounded intervals with $Q_k \subset Q_0 = L$ and $f^k(Q_k) = I_k = L$. Consequently, $L \subset f^k(L)$. By applying Theorem 1.2 to f^k, we conclude that f^k has a fixed point in L and, consequently, f has a k-periodic point in I. ∎

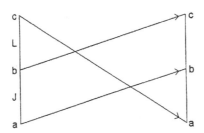

FIGURE 2.23
$f(a) = b$, $f(b) = c$, $f(c) = a$. Hence $J \subset f(L)$ and $L \subset f(J)$.

The next example [100] illustrates how the presence of period 3 for a continuous map on $\mathbb{R}$ implies the existence of periodic points of all periods.

Example 2.10
Contemplate the function (Fig. 2.24)

$$f(x) = \begin{cases} x + \frac{1}{2} & \text{for } 0 \le x < \frac{1}{2}, \\ 2 - 2x & \text{for } \frac{1}{2} \le x \le 1. \end{cases}$$

The map f has the periodic orbit $\{0, \frac{1}{2}, 1\}$. Moreover, f has the fixed point $\frac{2}{3}$. Let us write every $x \in [0, 1]$ in its binary expansion,

$$x = \sum_{j=1}^{\infty} \frac{a_j}{2^j} = 0.a_1 a_2 a_3 \ldots,$$

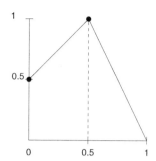

FIGURE 2.24

where each a_j is either 0 or 1. Now for $0 \le x < \frac{1}{2}$, $a_1 = 0$ and thus

$$f(x) = \frac{1}{2} + 0.0a_2a_3 \cdots = 0.1a_2a_3\ldots.$$

On the other hand if $\frac{1}{2} \le x \le 1$, $a_1 = 1$ and hence

$$f(x) = 2 - 2(0.1a_2a_3\ldots) = 2 - 1.a_2a_3a_4\ldots$$
$$= 1 - 0.a_2a_3a_4\ldots$$

$$\boxed{f(x) = 0.a_2'a_3'a_4'\ldots \quad \text{where } a_j' = 1 - a_j.}$$

One may show that

$$f^2(0.10a_3a_4\ldots) = 0.a_3a_4a_5\ldots \tag{2.13}$$

and

$$f^2(0.0a_2a_3\ldots) = 0.a_2'a_3'a_4'\ldots. \tag{2.14}$$

To find a point of period 2 we use (2.14) which implies that $a_2' = 1, a_3' = a_2, a_4' = a_3,\ldots$. Thus the point $0.\overline{01} = \frac{1}{3}$ is a point of period 2 and the periodic 2-cycle is $\left\{ \frac{1}{3}, \frac{5}{6} \right\}$.

To find a point of period 4, we let $a_2' = 1$ and $a_3' = 0$ in (2.13). Then

$$f^4(0.001a_4a_5\ldots) = f^2(0.10a_4'a_5'\ldots) = 0.a_4'a_5'a_6'\ldots.$$

Therefore $0.001a_4a_5\ldots$ is of period 4 if $a_4' = 0$, $a_5' = 0$, $a_6' = 1$. Hence the point $x = 0.\overline{001110} = \frac{2}{9}$ (Why?) and the periodic 4-cycle is given by $\left\{ \frac{2}{9}, \frac{13}{18}, \frac{5}{9}, \frac{8}{9} \right\}$. In general this procedure yields a point of period $2 + 2k$ of the form $x = 0.0a_1\ldots a_{2k}1a_1'\ldots a_{2k}'$, where $a_j = 0$ if j is odd and $a_j = 1$ if j is even, $1 \le j \le 2k$. (Problem 3) Now to find a point of period 5, we let

$a_3 = 1$ and $a_4 = 0$ and use (2.14). Then $f^5(0.0010a_5a_6\ldots) = 0.a_5a_6\ldots$. Hence $x = 0.0010a_5a_6a_7a_8\ldots$ is of period 5 if $a_5 = a_6 = 0$, $a_7 = 1$, $a_8 = 0$ and thus $x = 0.\overline{0010} = \frac{2}{15}$ (Why?). Hence the periodic 5-cycle is given by $\{\frac{2}{15}, \frac{19}{30}, \frac{11}{15}, \frac{8}{15}, \frac{14}{15}\}$. In general, a point $x = 0.\overline{00a_1\ldots a_{2k}}$ with $a_j = 1$ if j is odd and $a_j = 0$ if j is even, $a < j < 2k$, yields a point of period $3 + 2k$ (Problem 4). □

2.6.2 A Converse of Sharkovsky's Theorem

The question that we are going to address in this section is the following: given any positive integers k and r with $k \rhd r$, is there a continuous map that has a point of period r but no points of period k? The answer to this question is a definite yes. Here we give a simple proof of this result which is based on my paper [31].

THEOREM 2.10 (A Converse of Sharkovsky's Theorem)
  For any positive integer r, there exists a continuous map $f_r : I_r \to I_r$ on the closed interval I_r such that f_r has a point of prime period r but no points of prime periods s, for all positive integers s that precede r in the Sharkovsky's ordering, i.e., $s \rhd \ldots \rhd r$.

PROOF   In order to accomplish the proof, we have three cases to contemplate.

1. Odd periods

2. Periods of the form $2^n \times$ odd positive integers

3. Periods of powers of 2, i.e., 2^n

Case 1: Odd Periods.

(a) Let us construct a continuous map that has points of period 5 but no points of period 3. Define a map $f : [1,5] \to [1,5]$ as follows:

$$f(1) = 3, f(2) = 5, f(3) = 4, f(4) = 2, \text{ and } f(5) = 1.$$

On each interval $[n, n+1], 1 \le n \le 4$, we assume f to be linear (see Fig. 2.25).

Observe first that none of the points $1, 2, 3, 4, 5$ is a 3-periodic point; they all belong to the single 5-cycle: $1 \xrightarrow{f} 3 \xrightarrow{f} 4 \xrightarrow{f} 2 \xrightarrow{f} 5 \xrightarrow{f} 1$. Note also that

$$f^3([1,2]) = [2,5], f^3([2,3]) = [3,5], \text{ and } f^3([4,5]) = [1,4].$$

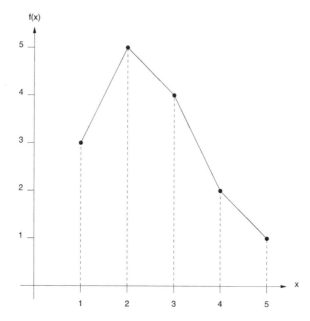

FIGURE 2.25
A map of period 5 but no points of period 3.

Hence, f^3 has no fixed points in the intervals $[1, 2], [2, 3]$, and $[4, 5]$. The situation with the interval $[3, 4]$ is much more involved since $f^3([3, 4]) = [1, 5]$. This implies by Theorem 1.1 that f^3 must have a fixed point $\bar{x}$ in the interval $[3, 4]$. We must show now that this fixed point of f^3 is really a fixed point of f and thus is not of prime period 3. Observe that $f(\bar{x}) \in [2, 4]$. So, if $f(\bar{x}) \in [2, 3]$, then $f^2(\bar{x}) \in [4, 5]$ and $f^3(\bar{x}) \in [1, 2]$. But, this is impossible since $f^3(\bar{x}) = \bar{x} \in [3, 4]$. Therefore, we conclude that $f(\bar{x}) \in [3, 4]$. Note that $f^2(\bar{x}) \in [2, 4]$. Again, if $f^2(\bar{x}) \in [2, 3]$, then $f^3(\bar{x}) \in [4, 5]$, yet another contradiction. Thus, the orbit of $\bar{x}, \{\bar{x}, f(\bar{x}), f^2(\bar{x})\}$ is a subset of the interval $[3, 4]$.

Now, on the interval $[3,4]$, $f(x) = 10 - 2x$ has the unique fixed point $x^* = \frac{10}{3}$. Moreover, on $[3,4]$, $f^3(x) = 30 - 8x$ also has the unique fixed point $\bar{x} = \frac{10}{3} = x^*$. Hence, f has no points of prime period 3.

(b) One may generalize the above construction in order to manufacture continuous maps that have points of period $2n + 1$ but no points of period $2n - 1$. Details will be given in the problems (Problems 3, 4, and 5).

Case 2: Periods of the Form $2^k(2n + 1)$.

(a) We begin by constructing a map that has points of period 2×5 but has no points of period 2×3. Consider first the map $f : [1,5] \to [1,5]$ as defined in Case 1(a). This map has points of period 5, but has no points of period 3. We will use this map to construct a new map $\tilde{f}$, called **the double** of f, as follows:

$$\tilde{f} : [1, 13] \twoheadrightarrow [1, 13],$$

$$\tilde{f}(x) = \begin{cases} f(x) + 8; \ 1 \le x \le 5 \\ x - 8; \quad\ \ 9 \le x \le 13. \end{cases}$$

For $5 < x < 9$, we connect the points $(5, 9)$ and $(9, 1)$ by a line (Fig. 2.26). The proof that the double map $\tilde{f}$ has a 10-cycle but no 6-cycle is left to the reader as Problem 6.

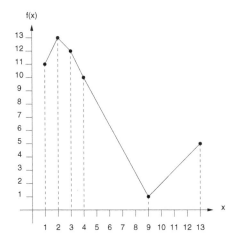

FIGURE 2.26
A 10-cycle but no 6-cycles.

(b) The general procedure for constructing the double $\tilde{f}$ of any map $f :$ $[1, 1 + h] \to [1, 1 + h]$ is as follows: $\tilde{f} : [1, 1 + 3h] \to [1, 1 + 3h]$, where

$$\tilde{f}(x) = \begin{cases} f(x) + 2h; \quad 1 \le x \le 1 + h \\ x - 2h; \quad 1 + 2h \le x \le 1 + 3h \end{cases}$$

and $\tilde{f}$ is linear for $1 + h < x < 1 + 2h$. So, if we want to construct a map with points of period $2(2n + 1)$, but no points of period $2(2n - 1), n = 3, 4, 5, \ldots$, we start with a map f that has points of period $(2n + 1)$, but no points of period $(2n - 1)$. Then, its double map $\tilde{f}$ will have the desired properties (Problem 7).

Case 3: Periods of the Form 2^n.

(a) It is easy to construct a map that has points of period $2^0 = 1$ (fixed points), but no points of prime period 2^1. Just pick any map $f(x) = ax + b$ with $a \neq \pm 1$. To construct a map that has points of period 2 but no points of period 2^2, we consider the map $f(x) = -x + b$. Then, $x = \frac{b}{2}$ is a fixed point of f. However, $f^2(x) = -(-x+b)+b = x$. Thus, every point, with the exception of $x^* = \frac{b}{2}$, is of prime period 2.

(b) To construct a map that has points of period 2^2, but no points of period 2^3, we use the double map $\tilde{f}$ of the map $f(x) = -x + 3$ (see Fig. 2.27). Map doubling may be used repeatedly to construct maps with 2^n-cycles, but no 2^{n+1}-cycles. ∎

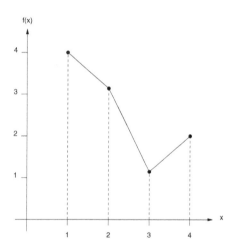

FIGURE 2.27
A 4-cycle but no 8-cycles.

Unresolved questions that remain to be settled are as follows:

1. Can we construct a continuous map that has points of period $2^n \times 3$, but has no points of any period of the form $2^{n-1} \times$ odd integer (see Problem 12)?

2. Can we construct a continuous map that has points of period 2^n for all $n \in \mathbb{Z}^+$, but no points of any other period [2] (see Problems 12 and 13)?

Exercise - (2.6)

In Problem 1–4, consider the map f in Example 2.10.

1. Find a point of period 6, and then find the corresponding orbit.

2. Find a point of period 7, and then find the corresponding orbit.

3. Show that the point $x = 0.\overline{0a_1 \ldots a_{2k} 1 a_1' \ldots a_{2k}'}$ is of period $2 + 2k$, if $a_j = 0$, if j is odd, and $a_j = 1$, if j is even $1 \le j \le 2k$.

4. Show that the point $x = 0.\overline{00a_1 \ldots a_{2k}}$, with $a_j = 1$, if j is odd, and $a_j = 0$ if j is even is of period $3 + 2k$, $1 \le j \le 2k$.

5. Show that the piecewise linear map $g : [1, 7] \to [1, 7]$, shown in Fig. 2.28, has a 7-cycle, but does not have a 5-cycle.

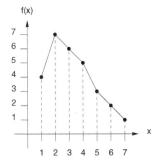

FIGURE 2.28
A 7-cycle but no 5-cycles.

6. Mimic Problem 1 to construct a map that has a 9-cycle, but not a 7-cycle.

7. Construct a map that has a $(2k+1)$-cycle, but has no $(2k-1)$-cycle for any $k > 3$.

8. Consider the map f, defined in Fig. 2.25, on the interval $I = [1, 5]$. Define a new function $\tilde{f}$ on $J = [1, 13]$ (called the **double** of f) by compressing the graph of f into the upper left square. Explicitly, we let

$$\tilde{f}(x) = \begin{cases} f(x) + 8 & \text{for } 1 \le x \le 5 \\ x - 8 & \text{for } 9 \le x \le 13. \end{cases}$$

Then connect the points (5, 9) and (9, 1) by a line. Show that the map $\tilde{f}$ (Fig. 2.26) has a 10-cycle, but not a 6-cycle.

9. Mimic Problem 4 to produce a map with a 14-cycle but not a 10-cycle.

10. Construct a map that has a $2(2n + 1)$-cycle but no $2(2n - 1)$-cycles.

11. Let f be a map defined on the interval $I = [1, 1 + h]$. Define $\tilde{f}$, "the double of f," on $[1, 1 + 3h]$, as follows:

$$\tilde{f}(x) = \begin{cases} f(x) + 2h & \text{for } 1 \le x \le 1 + h \\ x - 2h & \text{for } 1 + 2h \le x \le 1 + 3h. \end{cases}$$

and filling the rest of the graph as in Fig. 2.28. Prove that $\tilde{f}$ has a $2n$-periodic point at x if and only if f has an n-periodic point at x. Show that if f has points of period $2^k(2n + 1)$, then $\tilde{f}$ has points of period $2^{k+1}(2n + 1)$.

12. Construct a map that has an 8-cycle but no 16-cycle.

13. Construct a map that has a 2^k-cycle, but no 2^{k+1}-cycle, for $k > 3$.

14. (a) Construct a continuous map that has a point of period 2×3, but no points of odd periods.

     (b) Describe the procedure of constructing a map of period $2^n \times 3$, but has no points of period $2^{n-1} \times$ odd integer.

For another construction of the double map on the same interval: Let $I = [0, 1]$ and $f : I \to I$ be continuous. Define the double map $\tilde{f}$ by

$$\tilde{f}(x) = \begin{cases} \frac{2}{3} + \frac{1}{3}f(3x) & \text{for } 0 \le x \le \frac{1}{3} \\ [2 + f(1)](\frac{2}{3} - x) & \text{for } \frac{1}{3} \le x \le \frac{2}{3} \\ x - \frac{2}{3} & \text{for } \frac{2}{3} \le x \le 1. \end{cases}$$

15. Show that $\tilde{f}$ has a $2n$-periodic point at x, if and only if f has an n-periodic point at x.

16.* Use Problem 11 to construct a continuous map that has fixed points of period 2^n for all $n \in \mathbb{Z}^+$, but has no points of any other period.

     (Hint: Start with $f(x) = \frac{1}{3}$ on $[0, 1]$. Let $f_1 = \tilde{f}$ by its double map, $f_2 = \tilde{f}_1, \ldots, f_n = \tilde{f}_{n-1}$. Define $f_\infty(x) = \lim_{n \to \infty} f_n(x)$. Show that f_∞ is continuous and has points of period 2^n for all n and no other periods.)

17. Let f be a continuous map on the interval $[a, b]$. If there exists a point $x_0 \in [a, b]$ with $f^2(x_0) < x_0 < f(x_0)$, or $f(x_0) < x_0 < f^2(x_0)$, prove that f has a 2-cycle in $[a, b]$.

18. Prove that a homeomorphism of R cannot have periodic points with prime period greater than 2. Give an example of a homeomorphism that has a point of prime period 2.

19.* (Term Project) Generalize the Li-Yorke theorem (Theorem 2.9) as follows: Let J be an interval and let $f : J \to J$ be continuous. Assume there is a point $a \in J$ for which the points $b = f(a), c = f^2(a)$, and $d = f^3(a)$, satisfy $d \leq a < b < c$ $(d \geq a > b > c)$. Prove that for every $k = 1, 2, \ldots$ there is a periodic point in J having period k.

20.* (Term Project) (Li and Yorke) [62]. Under the assumption of Problem 14, show that there is an uncountable set $S \subset J$, containing no periodic points, which satisfies the following conditions:

(a) For every $x, y \in S$ with $x \neq y$,

$$\lim_{n \to \infty} \sup |F^n(x) - F^n(y)| > 0$$

and

$$\lim_{n \to \infty} \inf |F^n(x) - F^n(y)| = 0.$$

(b) For every $y \in S$ and periodic point $q \in J$,

$$\lim_{n \to \infty} \sup |F^n(x) - f^n(q)| > 0.$$

2.7 The Lorenz Map

Let us consider the Raleigh-Benard convection experiment, which led the MIT meteorologist Ed Lorenz in 1963 to introduce his now famous Lorenz equations. Consider a fluid contained between two rigid plates and subjected to gravity (see Fig. 2.29). The lower plate is kept at a higher temperature $T_0 + \Delta T$ than the temperature T_0 of the upper plate. The warm fluid near the bottom plate expands and rises, and the cool fluid above sinks, setting up a clockwise or counterclockwise current. An equilibrium state of this system occurs when the fluid is at rest and heat is transported upward through thermal conditions. Now, if the difference in temperature exceeds a critical value, the equilibrium loses its stability and convection rolls appear.

In his paper, "Deterministic non-periodic flow," Lorenz considered the Raleigh-Benard convection assuming that variations of the fluid occur in only two spatial dimensions. As a mathematical model to describe the Raleigh-Benard convection, he suggested the following highly idealized 3-dimensional system of first-order differential equations (now called the Lorenz equations):

Edward Norton Lorenz (1917 -)

Edward Lorenz is one of the earliest and most influential pioneers of chaos theory. Though his training was in Meteorology, he has a deep understanding of differential equations. In 1960 he began a project to simulate weather patterns modeled by a system of three differential equations (now named after him). To his total surprise, he discovered that each time he changes slightly the initial conditions, the simulated weather patterns changed dramatically. This led him to the discovery of one of the hallmarks of chaos, namely sensitive dependence on initial conditions, popularly known as the butterfly effect. Moreover, the orbits seemed to trace a strange distinctive shape, a kind of double spiral in three dimensions, like a butterfly with its two wings, the Lorenz strange attractor. Lorenz is currently a professor Emeritus of Meteorology at the Massachusetts Institute of Technology.

1. $\dot{x} = -\sigma x + \sigma y$

2. $\dot{y} = -xz + rx - y$

3. $\dot{z} = xy - bz.$

where the quantity x is proportional to the circulatory fluid flow velocity; the fluid circulates clockwise if $x > 0$ and counterclockwise if $x < 0$. The quantity y is proportional to the temperature difference between ascending and descending fluid elements, and z is proportional to the distortion of the vertical temperature profile from its equilibrium. The parameters of the system are the Prandtl number σ, the Raleigh number r, and b has no name but is related to the height of the fluid layer.

Lorenz used numerical methods to study the limiting behavior of the trajectories (orbits) of the system. He used the parameters $\sigma = 10, b = \frac{8}{3}, r = 28$.

He plotted $x(t)$ against $z(t)$ to get, to his surprise, a butterfly pattern (see Fig. 2.30). This is the Lorenz strange attractor, which we will comment on shortly.

Lorenz wrote that "the trajectory apparently leaves one spiral only after exceeding some critical distance from the center. Moreover, the extent to which this distance is exceeded appears to determine the number of circuits to be executed before changing spirals again. It, therefore, seems that some single feature of a given circuit should predict the same feature of the following circuit."

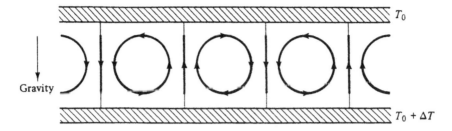

FIGURE 2.29
Raleigh-Benard convection.

Let us define what a strange attractor is. We say that a set A is a stable attractor if:

1. It is invariant; i.e., a trajectory $x(t)$ [or orbit $O^+(x)$] that starts in A stays indefinitely in A.

2. A attracts an open set of initial conditions; that is the basin of attraction of A is open.

3. A is minimal; there is no proper subset of A that satisfies the above two conditions.

   If, in addition, A possesses sensitive dependence on initial conditions, which means roughly that points that are arbitrarily close initially become (exponentially) separated at the attractor for a sufficient time, then A is said to be a strange attractor.

Lorenz found an ingenious method to study his model using only one-dimensional maps. Consider the surface on which $\dot{z} = 0$, given by $xy - bz = 0$. The maximum of z for every trajectory lies on this surface; let $z(n)$ denote the nth intersection of trajectory with this surface (Fig. 2.31(a)). Plotting the next maximum $z(n+1)$ as a function f of the current maximum $z(n)$ we get the tent-like curve depicted in Fig. 2.31(b). The map $z(n+1) = f(z(n))$ is called the Lorenz map.

Note that the graph of Lorenz map is reminiscent of an old friend, the tent map T defined as

$$
T(x) = \begin{cases} 2x & \text{if } 0 \le x \le \tfrac{1}{2} \\ 2(1-x) & \text{if } \tfrac{1}{2} \le x \le 1. \end{cases}
$$

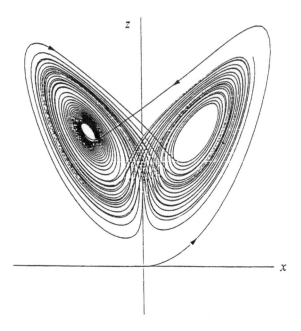

FIGURE 2.30
A Lorenz attractor.

2.8   Period-Doubling in the Real World

Period-doubling has been observed in real experiments and the Feigenbaum number which has been extracted from the experimental bifurcation points shows quite reasonable agreement with $\delta = 4.6692\ldots$ obtained from the logistic map in section 1.9.

The following table [35] shows the experimental period-doubling scenario for seven different materials.

Let us now explain one of the experiments that was performed by Libchaber [64] and his coworkers. A box of liquid Mercury was heated from below and a temperature gradient established through the thickness of Mercury. A dimensionless measure of the temperature gradient is the Rayleigh number R_a. For R_a less than a critical value, R_c, the fluid remains motionless and the heat flow up through the mecury via conduction. For $R_a > R_c$, convection occurred and cylinderical rolls formed (Fig. 2.29). The rolls were then stabilized by the application of an external DC magnetic field, the rolls tending to align along the magnetic field direction. For R_a slightly above R_c, the temperature at a fixed point on a roll was observed to be constant. As R_a was further increased, instability occured with a wave propagating along the roll

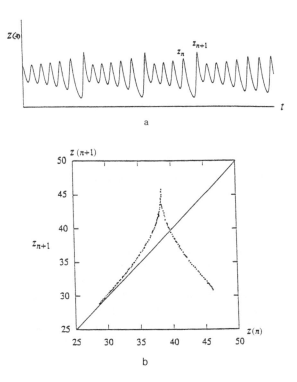

FIGURE 2.31
(a) Lorenz examined the behavior of successive maxima of the 3-coordinate of the trajectory; this is the vertical direction z in Fig. 2.30. (b) The Lorenz Map.

TABLE 2.2
Observed period-doubing and Feigenbaum constant δ.

Experiment	Period-doubling observed	δ
Water	4	4.3 ± 0.8
Mercury	4	4.4 ± 0.1
Diode	4	4.5 ± 0.6
Transistor	4	4.7 ± 0.3
Josephson	4	4.5 ± 0.3
Laser feedback	3	4.3 ± 0.3
Helium	3	4.8 ± 0.6

leading to a temperature variation at the point being monitored. As R_a was further increased, the recorded temperature variations, as a function of time, indicated a series of period-doubling from which the Feigenbaum number was extracted.

In all the performed experiments, scientists observed that it was difficult to obtain accurate data for more than about four period-doublings. However, it is remarkable that the observed Feigenbaum constant δ is very much in agreement with that obtained from the theoretical logistic map.

2.9 Poincaré Section/Map

2.9.1

In Section 1.3.2, we explained how a Poincaré map may be used to show the existence of a periodic solution of a system of two differential equations. In this section we provide a concrete example of how this scheme works.

Example 2.11

Consider the differential system

$$\frac{dx}{dt} = y + x(1 - x^2 - y^2)$$
$$\frac{dy}{dt} = -x + y(1 - x^2 - y^2)$$

or in polar coordinates

$$\frac{dr}{dt} = r(1 - r^2), \quad \frac{d\theta}{dt} = 1.$$

Choose the line L as the positive x-axis. Let r_0 be an initial point on L. Then solving the second equation yields $\theta(t) = t$. Thus at $t = 2\pi$, $\theta(2\pi) = 2\pi$, the first return of $(r_0, 0)$ to the x-axis (L) occurs at $t = 2\pi$. Let P denotes the Poincaré map, then $P(r_0) = r_1$, where r_1 satisfies

$$\int_{r_0}^{r_1} \frac{dr}{r(1 - r^2)} = \int_0^{2\pi} dt = 2\pi.$$

Using partial fractions

$$\int_{r_0}^{r_1} \frac{dr}{r} + \frac{1}{2} \int_{r_0}^{r_1} \frac{dr}{1-r} - \int_{r_0}^{r_1} \frac{dr}{1+r} = 2\pi$$

$$\ln \frac{r_1}{\sqrt{1-r_1^2}} \quad \ln \frac{r_0}{\sqrt{1-r_0^2}} - 2\pi$$

$$\ln \frac{r_1\sqrt{1-r_0^2}}{r_0\sqrt{1-r_1^2}} = 2\pi$$

$$\frac{r_1^2}{1-r_1^2} = \frac{r_0^2}{1-r_0^2} e^{4\pi}.$$

Hence

$$r_1^2 = \frac{r_0^2 e^{4\pi}}{1 - r_0^2 + r_0^2 e^{4\pi}} = \frac{1}{1 + (r_0^{-2} - 1)e^{-4\pi}}$$

and

$$r_1 = [1 + (r_0^{-2} - 1)e^{-4\pi}]^{-\frac{1}{2}}.$$

The Poincaré map is thus given by

$$P(r) = [1 + e^{-4\pi}(r^{-2} - 1)]^{-\frac{1}{2}}.$$

This map has two fixed points $r_1^* = 0$, $r_2^* = 1$. The fixed point $r_2^* = 1$ is globally asymptotically stable (Fig. 2.32) (Problem 1).

From this we conclude that the differential system has a (globally) asymptotically stable limit cycle (excluding the origin); see Fig. 2.33. ∐

2.9.2 Belousov-Zhabotinskii Chemical Reaction

One of the most interesting application of Poincaré sections is to the Belousov-Zhabotinskii reaction (BZ). In this experiment, malonic acid is oxidized in an acidic medium by bromate ions. Several chemists, Rous, Simoyi, Wolf, and Swinney [88, 96] conducted an experiment on the BZ reaction in a continuous flow stirred tank reactor.

Fresh chemicals are pumped through the reactor at a constant rate to replenish the reactants. The flow rate represents a control parameter. The scientists measured the concentration of bromide ions $B(t)$ at time periods. From the collected data, a time series graph is plotted (Fig. 2.34).

Roux et al (1983) suspected the presence of chaos as the time series show an oscillatory chaotic motion. To verify their conjecture, they used an effective method, known as attractor reconstruction where delay is introduced. A two dimensional vector $X(t) = (B(t), B(t+T))$ is introduced, for some delay

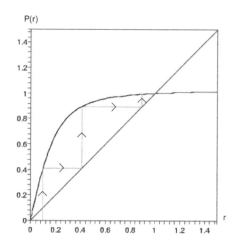

FIGURE 2.32
The Poincaré map shows the fixed point $r_2^* = 1$ is globally asymptotically stable.

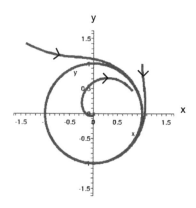

FIGURE 2.33
A limit cycle.

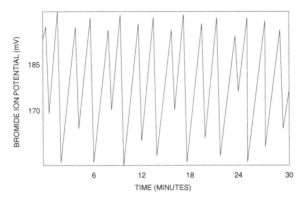

FIGURE 2.34

Time series.

$T > 0$. Plotting the orbit of the vector $X(t)$ in a two-dimensional phase space, for $T = 8.8$, the resulting figure looks similar to the Rössler attractor[4] (Fig. 2.35). Consider a line L crossing the orbit of $X(t)$ (dashed line). Let $x(1), x(2), \ldots, x(n), \ldots$ denote successive values of $B(t + T)$ at points where the orbit of $X(t)$ crosses the dashed line L. Plotting x_{n+1} versus x_n yields the graph shown in Figure 2.36. The graph looks like the logistic map and thus chaos occurs through period-doubling scenario. Coffman et al [18] verified the period-doubling scenario in the laboratory (Figure 2.37).

Exercises - (2.9)

1. Show that the fixed point $r_2^* = 1$ in Example 2.9 is globally asymptotically stable on $(0, \infty)$.

2. Consider the differential system

$$\frac{dr}{dt} = br(1 - r), \quad \frac{d\theta}{dt} = 1.$$

   (a) Find the Poincaré map associated with this system.

   (b) For which values of b is the orbit at $r_0 = 1$ asymptotically stable?

---

[4]See Chapter 3

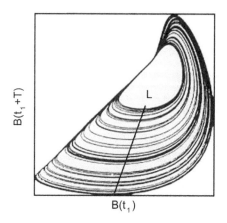

FIGURE 2.35
The Rössler attractor.

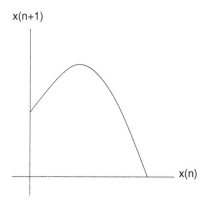

FIGURE 2.36
$x(n+1)$ versus $x(n)$, where $x(n)$'s are the successive values of $B(t+T)$ where
the orbit of $X(t)$ crosses the dashed line L.

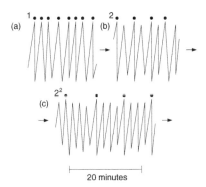

FIGURE 2.37
Period-doubling scenario.

Appendix

Proof of Sharkovsky's Theorem (Theorem 2.8)

We first need the following lemma.

LEMMA 2.8
Let f be continuous on the interval $[a, b]$, and let $I_0, I_1, \ldots, I_{k-1}$ be closed subintervals of $[a, b]$. If $f(I_j) \supset I_{j+1}, j = 0, 1, \ldots, k-2, f(I_{k-1}) \supset I_0$, then the equation $f^k(x) = x$ has at least one solution $x_0 \in I_0$ such that

$$f^j(x_0) \in I_j, j = 0, 1, \ldots, k-1$$

PROOF Let us use the notation $I_i \to I_j$ or $I_j \leftarrow I_i$ to denote $f(I_i) \supset I_j$. Hence the assumption in the lemma can be written as

$$I_0 \to I_1 \to I_2 \to \ldots \to I_{k-1} \to I_0$$

Now, for any $j, 0 \leq j \leq k-1$, $I_{j+1} = [f(p), f(q)]$ for some $p, q \in I_j$. If $p < q$, we let r be the largest number in $[p, q]$ such that $f(r) = f(p)$ and let s be the smallest number in $[p, q]$ with $f(s) = f(q)$ and $s > r$. We claim that $f([r, s]) = I_{j+1}$. We observe that by the intermediate value theorem we have $f([r, s]) \supset I_{j+1}$. Assume now there exists t with $r < t < s$ such that $f(t) \notin I_{j+1}$. Without loss of generality, suppose that $f(t) > f(q)$. Applying the intermediate value theorem again yields $f([r, t]) \supset I_{j+1}$. Hence there is $x \in [r, t)$ such that $f(x) = f(q)$, which contradicts our assumption that s is the smallest number in $[p, q]$ with $f(p) = f(q)$. The case where $p < q$ is similar.

Let $I_j^* = [r, s]$. Then we have $I_j^* \supset I_j$ and $f(I_j^*) = I_{j+1}$. Summarizing, there exist $I_j^* \supset I_j$ such that $f(I_j^*) = I_{j+1}^*$. This implies that there exists $I_{k-1}^* \supset I_{k_1}$ with $f(I_{k-1}^*) = I_0$, $I_{k-2}^* \supset I_{k-2}$ with $f(I_{k-2}^*) = I_{k-1}^*, \ldots$, and $I_0^* \supset I_0$ such that $f(I_0^*) = I_1^*$. This implies that $f^j(I_0^*) = I_j^*$, for $j = 0, 1, \ldots, k - 2$, and $f^k(I_0^*) \supset I_0^*$. Hence, by Theorem 1.2, f^k has a fixed point $x_0 \in I_0^* \supset I_0$. This completes the proof of the lemma. ∎

PROOF OF THEOREM 2.8   Assume that f has a periodic point x_0 of prime odd period k. Reorder the points in the orbit of x_0 and write it as $x_1, x_2, x_3, \ldots, x_k$ with $x_i < x_{i+1}$, $i = 1, 2, \ldots, k - 1$. Observe that $f(x_k)$ must be less than x_k. Let j be the largest index for which $f(x_j) > x_j$. Let $I_1 = [x_j, x_{j+1}]$. Since $f(x_{j+1}) \leq x_{j+1}$, we have $f(x_{j+1}) \leq x_j$. Consequently, $f(I_1) \supset I_1$ and thus $I_1 \to I_1$. Since x_0 is not of period 2, $f(I_1)$ must contain at least one other interval of the form $[x_i, x_{i+1}]$.

Let U_2 denote the union of all intervals of the form $[x_i, x_{i+1}]$ that are covered by $f(I_1)$. Then $U_2 \supset I_1$ and $U_2 \neq I_1$. Moreover, if $I_2 = [x_i, x_{i+1}]$ is any interval in U_2, then $I_1 \to I_2$.

Let U_3 denote all the intervals of the form $[x_i, x_{i+1}]$ that are covered by the image of some interval in U_2. Repeating this process inductively, we let U_{r+1} be the union of all intervals that are covered by the image of some interval in U_r. Observe that if I_{r+1} is any interval in U_{r+1}, there is a sequence of intervals $I_2, I_3, \ldots, I_r$ with $I_i \supset U_i$ such that $I_1 \to I_2 \to \cdots \to I_r \to I_{r+1}$.

Note that the sequence U_r forms an increasing union of intervals. Since the orbit $\{x_1, x_2, \ldots, x_k\}$ is finite, there exists an s such that $U_{s+1} = U_s$. For this s, U_s contains all intervals of the form $[x_i, x_{i_1}]$, for otherwise x_0 would have a period less than k. Observe that there is at least one interval $[x_i, x_{i+1}] \neq I_1$ in some U_r whose image covers I_1. This is clear since k is odd and thus there are more x_i's on one side of I_1 than on the other side. Hence some x_i's must change sides under f, and some must not. Thus, there is at least one interval whose image covers I_1. This implies that there is a chain of the form (see Fig. 2.38) $I_1 \to I_2 \to \cdots \to I_s \to I_1$ where I_i is of the form $[x_j, x_{j+1}]$ for some j, and $I_2 \neq I_1$. Moreover, we assume that s is the smallest integer for which this chain exists, i.e., this chain is the shortest nontrivial path from I_1 to I_1.

If $s < k - 1$, then one of loops $I_1 \to I_2 \to \cdots \to I_s \to I_1$ or $I_1 \to I_2 \to \cdots \to I_s \to I_1 \to I_1$ gives a fixed point of f^m with m odd and $m < s$. This point must have a prime period less than s since $I_1 \cap I_2$ consists of only one point, and that point has period greater than m. Therefore $s = n - 1$.

Since s is the smallest integer that works, we cannot have $I_\ell \to I_j$ for any $j > \ell + 1$. This implies that the orbit of x_0 should be ordered in $\mathbb{R}$ in one two possible ways as depicted in Figs. 2.39 and 2.40.

Thus we can extend the diagram in Fig. 2.38 to that shown in Fig. 2.41.

This proves Sharkovsky's theorem for the case when the period k is odd.

Note that periods larger than k are given by cycles of the form $I_1 \to I_2 \to \cdots \to I_{k-1} \to I_1 \to \cdots \to I_1$. The smaller even periods are given by cycles of

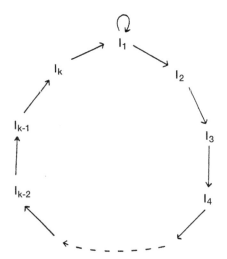

FIGURE 2.38
A chain $I_1 \to I_2 \to \ldots \to I_k \to I_1$.

the form

$$I_{k-1} \to I_{k-2} \to I_{k-1}$$
$$I_{k-1} \to I_{k-4} \to I_{k-3} \to I_{k-2} \to I_{k-1}$$

etc.

Observe that if k is even, then f must have a periodic point of period 2. This follows from the above argument provided we can guarantee that some x_i's change sides under f and some do not. For if this is not true, then all of the x_i's must change sides and hence $f[x_1, x_j] \supset [x_{j+1}, x_j]$ and $f[x_{j+1}, x_k] \supset [x_1, x_j]$. But then we must have a 2-periodic point in $[x_1, x_j]$.

Now we prove the theorem for $k = 2^m$. Let $n = 2^\ell$ with $\ell < m$. Consider $g = f^{k/2}$. By assumption g has a periodic point of period $2^{m-\ell+1}$. Thus g has a point z of period 2. The point z has period 2^ℓ under f. Finally, we assume that $k = p.2^m$, where p is odd. This will be left to the reader as exercises.

∎

Exercises

1. Prove that if f has period $p.2^m$ with p odd, then f has period $q.2^m$ with q odd and $q > p$.

2. Prove that if f has period $p.2^m$ with p odd, then f has period 2^ℓ, $\ell \leq m$.

3. Prove that if f has period $p.2^m$ with p odd, then f has period $q.2^m$ with q even.

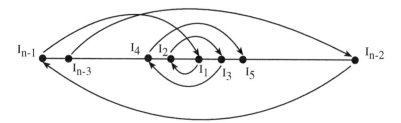

FIGURE 2.39

One possible ordering of I_j's.

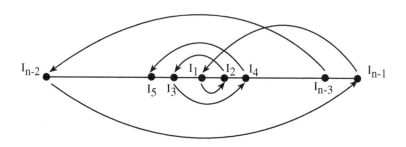

FIGURE 2.40

Another possible ordering of I_j's.

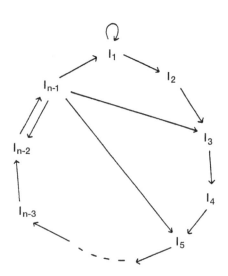

FIGURE 2.41

A chain for odd periods.

3

Chaos in One Dimension

The "butterfly effect": Does the flap of a butterfly's wings in Brazil set off a tornado in Texas?

Edward Lorenz

3.1   Introduction

The origin of the word chaos ($\chi\alpha\prime$ os) is the Greek verb *chasken,* meaning "to yawn" or "to gape open," referring either to the primeval emptiness of the universe before things came into being, or the abyss of Tartarus, the underworld (*Encyclopedia Britannica,* Vol. 5, p. 276). In modern dictionaries, chaos is defined as "total disorder and confusion."

The study of what we now call "chaotic systems" is due to Henri Poincaré and certainly to the ergodic theorists Birkhoff [11] and Von Neuman [106] in the 1930s. The expression "chaos" became popularized through the paper of Li and Yorke [62], "Period three implies chaos." Examples of chaotic systems include turbulent flow of fluids, population dynamics, irregularities in heartbeat, plasma physics, economic systems, weather forecasting, etc. These systems share the property of having a high degree of sensitivity to initial conditions. In other words, a very small change in initial values (due to error in measurement, noise, etc.) will multiply in such a way that the new computed system bears no resemblance to the one predicted.

Meteorologist and mathematician Edward Lorenz [65] introduced one of the most interesting examples of chaotic systems. In his study of weather forecasting, he concluded that weather is unpredictable, although it is deterministic. Thus, long-term weather forecasting would always elude science. This is due again to the fact that weather patterns are sensitive to initial conditions. Lorenz called this magnification of errors in weather forecasting the "butterfly effect." The metaphor says the flapping of a butterfly's wings in Brazil may cause a tornado in Texas several weeks later.

Another simple example of a chaotic system is the ball in a two-well potential as shown in Fig. 3.1. If the base vibrates with periodic motions of sufficiently large amplitude, the ball will jump from one well to the other in

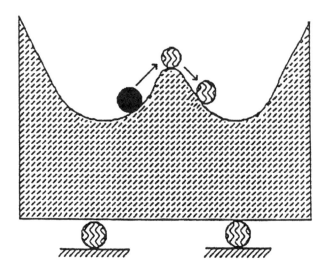

FIGURE 3.1
A pinball machine.

a chaotic manner. A popular model of such a mechanical system is a pinball machine where the ball's movement is again totally unpredictable.

Finally, it has been recognized by Sharkovsky [93], Li and Yorke [62], and many others that there is a hidden, self-organizing order in chaotic systems. A certain degree of order in chaotic systems has led to various definitions of chaos in literature. In this book, we adopt Devaney's definition of chaos (see Section 3.4).

In Section 3.9 we present a comparative study of the various notions of chaos that are used in the literature.

Devaney's definition of chaos has three components: transitivity, sensitive dependence on initial conditions, and the density of the set of periodic points. In the sequel we study these three important concepts in detail.

3.2   Density of the Set of Periodic Points

One of the building blocks of chaos is the abundance of periodic points in the system. This can be described mathematically as the presence of a <u>dense</u> set of periodic points. So what is a dense set?

DEFINITION 3.1   *Let I be an interval in* $\mathbb{R}$*. Then a set A is said to be*

<u>dense</u> *in I if for any x ∈ I any open interval containing x must intersect A. In other words, for each δ > 0, the open interval J = (x − δ, x + δ) contans a point in A.*

Example 3.1

The set of rational numbers Q is dense in $\mathbb{R}$. ▯

SOLUTION To prove this we consider $x \in \mathbb{R}$ and we write it in its decimal expansion

$$x = \sum_{n=0}^{\infty} \frac{d_n}{10^n}$$

where $d_n \in \{0, 1, \ldots, 9\}$. Let $\delta > 0$. Then there exists a positive integer m such that $10^{-m} < \delta$. Consider the rational number $y = \sum_{n=0}^{m} \frac{d_n}{10^n}$. Then

$$|x - y| = \sum_{n=m+1}^{\infty} \frac{d_n}{10^n} \le \sum_{n=m+1}^{\infty} \frac{9}{10^n} = \frac{\frac{9}{10^{m+2}}}{1 - \frac{1}{10}} = \frac{1}{10^m}.$$

Hence $|x - y| < \delta$. This proves that Q is dense in $\mathbb{R}$.

Similarly, one may show that for any interval I in $\mathbb{R}$, the set $Q \cap I$ is dense in I. ∎

In the next example we revisit the tent map and study the density of the set of periodic points in a greater detail.

Example 3.2

(The Tent Map Revisited). Think about the tent map T that we previously encountered, which is defined as

$$T(x) = \begin{cases} 2x; & 0 \le x \le \frac{1}{2} \\ 2(1-x); & \frac{1}{2} < x \le 1. \end{cases}$$

Show that the set of periodic points of T is dense in the closed interval $[0, 1]$.
▯

SOLUTION Before formalizing our investigation, let us do some exploration. Recall that the fixed points of T are $x_1^* = 0$ and $x_2^* = \frac{2}{3}$. Consider now the following rational numbers:

$$\tfrac{1}{3} \xrightarrow{T} \tfrac{2}{3}; \qquad\qquad\qquad \tfrac{1}{3} \text{ is eventually fixed.}$$

$$\tfrac{1}{4} \xrightarrow{T} \tfrac{1}{2} \xrightarrow{T} 1 \xrightarrow{T} 0; \ \tfrac{1}{4} \text{ is eventually fixed.}$$

$$\tfrac{3}{4} \xrightarrow{T} \tfrac{1}{2} \xrightarrow{T} 1 \xrightarrow{T} 0; \ \tfrac{3}{4} \text{ is eventually fixed.}$$

$$\tfrac{1}{5} \xrightarrow{T} \tfrac{2}{5} \xrightarrow{T} \tfrac{4}{5} \xrightarrow{T} \tfrac{2}{5}; \tfrac{1}{5} \text{ is eventually 2-periodic.}$$

$$\tfrac{2}{5} \xrightarrow{T} \tfrac{4}{5}; \qquad\qquad\qquad \tfrac{2}{5} \text{ is of period 2.}$$

$$\tfrac{3}{5} \xrightarrow{T} \tfrac{4}{5} \xrightarrow{T} \tfrac{2}{5}; \qquad \tfrac{3}{5} \text{ is eventually 2-periodic.}$$

$$\tfrac{4}{5} \xrightarrow{T} \tfrac{2}{5}; \qquad\qquad\qquad \tfrac{4}{5} \text{ is of period 2.}$$

Clearly, all points of the form $\frac{r}{2^k}$ are eventually fixed points. Moreover, points of the form $\frac{r}{s}$ with s odd are eventually periodic when r is also odd, and periodic if r is even. These observations will be established rigorously by the following two lemmas. However, our main objective here is to show that the set of periodic points is dense. This is done in Theorem 3.1. ∎

LEMMA 3.1
A point $b \in (0, 1)$ is eventually periodic under T if and only if it is rational.

PROOF   Let $b = \frac{r}{s}$ be in its reduced form. First, assume that $s = 2k + 1$ be an odd integer. Then $T^n \left(\frac{r}{s} \right) = \frac{\text{even integer}}{s}$, for all $n \in \mathbb{Z}^+$. Moreover, there are exactly k numbers in the interval $[0, 1]$ of the form $\frac{\text{even integer}}{s}$, namely, $\frac{2}{s}, \frac{4}{s}, \dots, \frac{2k}{s}$. Hence the orbit of the point b has at most k elements, and consequently b is eventually periodic. On the other hand, if $s = 2k$ is an even integer, then for some positive integer m either:

1. $T^m(b) = \frac{\text{integer}}{\text{odd integer}}$, which we have discussed above, or

2. $T^m(b) = 1$ and hence $T^{m+n}(b) = 0$ for all $n \in \mathbb{Z}^+$; b is eventually fixed.

Thus, in either case, b is eventually periodic. Conversely, assume that b is an eventually periodic point of T.

Generally, $T^n(b) = t_n \pm 2^n b$, for some integer t_n. Since b is eventually periodic, $T^n(b) = T^{n+k}(b)$ for some positive integer k. Thus, $t_{n+k} \pm 2^{n+k} b = t_n \pm 2^n b$, or

$$b = \frac{t_{n+k} - t_n}{\pm 2^n \mp 2^{n+k}}$$

which shows that b is rational. ∎

LEMMA 3.2
Let $b = \frac{r}{s}$ be a rational number in the interval $(0,1)$. Then, b is a periodic point of the tent map T if and only if r is an even integer and s is an odd integer.

PROOF   Let $b = \frac{r}{s} \in (0,1)$ with r being an even integer and s an odd integer. It follows from Theorem 3.1 that the point b is an eventually periodic point of the tent map T. Therefore, there exists a least nonnegative integer m and a least positive integer $n > m$ such that $T^m(b) = T^n(b)$. If $m = 0$, we are done and b is an n-periodic point of T. But, suppose that $m > 0$. Then, from the definition of T,

$$T^{m-1}\left(\frac{r}{s}\right) = \frac{\text{even integer}}{s}.$$

Thus,

$$T^m\left(\frac{r}{s}\right) = \begin{cases} \frac{2(\text{EI})}{s} = \frac{4(\text{I})}{s}, & \text{if } 0 \le T^{m-1}\left(\frac{r}{s}\right) \le \frac{1}{2} \\ 2\left(1 - \frac{\text{EI}}{s}\right) = \frac{4(\text{I})}{s} + 2, & \text{if } \frac{1}{2} < T^{m-1}\left(\frac{r}{s}\right) \le 1, \end{cases}$$

where EI is an even integer and I is an integer. Hence, for $T^m(\frac{r}{s})$ to be equal to $T^n(\frac{r}{s})$, we must either have both $T^{m-1}(\frac{r}{s})$ and $T^{n-1}(\frac{r}{s})$ in the interval $[0, \frac{1}{2}]$ or have both in the interval $(\frac{1}{2}, 1]$.
   Without loss of generality, assume that both $T^{m-1}(\frac{r}{s})$ and $T^{n-1}(\frac{r}{s})$ are in the interval $[0, \frac{1}{2}]$. Then,

$$2T^{m-1}\left(\frac{r}{s}\right) = T^m\left(\frac{r}{s}\right) = T^n\left(\frac{r}{s}\right) = 2T^{n-1}\left(\frac{r}{s}\right).$$

Consequently, $T^{m-1}(\frac{r}{s}) = T^{n-1}(\frac{r}{s})$, which contradicts the minimality of m and n. Therefore, $m = 0$ and b is an n-periodic point of the tent map T.
   The converse is left to the reader as Problem 1. ∎

THEOREM 3.1
The set of periodic points of the tent map T is dense in the closed interval $[0,1]$.

PROOF   Let $J = (a,b)$ be an open subinterval of $[0,1]$ where $t = b - a$. Choose an odd integer s such that $s > \frac{2}{t}$. Consider now that the set $A = \left\{\frac{1}{s}, \frac{2}{s}, \ldots, \frac{(s-1)}{s}\right\}$. We observe that for any two successive numbers

$\frac{r}{s}, \frac{r+1}{s}$ in A, $\frac{r+1}{s} - \frac{r}{s} = \frac{1}{s} < \frac{t}{2}$. This implies that there are two successive numbers $\frac{m}{s}$ and $\frac{(m+1)}{s}$ in A that belong to the interval J. Now, either m or $m+1$ is an even integer. Thus, the interval J contains a point c of the form $c = \frac{\text{even integer}}{\text{odd integer}}$. Now, by virtue of Lemma 3.2, the point c is a periodic point of T. Consequently, the set of periodic points of T is dense in $[0,1]$. ∎

3.3 Transitivity

In this section we study topological transitivity of maps on $\mathbb{R}$ or on an interval I. We begin our exposition by stating the definition.

DEFINITION 3.2 *Let f be a map on an interval I (or $\mathbb{R}$). Then f is said to be topologically transitive if for any pair of nonempty open intervals J_1 and J_2 in I there exists a positive integer k such that $f^k(J_1) \cap J_2 \neq \emptyset$.*

Equivalently, one may replace the interval J_1 and J_2 be open subsets U_1 and U_2 of I. Note that an open set is just the union of open intervals.

Intuitively, under a transitive map, a point in I wanders all over I, and its orbit gets as close as we wish to every other point in I.

In many instances, it is easier to use the following criterion to prove that the given map is transitive.

THEOREM 3.2

If the map $f : I \to I$ on the interval I has a dense orbit, then it is topologically transitive. The converse is true if I is a closed interval.

PROOF Suppose that the point $a \in I$ has a dense orbit. Let U and V be any pair of nonempty open sets in I. There are positive integers r, s such that $f^r(a) \in U$ and $f^s(a) \in V$. Let $b = f^r(a)$ and $c = f^s(a)$. Without loss of generality, assume that $k = s-r \geq 0$. Then $f^k(b) = f^{s-r}(f^r(a)) = f^s(a) = c$. Consequently, $f^k(U) \cap V \neq \phi$. The proof of the converse is beyond the scope of this book. ∎

In the following example, the doubling map D will play an important role in our discussion of chaos.

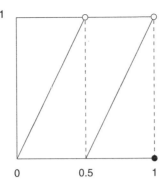

FIGURE 3.2
The doubling map D.

Example 3.3
The doubling map $D : [0,1] \to [0,1]$ is defined as

$$D(x) = \begin{cases} 2x & \text{for } 0 \le x < \frac{1}{2}, \\ 2x - 1 & \text{for } \frac{1}{2} \le x < 1, \\ 0 & \text{for } x = 1. \end{cases}$$

Equivalently, the doubling map D may be defined in the compact form

$$\boxed{D(x) = 2x \pmod{1},}$$

where " (mod 1) " stands for modulo 1. This means we drop the integral part leaving only the fractional part. For example, $D(0.8) = 1.6 \pmod 1 = 0.6$
⬚

We are going to show that the map D is transitive and has a dense set of periodic points.

THEOREM 3.3
Let $D : [0,1] \to [0,1]$ be the doubling map. Then Per_D is dense in $[0,1]$ and the map D is topologically transitive.

PROOF   We first show that Per_D is dense in $[0,1]$. Let $x \in [0,1]$ be written in its binary expansion, $x = \sum\limits_{j=1}^{\infty} \frac{a_j}{2^j}$ where a_j is either 0 or 1. Then

$$D(x) = a_1 + \sum_{j=2}^{\infty} \frac{a_j}{2^{j-1}} \pmod 1 = \sum_{j=2}^{\infty} \frac{a_j}{2^{j-1}} = \sum_{r=1}^{\infty} \frac{a_{r+1}}{2^r} \quad (\text{since } a_1 = 0 \text{ or } 1 \text{ and}$$
thus omitted under "(mod 1)").

Note that the nth iterate simply shifts the expansion by n places:

$$D^n(x) = D^n \left(\sum_{j=1}^{\infty} \frac{a_j}{2^j} \right) = \sum_{r=1}^{\infty} \frac{a_{r+n}}{2^r}.$$

Moreover, if a point x has an expansion that repeats every n places, $a_{j+n} = a_j$ for all j, then $D^n(x) = x$ and, consequently, x is a periodic point of period n. Now given a point $x = \sum_{j=1}^{\infty} \frac{a_j}{2^j} \in [0,1]$ and $\delta > 0$, there exists a positive integer N such that $\frac{1}{2^N} < \delta$. Consider the N-periodic point $y = \sum_{j=1}^{\infty} \frac{b_j}{2^j}$ in $[0,1]$, where $b_1, b_2, \ldots, b_N$ repeats and $b_1 = a_1, b_2 = b_2, \ldots, b_N = a_N$. Then

$$|x - y| = \left| \sum_{j=N+1}^{\infty} \frac{a_j}{2^j} - \sum_{j=N+1}^{\infty} \frac{a_j}{2^j} \right| \leq \sum_{j=N+1}^{\infty} \frac{1}{2^j}$$

$$= \frac{1}{2^N} < \delta.$$

Hence Per_D is dense in $[0,1]$.

Next, we show that D is topologically transitive. To show this, by virtue of Theorem 3.2, it suffices to produce a dense orbit. Consider the point $z = \sum_{j=1}^{\infty} \frac{c_j}{2^j}$, where c_j's appear as follows

$$\underbrace{01}_{1-\text{block}}, \quad \underbrace{00 \ 01 \ 10 \ 11}_{2-\text{block}}, \quad \underbrace{000 \ 001 \ 010 \ 100 \ 011 \ 101 \ 110 \ 111}_{3-\text{block}}.$$

This sequence consists of all the blocks of 0's and 1's of length 1 (there are only 2), of length 2 (these are four 2-blocks), etc. We claim that the orbit $O(z)$ is dense in $[0,1]$. Now let $x = \sum_{j=1}^{\infty} \frac{a_j}{2^j}$ be an arbitrary point in $[0,1]$ and let $\delta > 0$ be given. Then, there exists a positive integer N such that $\frac{1}{2^N} < \delta$. Now the string $a_1, a_2, \ldots, a_N$ must appear as one of the N-blocks in the binary expansion of z. Hence there is a positive integer k such that $D^k(z) = \sum_{r=1}^{\infty} \frac{c_{r+k}}{2^r}$ with $a_1 = c_{k+1}, a_2 = c_{k+2}, \ldots, a_N = c_{k+N}$. Now

$$|D^k(z) - x| = \left| \sum_{j=N+1}^{\infty} \frac{a_j}{2^j} - \sum_{r=N+1}^{\infty} \frac{c_{r+k}}{2^r} \right| \leq \sum_{j=N+1}^{\infty} \frac{1}{2^j}$$

$$= \frac{1}{2^N} < \delta.$$

Hence $O(z)$ is dense in $[0,1]$. ∎

Exercises - (3.2 and 3.3)

In Problems 1 and 2, consider the Baker map $B : [0, 1] \to [0, 1]$ defined as

$$B(x) = \begin{cases} 2x & \text{for } 0 \le x < \frac{1}{2}, \\ 2x - 1 & \text{for } \frac{1}{2} \le x \le 1. \end{cases}$$

(Notice the difference between the Baker map and the doubling map.)

1. (a) Prove that $x \in (0, 1)$ is a periodic point of the map B if $x = \dfrac{r}{(2^n - 1)}$, for $r = 0, 1, 2, \ldots, 2^n - 2$.

   (b) Prove that Per_B is dense in $[0, 1]$.

2. Show that B is transitive.

3. Show that the tripling map $f(x) = 3x \pmod 1$ is topologically transitive and the set of periodic points Per_f is dense in $[0, 1]$ (use a ternary expansion).

4. Consider the map $f : [0, 1) \to [0, 1)$ defined by $f(x) = 10x \mod 1$.

   (a) Show that the set of periodic points of f is dense in $[0, 1)$.

   (b) Show that f is transitive.

   In Problems 5 and 6 consider the <u>double angle-map</u> $g : S^1 \to S^1$ on the unit circle S^1 defined as $g(\theta) = 2\theta$, where θ represents the point $e^{i\theta}$ on S^1 (Fig. 3.3).

5. Show that g is topologically transitive.

6. (a) Show that a point $\theta \in S^1$ is n-periodic if it is of the form

$$\theta = \frac{2k\pi}{2^n - 1}, \quad \text{for some positive integer } k.$$

   (b) Show that Per_g is dense in S^1. In S^1 an open interval $(x - \delta, x + \delta)$ is now replaced by an open arc $(\theta - \delta, \theta + \delta)$.

7. Consider the triple angle map $h : S^1 \to S^1$ on the unit circle S^1 defined as $h(\theta) = 3\theta$.

   (a) Show that the set of periodic points Per_h is dense in S^1.

   (b) Show that h is topologically transitive.

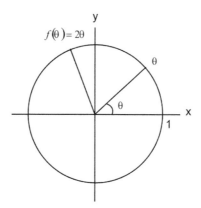

FIGURE 3.3

The double angle-map.

8. Let $f : I \to I$ be a continuous map on an interval I. Prove that f is topologically transitive and Per$_f$ is dense in I if and only if for each pair of open intervals J_1 and J_2 in I, there exists a periodic point $p \in J_1$ and a nonnegative integer k such that $f^k(p) \in J_2$.

9.* (Term Project)[73]

Let $f : I \to I$ be a one-to-one continuous map on the closed bounded interval I.

(a) Show that either f is strictly increasing or f is strictly decreasing on I.

(b) Show that f is not transitive on I.

(c) Show that if f is strictly increasing, then every periodic point is a fixed point. Moreover, if f has a non-fixed point, then the set of periodic points Per$_f$ is not dense in I.

(d) Show that if f striclty decreasing, then there is exactly one fixed point and all other periodic points are of period 2. Moreover, if $f^2(x) \neq x$ for some $x \in I$, then the set of periodic points Per$_f$ is not dense in I.

10.* [102] Let $f : I \to I$ be a continuous map on an interval I. Show that the following statements are equivalent.

(a) f is transitive and its set of periodic point Per$_f$ is dense in I.

(b) Given any two open subintervals J_1 and J_2 in I, there exists a periodic point $p \in J_1$ and a nonnegative integer k such that $f^k(p) \in J_2$.

3.4 Sensitive Dependence

If you look back at the introduction of this chapter you will remember that one of the main characteristics of a chaotic system is its sensitive dependence on initial conditions or, metaphorically speaking, the butterfly effect. In other words, a small error in the initial data will be magnified significantly by iteration. Therefore, in such a system, computer calculations may be misleading. Below is the formal definition of sensitivity.

DEFINITION 3.3 *A map of an interval I is said to possess **sensitive dependence** on initial conditions if there exists $\nu > 0$ such that for any $x_0 \in I$ and $\delta > 0$, there exists $y_0 \in (x_0 - \delta, x_0 + \delta)$ and a positive integer k such that*

$$\left| f^k(x_0) - f^k(y_0) \right| \geq \nu.$$

The number ν will be called the sensitivity constant of f.

The simplest function with sensitive dependence is the linear map $f(x) = cx$, $c > 1$. For the initial points x_0, and $x_0 + \delta$, we have

$$f^n(x_0 + \delta) - f^n(x_0) = c^n(x_0 + \delta) - c^n x_0$$
$$= c^n \delta.$$

Hence, $|f^n(x_0 + \delta) - f^n(x_0)|$ will increase to ∞ as n goes to ∞, regardless of how small δ is. However, this linear map is not an interesting example because it does not possess any of the other properties of chaos.

A more interesting example is provided by the quadratic map $F_4(x) = 4x(1 - x)$. In Fig. 3.4, we let $x_0 = 0.09$ and $x_0 + \delta = 0.11$. We observe that after each iteration the error almost doubles.

A similar phenomenon is observed in the tent map T.

We now examine this phenomenon in the doubling map that we discussed in Example 3.3.

Example 3.4
(The Doubling Map Revisited). Consider the doubling map $D : [0,1] \to [0,1]$. We will show that D has sensitive dependence on initial conditions. Let $I_1 = [0, \frac{1}{2})$ and $I_2 = [\frac{1}{2}, 1)$. Then for any two points $x, y \in [0,1)$, either (i) $x, y \in I$, or (ii) $x \in I_1$, $y \in I_2$. □

SOLUTION Case (i): If $x, y \in I_1$, then $|D(x) - D(y)| = |2x - 2y| = 2|x - y|$ on the other hand if $x, y \in I_2$, then $|D(x) - D(y)| = |2x - 1 - 2y + 1| = 2|x - y|$.

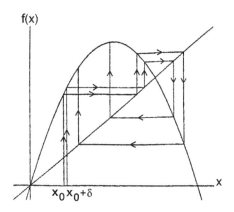

FIGURE 3.4
The error almost doubles after each iteration.

Case (ii): Suppose now that $x \in I_1$ and $y \in I_2$. Then $|D(x) - D(y)| = |2x - 2y + 1| \geq 1 - 2|x - y|$. Note that if $|x - y| < \frac{1}{4}$, then $-2|x - y| > -\frac{1}{2}$. Hence

$$|D(x) - D(y)| \geq 1 - \frac{1}{2} = \frac{1}{2}.$$

Select $\nu = \frac{1}{4}$ where ν is the sensitivity constant of D. If $x_0 \in I$ and $\delta > 0$, we pick any point $y_0 \in (x_0 - \delta, x_0 + \delta)$. If $x_0, y_0 \in I_1$ or I_2, then the distance between $x_j = D^j(x_0)$ and $y_k = D^k(y_0)$ will at least double the distance between y_{k-1} and x_{k-1}. Hence

$$|y_k - x_k| \geq 2|y_{k-1} - x_{k-1}|$$
$$\geq 2^k |y_0 - x_0|.$$

Hence, eventually $|x_j - y_j| \geq \frac{1}{4} = \nu$ for some positive integer j. On the other hand if $x_0 \in I_1$ and $y_0 \in I_2$ and $|x_0 - y_0| < \frac{1}{4}$, then $|D(x_0) - D(y_0)| > \frac{1}{4} = \nu$. Finally if $x_0 = 0$ or $x_0 = 1$, the above argument will carry over in a straightforward manner. ∎

This exponential stretching exhibited by the preceding map may be expressed by the **Liapunov exponent** λ. Roughly speaking, the Liapunov exponent $\lambda(x)$ at a point x measures the growth in error per iteration or the average loss of information during successive iterates of points near x.

How do we define this formally? We begin by considering a point x_0 and a neighboring point $x_0 + \delta$. Then the error e_n is defined as

$$e_n = |f^n(x_0 + \delta) - f(x_0)|,$$

and the relative error

$$\left| \frac{e_n}{\delta} \right| = \frac{|f^n(x_0 + \delta) - f^n(x_0)|}{\delta}.$$

If the map f possesses sensitive dependence on initial conditions, we expect the relative error $\frac{e_n}{\delta}$ to grow exponentially with n, and thus

$$e^{n\tilde{\lambda}} = \lim_{\delta \to 0} \frac{e_n}{\delta} = \lim_{\delta \to 0} \frac{|f^n(x_0 + \delta) - f^n(x_0)|}{\delta}, \quad \text{for some } \tilde{\lambda} > 0.$$

Hence

$$e^{\tilde{\lambda} n} = \left| \frac{d}{dx} f^n(x_0) \right| = |f'(x_0)f'(x(1)) \ldots f'(x(n-1))|$$

and

$$\tilde{\lambda} = \frac{1}{n} \sum_{k=0}^{n-1} \ln f'(x(k)).$$

This motivates us to define the Liapunov exponent $\lambda(x_0)$ for a map f as

$$\lambda(x_0) = \lim_{n \to \infty} \frac{1}{n} \ln |[f^n(x_0)]'|. \tag{3.1}$$

We can easily verify that

$$\ln |[f^n(x_0)]'| = \sum_{k=0}^{n-1} \ln |f'(x(k))|, \tag{3.2}$$

where $x(k) = f^k(x_0)$. Thus, Equation 3.1 becomes

$$\lambda(x_0) = \lim_{n \to \infty} \frac{1}{n} \sum_{k=0}^{n-1} \ln |f'(x(k))|. \tag{3.3}$$

Formula (3.3) tells us that the Liapunov exponent (the rate of convergence of two orbits) is the rate of change of the natural logarithm of the absolute value of the derivatives of the map evaluated at the orbit points. Note that if the application of the map to two nearby points leads to two points further apart, then the absolute value of the derivative of the map is greater than 1 when evaluated at these orbit points, and hence its logarithm is positive. If the orbit points continue to diverge, then the rate of change of the logarithm of the absolute values of the derivatives is positive, and hence the presence of sensitive dependence on initial conditions.

As we will see in the examples that follow, if the Liapunov exponent λ is positive, then sensitive dependence exists. Moreover, as the Liapunov exponent becomes larger, the magnification of error becomes greater.

Example 3.5
Find the Liapunov exponent of the tent map

$$T(x) = \begin{cases} 2x & \text{for } 0 \le x \le \frac{1}{2} \\ 2(1-x) & \text{for } \frac{1}{2} < x \le 1. \end{cases} \qquad \square$$

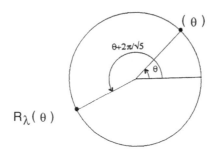

FIGURE 3.5

Failure of the rotation map $R_\lambda(\theta) = \theta + 2\pi\lambda$ to be sensitive dependent.

SOLUTION   Let $x_0 \in (0,1)$. Then,

$$T(x(k)) = \begin{cases} 2x(k) & \text{if } 0 \leq x(k) \leq \frac{1}{2} \\ 2(1 - x(k)) & \text{if } \frac{1}{2} < x(k) \leq 1. \end{cases}$$

Hence, $|T'(x(k))| = 2$.

By Equation (3.3), we have

$$\lambda(x_0) = \lim_{n\to\infty} \frac{1}{n} \sum_{k=0}^{n-1} \ln 2 \approx 0.6931.$$

This implies that the tent map possesses sensitive dependence.   ∎

Example 3.6

Consider the rotation map on the unit circle S^1 defined by $R_\lambda(\theta) = \theta + 2\pi\lambda$ for some $\lambda \in R$. Hence $|R_\lambda'(x)| = 1$ and

$$\lambda(x_0) = \lim_{n\to\infty} \frac{1}{n} \sum_{k=0}^{n-1} \ln 1 = 0.$$

This clearly indicates that the rotation map R_λ fails to be sensitive (see Fig. 3.5).   ∎

A Numerical Scheme to Compute Liapunov Exponents

It is often the case that one may not be able to exactly compute Liapunov exponents. In this case, one resorts to numerical schemes. We will illustrate the scheme for the logistic map $F_\mu(x) = \mu x(1-x)$. For a fixed value of μ, start with an initial point say 0.5. Discard the first 400 (transient) iterates. Then

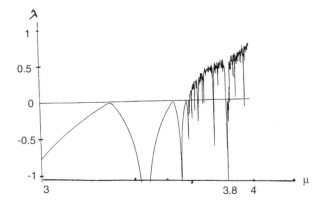

FIGURE 3.6

The graph of the Liapunov exponent λ of the logistic map F_μ as a function of μ.

compute an additional 100 iterates. The Liapunov exponent is now estimated by the formula

$$\lambda(0.5) = \frac{1}{500} \sum_{k=401}^{500} \ln |\mu - 2\mu x(k)|.$$

Starting with $\mu = 3$ and increasing μ by $\frac{1}{1000}$, we end up with Fig. 3.6.

The negative spikes corresponds to the 2^n-cycles where we have stable cycles that do not possess sensitive dependence. Note also that λ remains negative for $3 < \mu < \mu_\infty \approx 3.57$, and approaches zero at the period-doubling bifurcation. As λ increases toward 4, it oscillates between positive and negative values. The positive values of λ increase to $\ln 2$ as we get closer and closer to $\lambda = 4$, which demonstrates that F_μ is increasingly sensitive to initial conditions.

REMARK 3.1   As was shown in the case of the tent map T, it is easy to compute Liapunov exponents if the derivative of the considered map is constant. Another instance is when the point we are considering is either periodic or eventually periodic. For if $\{x(0), x(1), \dots, x(k-1)\}$ is a k-cycle of a point $x_0 = x(0)$ under a map f, then, $\lambda(x_0) = \frac{1}{k} \sum_{j=0}^{k-1} \ln |f'(x(j))|$. Similarly, if a point y_0 is eventually periodic to a k-cycle $\{x(0), x(1), \dots, x(k-1)\}$, then $\lambda(y_0) = \frac{1}{k} \sum_{j=0}^{k-1} \ln |f'(x(j))|$ (Problem 8). The latter phenomenon may be extended to asymptotically periodic points. A point y_0 is asymptotically periodic if its orbit converges to a periodic orbit $\{x(0), x(1), \ldots, x(k-1)\}$ for some $k \in \mathbb{Z}^+$. In other words, $\lim\limits_{n \to \infty} |y(n) - x(n)| = 0$. ∎

THEOREM 3.4

Suppose that a point y_0 is asymptotically periodic to a periodic point x_0 such that $f'(y(k)) \neq 0$ for all k. Then $\lambda(y_0) = \lambda(x_0)$, provided both Liapunov exponents exist.

PROOF   This is left to the reader as Problem 13. ∎

In reference [2] a chaotic orbit of a map f is defined to be a bounded orbit with a positive Liapunov exponent, but not asymptotically periodic. Although this is a plausible definition, it is not universally accepted yet and we will not adopt it in this book.

Exercises - (3.4)

A rough estimate of a Liapunov exponent $\lambda(x_0)$ of a map f maybe obtained by the formula

$$\lambda(x_0) \approx \frac{1}{n} \ln \left| \frac{e_n}{e_0} \right|, \tag{3.4}$$

where e_0 is the initial error in x_0 and $e_n = |f^n(x_0 + e_0) - f^n(x_0)|$ is the error after n iterations.

In Problems 1–5, use Formula (3.4) to approximate the Liapunov exponent.

1. $f(x) = 4x(1-x)$ on $[0, 1]$, $n = 5, 6, 7$, $x_0 = 0.1$, $e_0 = 0.01$

2. $f(x) = 4x^3 - 3x$ on $[-1, 1]$, $n = 5, 6, 7$, $x_0 = 0.1$, $e_0 = 0.01$

3. $f(x) = \sin x$ on $[0, 2\pi]$, $n = 5, 6, 7$, $x_0 = 0.3$, $e_0 = 0.01$

4. $f(x) = 8x^4 - 8x^2$ on $[-1, 1]$, $n = 5, 6, 7$, $x_0 = 0.1$, $e_0 = 0.01$

5. Let

$$f(x) = \begin{cases} 3x & \text{for } 0 \leq x \leq \frac{1}{3} \\ 2 - 3x & \text{for } \frac{1}{3} < x \leq \frac{2}{3} \\ 3 - 3x & \text{for } \frac{2}{3} < x \leq 1. \end{cases}$$

   (a) Find the Liapunov exponent of f.

   (b) Show directly that f possesses sensitive dependence on initial conditions.

6. (a) Find the Liapunov exponent of the Baker map

$$B(x) = \begin{cases} 2x, & 0 \le x < \frac{1}{2} \\ 2x - 1, & \frac{1}{2} \le x \le 1. \end{cases}$$

   (b) Show directly that B has sensitive dependence on initial conditions.

7. Consider the generalized Baker map

$$B_\mu(x) = \begin{cases} 2\mu x & \text{if } 0 \le x < \frac{1}{2} \\ \mu(2x - 1) & \text{if } \frac{1}{2} \le x \le 1, \end{cases}$$

where $m > 0$.

   (a) Determine the Liapunov exponent of B_μ.

   (b) Determine the values of μ for which B_μ has sensitive dependence.

8.* Suppose that y is an eventually periodic point of a map f where it joins a k-cycle $\{x_1, x_2, \ldots, x_k\}$ of a point x. Show that

$$\lambda(y) = \lambda(x) = \frac{1}{k} \sum_{j=1}^{k} |f'(x_j)|.$$

9. Show that for $1 < \mu < 3$ and $\mu \ne 2$, the Liapunov exponent of the logistic map $F_\mu(x) = \mu x(1 - x)$ is given by $\lambda(x) = \ln|2 - \mu|$.

10. Define the Liapunov number $L(x_0)$ of a point $x_0 = x(0)$ as

$$L(x_0) = \lim_{n \to \infty} \left(|f'(x_0)||f'(x(1))| \ldots |f'(x(n))| \right)^{\frac{1}{n}}.$$

   Show that if L is the Liapunov number of x_0 under f, then L^k is the Liapunov number of x_0 under f^k.

11. Show that if $\lim_{n \to \infty} x_n = x$, then

$$\lim_{n \to \infty} \frac{1}{n} \sum_{i=1}^{n} x_i = x.$$

12. Show that any asymptotically periodic point of the tent map T must be eventually periodic.

13. Prove Theorem 3.4. (Hint: Use Problem 12.)

14. A map f is said to be expansive if there exists $\delta > 0$ such that for any $x, y \in X$ with $x \ne y$, there exists $k \in \mathbb{Z}^+$ with $d(f^k(x), f^k(y)) > \delta$.

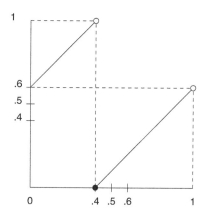

FIGURE 3.7
Graph of $f_{.4}$.

    (a) Show that the map $f(x) = cx, c > 1$ is expansive.

    (b) Show that neither the tent map T nor the logistic map F_4 is expansive.

15. (Senior Project/Master's Thesis) [73] For each $\alpha \in \mathbb{R}$, defined the function $f_\alpha : [0, 1) \to [0, 1)$ by $f_\alpha(x) = \text{frac}(x - \alpha)$. To explain frac, we write a given number x as the sum of an integer and a number $\text{frac}(x)$ in $[0, 1)$, which is the fractional part of x. For instance, $f_{.4}(0) = \text{frac}(0 - \cdot 4) = \text{frac}(-1 + \cdot 6) = \cdot 6$; $f_{.4}(\cdot 5) = \text{frac}(\cdot 5 - \cdot 4) = \cdot 1$ (Fig. 3.7).

    (a) Show that

$$f_\alpha(x) = \begin{cases} x + 1 - \text{frac}(\alpha) & \text{if } 0 \le x < \text{frac}(\alpha), \\ x - \text{frac}(\alpha) & \text{if } \text{frac}(\alpha) \le x < 1. \end{cases}$$

    (b) Show that $f_\alpha \circ f_\beta = f_{\alpha + \beta}$, for any $\alpha, \beta \in \mathbb{R}$.

    (c) Show that f_α is one-to-one.

    (d) Show that if α is not an integer, then f_α has a discontinuity at $\text{frac}(\alpha)$, but f_α is continuous and increasing on each of the intervals $[0, \text{frac}(\alpha))$ and $[\text{frac}(\alpha), 1)$.

    (e) Show that f_α has a periodic point if and only if α is rational and if $\alpha = \pm\frac{p}{q}$ in reduced form, p, q are positive integers, then every point in $[0, 1)$ is a periodic point of period q.

    (f) If α is irrational, show that f_α is transitive and possesses sensitive dependence on initial conditions but has no periodic points.

3.5 Definition of Chaos

We are now ready to define chaos according to Devaney [25].

DEFINITION 3.4 *A map $f : I \to I$, where I is an interval, is said to be chaotic if:*

1. *f is transitive*

2. *The set of periodic points P is dense in I*

3. *f has sensitive dependence on initial conditions*

REMARK 3.2 Recently, Banks et al. [4] showed that conditions (1) and (2) in Definition 3.4 imply Condition (3) of sensitive dependence on initial conditions. However, no other two conditions imply the third (see Problems 5 and 6). ∎

We start this section by establishing Banks et al. result [4].

THEOREM 3.5
Let $f : I \to I$ be a continuous map on an interval I. If f is transitive and its set of periodic points is dense, then f possesses sensitive dependence on initial conditions, i.e., f is chaotic.

To simplify the proof of this theorem, we first present the following lemma.

LEMMA 3.3
Let $f : I \to I$ be a continuous map on an interval I such that f has at least two periodic points with nonoverlapping orbits. Then, there exists $\varepsilon_0 > 0$ such that for every $x \in I$ there is a periodic point $p \in I$ with $d(x, f^n(p)) \geq \varepsilon_0$, for all $n \in \mathbb{Z}^+$.

PROOF Choose two periodic points p_0 and q_0 such that $O(p_0) \cap O(q_0) = \emptyset$. If $O(p_0) = \{p_0, p_1, \ldots, p_{n-1}\}$, and $O(q_0) = \{q_0, q_1, \ldots, q_{m-1}\}$, we let $\varepsilon_0 = \frac{1}{2} \min \{d(p_i, q_j) : i = 0, 1, \ldots, n-1, j = 0, 1, \ldots, m-1\}$. Let $x \in I$ be an arbitrary point. Then by the triangle inequality we have for any $r, s \in \mathbb{Z}^+, 2\varepsilon_0 \leq d(f^r(p_0), f^s(q_0)) \leq d(f^r(p_0), x) + d(x, f^s(q_0))$. So, if $d(f^r(p_0), x) \leq \varepsilon_0$, then $d(x, f^s(q_0)) \geq \varepsilon_0$ for all $s \in \mathbb{Z}^+$ and if $d(f^s(q_0), x) \leq \varepsilon_0$; then $d(x, f^r(p_0)) \geq \varepsilon_0$ for all $r \in \mathbb{Z}^+$. This completes the proof of the lemma. ∎

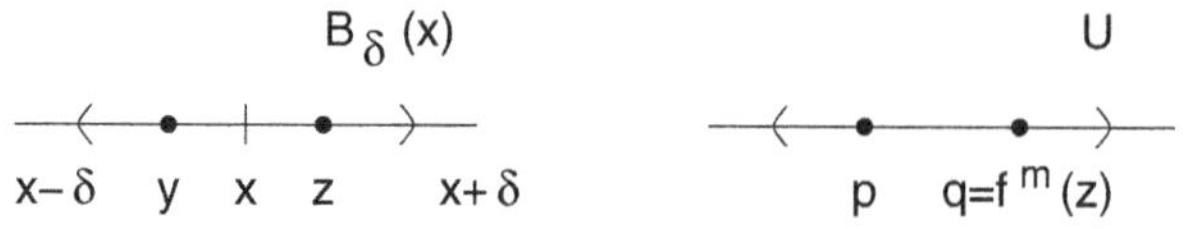

$$B_\delta\ (x) \qquad\qquad\qquad\qquad U$$

$$\text{x}-\delta \quad \text{y} \quad \text{x} \quad \text{z} \qquad \text{x}+\delta \qquad\qquad \text{p} \qquad \text{q}=\text{f}^{\,m}(\text{z})$$

FIGURE 3.8
There exists $z \in B_\delta(x)$ such that $f^m(z) \in U$.

PROOF OF THEOREM 3.5   We let our final ε be $\varepsilon = \frac{1}{4}\varepsilon_0$, where ε_0, as in the preceding lemma. Let $x \in I$ and $\delta > 0$ be given, with $\delta < \varepsilon$ (without loss of generality). By the density of the set of periodic points, there exists a k-periodic point $y \in I$ such that $|x - y| < \delta$. By Lemma 3.3 there exists a periodic point p such that

$$|x - f^n(p)| \geq \varepsilon_0 = 4\varepsilon, \quad \text{for all } n \in \mathbb{Z}^+. \tag{3.5}$$

Let

$$U = \bigcap_{i=0}^{k-1} f^{-i}(B_\varepsilon(f^i(p))). \tag{3.6}$$

It is easy to show that

$$U = \{z : |f^i(z) - f^i(p)| < \varepsilon \text{ for } 0 \leq i \leq k-1\}. \tag{3.7}$$

Then, U is open[1] and nonempty since it contains at least the point p (Problem 1). Since f is transitive there exists $z \in B_\delta(x)$ such that $f^m(z) \in U$ for some $m \in \mathbb{Z}^+$ (see Fig. 3.8). Let $r \in \mathbb{Z}^+$ such that $\frac{m}{k} < r < \frac{m}{k} + 1$ or $0 < kr - m < k$.

Now,

$$|x - f^{kr-m}(p)| \leq |x - y| + |y - f^{kr}(z)| + |f^{kr}(z) - f^{kr-m}(p)|. \tag{3.8}$$

Recall that $|x - y| < \delta < \varepsilon$. Furthermore, since

$$f^{kr}(z) = f^{kr-m}(q) \in B_\varepsilon(f^{kr-m}(p)), \quad \text{for some } q \in f^m(z) \text{ or } z = f^{-m}(q),$$

it follows that $|f^{kr}(z) - f^{kr-m}(p)| < \varepsilon$. Hence, from Equation (3.8) we get

$$4\varepsilon \leq |x - f^{kr-m}(p)| < 2\varepsilon + |y - f^{kr}(z)|.$$

This implies by (3.5) that

$$|y - f^{kr}(z)| > 2\varepsilon \text{ and since } y \text{ is of period } k, \ |f^{kr}(y) - f^{kr}(z)| > 2\varepsilon$$

---

[1]A set U is open if it is the union of open intervals.

since y is k-periodic. Therefore, by the triangle inequality,

$$2\varepsilon < |f^{kr}(y) - f^{kr}(z)| \le |f^{kr}(y) - f^{kr}(x)| + |f^{kr}(x) - f^{kr}(z)|.$$

Therefore, either $|f^{kr}(y) - f^{kr}(x)| > \varepsilon$ or $|f^{kr}(x) - f^{kr}(z)| > \varepsilon$. This establishes sensitive dependence on initial conditions.

REMARK 3.3 More recently, Vellekoop and Berglund [104] showed that for continuous maps on intervals in $\mathbb{R}$, transitivity implies that the set of periodic points is dense. It follows from Theorem 3.5 that in this case **transitivity implies chaos**. The proof of this result will be facilitated by first establishing the following lemma. ∎

LEMMA 3.4

Let $f : J \to J$ be a continuous map on an interval J (not necessarily finite) in $\mathbb{R}$. Suppose that there exists a subinterval I of J such that I contains no periodic points of f. If $x, f^m(x)$, and $f^n(x)$ are all in I, with $0 < m < n$, then either $x < f^m(x) < f^n(x)$ or $x > f^m(x) > f^n(x)$.

PROOF Let m and n be integers such that $0 < m < n$, and let $I \subset J$ be an interval with no periodic points of f. Suppose that for some $x \in I$, we have $x < f^m(x)$, $f^m(x) > f^n(x)$ and $f^m(x), f^n(x) \in I$. Define a new function $g = f^m$. Then, from the preceding sentence, we have $x < g(x)$. Claim that for all $k > 1$,

$$x < g(x) \le g^k(x). \tag{3.9}$$

Now, if $g^2(x) < g(x)$, then the function $h(z) = g(z) - z$ is positive at $z = x$ and negative at $z = g(x)$. By the intermediate value theorem, this implies that $h(y) = g(y) - y = 0$ for some y between x and $g(x)$. This means that $g(y) = f^m(y) = y$ and y is thus a periodic point of f. But $y \in I$ since x and $g(x)$ are in I; clearly we have a contradiction. This proves that $x < g(x) < g^2(x)$. By mathematical induction, we may complete the proof of the claim (Problem 2).

Now, $x < g^k(x)$ for all $k \in \mathbb{Z}^+$ and in particular for $k = n - m$ we have $x < g^{n-m}(x) = f^{(n-m)m}(x)$. By letting $h = f^{n-m}$, we have

$$x < h^m(x). \tag{3.10}$$

Furthermore, from the first line in the proof, $f^{n-m}(f^m(x)) = f^n(x) < f^m(x)$ or $h(f^m(x)) < f^m(x)$. By an argument similar to that used for g we can show that

$$h^m(f^m(x)) < f^m(x). \tag{3.11}$$

Using Eqs. (3.10) and (3.11), it is easy to see that the function $p(y) = h^m(y) - y$ is positive at $y = x$ and negative at $y = f^m(x)$. Thus, by the intermediate value theorem, there exists $z \in I$ between x and $f^m(x)$ such that $h^m(z) = f^{(n-m)m}(z) = z$, a contradiction. This proves that $x < f^m(x) < f^n(x)$. The second part of the proof is left to you as Problem 3. ∎

THEOREM 3.6
Let $f : J \to J$ be a continuous map on an interval J (not necessarily finite) in $\mathbb{R}$. If f is transitive, then the set of periodic points in J is dense in J, that is, f is chaotic.

PROOF   Suppose that f is transitive and assume there exists a subinterval I of J, which is void of periodic points of f. Let $x \in I$ such that x is not an end point of I, $U \subset I$ be an open interval containing x, and $V \subset I \backslash U$ be an open interval in I disjoint from U. By the transitivity of f, there exists $m \in \mathbb{Z}^+$ such that for some $y \in U, f^m(y) \in V$. Since f^m is continuous, there exists an open interval $\tilde{U}$ containing y with $f^m(\tilde{U}) \subset V$. Therefore, $f^m(\tilde{U}) \cap \tilde{U} = \emptyset$. Since $f^m(\tilde{U})$ may not be open, we choose an open interval $\tilde{V} \subset f^m(\tilde{U})$ such that $f^m(y) \in \tilde{V}$. Using the transitivity of f again, we have $f^k(\tilde{V}) \cup \tilde{U} \neq \emptyset$ for some $k \in \mathbb{Z}^+$. Thus, there exists $z \in \tilde{U}$ with $f^{m+k}(z) = f^n(z) \in \tilde{U}$. Obviously, this is a flagrant violation of Lemma 3.4 since $z, f^n(z) \in \tilde{U}$ while $f^m(z) \notin \tilde{U}$. This completes the proof of the theorem. ∎

We caution you that the preceding theorem fails to hold for nonintervals or higher dimensional spaces or even for the unit circle S^1, as may be seen in the following example.

Example 3.7
(Irrational Rotation of the Circle).   Consider the rotation map $R_\lambda : S^1 \to S^1$ defined by $R_\lambda(\theta) = \theta + 2\pi\lambda$, where λ is an irrational number (see Fig. 3.5). Show that R_λ is transitive, but the set of periodic points is not dense. ☐

SOLUTION   Let $\theta \in S^1$. Then, $R_\lambda^m(\theta) \neq R_\lambda^n(\theta)$ if $m \neq n$. Otherwise, $\theta + 2\pi m\lambda = \theta + 2\pi n\lambda$, which implies that $2\pi(m-n)\lambda = 1$. Thus, $(m-n)\lambda \in \mathbb{Z}$. But, since λ is irrational, we must have $m = n$. Hence, the orbit $O^+(\theta)$ is an infinite set in S^1. Furthermore, since $O^+(\theta)$ is a bounded sequence, it must have a convergent subsequence. Therefore, for $\varepsilon > 0$ there exist positive integers r and s with $|R_\lambda^r(\theta) - R_\lambda^s(\theta)| < \varepsilon$. Without loss of generality, we may assume that $m = r - s > 0$. Since R_λ preserves arc length in S^1, it follows that

$$|R_\lambda^m(\theta) - \theta| = |R_\lambda^s(R_\lambda^m(\theta)) - R_\lambda^s(\theta)|$$
$$= |R_\lambda^r(i\theta) - R_\lambda^s(\theta)| < \varepsilon.$$

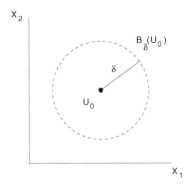

FIGURE 3.9
An open ball centered at U_0.

Now, under R_λ^m, the arc of length less than ε connecting θ to $R_\lambda^m(\theta)$ is mapped to the arc, of less than ε, connecting $R_\lambda^m(\theta)$ to $R_\lambda^{2m}(\theta)$. This arc, in turn, is mapped to the arc of length less than ε joining $R_\lambda^{2m}(\theta)$ to $R_\lambda^{3m}(\theta)$, etc. So the points $\theta, R_\lambda^m(\theta), R_\lambda^{2m}(\theta), \dots$ partition S^1 into arcs of length less than ε. But since ε was arbitrarily chosen, it follows that $O^+(\theta)$ intersects every open arc in S^1; and thus $O^+(\theta)$ is dense in S^1.

Observe that R_λ has no periodic points in S^1. ∎

REMARK 3.4   Theorem 3.5 is valid for more general spaces such as $\mathbb{R}^n$, in which the distance between two points (vectors) in the Euclidean distance

$$\|U - V\| = \sqrt{(u_1 - v_1)^2 + \cdots + (u_n - v_n)^2}$$

where $U = (u_1, u_2, \dots, u_n)$, $V = (v_1, v_2, \dots, v_n)$. The interval (a, b) is replaced by the open ball $B_\delta(U_0) = \{V \mid \|U_0 - V\| < \delta\}$ (Fig. 3.9).

One may go further in abstraction and define a distance d between two points in a space X as a function which satisfies the following properties (abstracted from the absolute value $|\;|$ function)

1. $d(x, y) \geq 0$ and $d(x, y) = 0$ if and only if $x = y$,

2. $d(x, y) = d(y, x)$ (symmetry),

3. $d(x, y) \leq d(x, z) + d(z, y)$ (the triangle inequality).

A space X with the metric d is called a metric space, and is denoted sometimes as (X, d).

One may show that $(\mathbb{R}^n, \|\|)$ is a metric space. In the next section we will encounter metric spaces that are not Euclidean spaces $\mathbb{R}^n$ and one may define an open ball in a metric space (X, d) as

$$B_\delta(x_0) = \{y_0 : d(x_0, y_0) < \delta\}.$$

A set U is open in X if it is the union of open balls, and a set G is closed if it is the complement of an open set.

The main point I wish to make here is that Theorem 3.5 is valid for general metric spaces, while Theorem 3.6 is valid only on intervals in $\mathbb{R}$ as illustrated in Example 3.7. It should be noted that the proof of Theorem 3.5, in the setting of metric spaces, is not much different from the proof given here. ∎

Exercises - (3.5)

1. Show that the set U defined in Equation (3.6) is open and nonempty.

2. Prove Statement (3.9).

3. Prove the second part of Lemma 3.4.

4. Give an example of a continuous function f, on an interval I, such that the set of periodic points of f is dense in I, but f does not have sensitive dependence on initial conditions.

5. Let
$$X = S^1 \setminus \left\{ \frac{2\pi p}{q} : p, q \in \mathbb{Z}, q \neq 0 \right\}.$$

   Define $f : I \to I$ by $f(\theta) = 2\theta$. Show that f is transitive on I, but I has no periodic points of f.

6. Let $Y = S^1 \times [0,1]$. Define $g : Y \to Y$ by $g(\theta, t) = (2\theta, t)$

   (a) Show that g possesses sensitive dependence on initial conditions.

   (b) Show that the set of periodic points of g is dense in Y.

   (c) Show that g is not transitive on Y.

7. Let
$$f(x) = \begin{cases} \frac{3}{2}x & 0 \leq x < \frac{1}{2} \\ \frac{3}{2}(1-x) & \frac{1}{2} \leq x \leq \frac{3}{4} \end{cases}$$

   be defined on the interval $I = [0, \frac{3}{4}]$.

   (a) Show that f has sensitive dependence on initial conditions.

   (b) Show that the set of periodic points is not dense in I.

8. Let $0 < \beta < 0.5$, and $f_\beta : [0,1] \to [0,1]$ be defined by

$$f_\beta(x) = \begin{cases} 2(\beta - x) & \text{for } 0 \le x \le \beta \\ 2(x - \beta) & \text{for } \beta \le x \le 0.5 + \beta \\ 2(0.5 + \beta - x) + 1 & \text{for } 0.5 + \beta \le x \le 1. \end{cases}$$

(a) Draw the graph of f_β.

(b) Find the intervals on which f_β is chaotic and describe its dynamics.

9. Define a function f on $\mathbb{R}^+$ as follows:

$$f(x) = \begin{cases} 3x & 0 \le x < \frac{1}{3} \\ -3x + 2 & \frac{1}{3} \le x \le \frac{2}{3} \\ 3x - 2 & \frac{2}{3} \le x < 1 \\ f(x-1) + 1 & x \ge 1. \end{cases}$$

(a) Show that f has sensitive dependence on initial conditions.

(b) Show that the set of periodic points of f is dense in I.

(c) Show that f is not transitive on $\mathbb{R}$.

10. Let $f : I \to I$ be a continuous map on an interval I.

(a) Suppose that for any nontrivial closed intervals U and V contained in I, we can find n so that $f^n(U) \supset V$. Prove that f is chaotic on I.

(b) Is the converse of part (a) true?

11. Suppose that $f : X \to X$ is topologically transitive on a metric space X. Prove that either X is an infinite set or X consists of the orbit of a single periodic point.

12. (a) Give the definition of chaos in metric spaces.

(b) Prove Theorem 3.5 from metric spaces.

13.* (Conjecture: Senior Project/Master's Thesis) Let $f : X \to X$ be a continuous map on a metric space X (an interval I) which is chaotic. Show that if X is connected, then f^m is chaotic for all $m \in \mathbb{Z}^+$.

14.* [103] Let $f : X \to X$ be a continuous map on metric space X. Show that f is chaotic if and only if for any given nonempty open sets U and V in X there exists a periodic point $p \in U$ and a nonnegative integer k such that $f^k(p) \in V$.

15. (Term Project) [102] If X is not connected, the conjecture in Problem 13 fails. Consider the space $S^1 \times \{0,1\}$, where S^1 is the unit circle.

   (a) The "switch" map $\sigma : \{0,1\} \to \{0,1\}$ is defined as $\sigma(0) = 1$ and $\sigma(1) = 0$. Show that σ^n is transitive for all $n \in \mathbb{Z}^+$.

   (b) Let g be the double angle map on S^1. Define

$$f = g \times \sigma : S^1 \times \{0,1\} \to S^1 \times \{0,1\}$$

   as

$$f(\theta, i) = (g(\theta), \sigma(i)) = \begin{cases} (2\theta, 1) & \text{if } i = 0, \\ (2\theta, 0) & \text{if } i = 1. \end{cases}$$

   Show that f is chaotic.

   (c) Show that f^2 is not chaotic by showing that it is not transitive.

3.6   Cantor Sets

Cantor sets play an important role in both analysis and in chaos theory. In this section we will present a brief description of Cantor sets and their topological properties. We begin our task by presenting some notions from topology.

DEFINITION 3.5

(i) Let A be a subset of $\mathbb{R}$. Then $x_0 \in \mathbb{R}$ is said to be a *limit point* of A if for every $\delta > 0$, there exists $a \in A \cap (x_0 - \delta, x_0 + \delta)$, $a \neq x_0$.

   A is said to be <u>perfect</u> if every point in A is a limit point of A.

(ii) A subset A of $\mathbb{R}$ is connected if it is not the union of two nonempty open subset of $\mathbb{R}$. Equivalently, $A \subseteq \mathbb{R}$ is connected if and only if it is an interval.

   A is said to be totally disconnected if the only nonempty connected subset of A are the one-point sets.

(iii) A subset A of $\mathbb{R}$ is a Cantor set if it is closed, bounded, perfect, and totally disconnected.

Example 3.8

(The Cantor Middle-Third Set). We begin with a bounded closed interval $S_0 = [0,1]$. Remove its open middle third $(\frac{1}{3}, \frac{2}{3})$ and denote the remaining set as S_1 so that $S_1 = [0, \frac{1}{3}] \cup [\frac{2}{3}, 1]$. Next remove the open middle third of

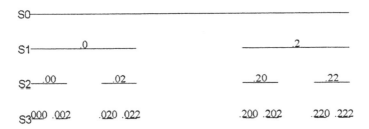

FIGURE 3.10
The construction of the Cantor middle-third set.

each of the subintervals of S_1 to obtain $S_2 = [0, \frac{1}{9}] \cup [\frac{2}{9}, \frac{1}{3}] \cup [\frac{2}{3}, \frac{7}{9}] \cup [\frac{8}{9}, 1]$. Continue this inductively (Fig. 3.10) to obtain a nested sequence of closed intervals $\{S_n\}$, where $S_{n+1} \subset S_n$. ⬜

The Cantor set is defined as

$$K = \bigcap_{n=1}^{\infty} S_n. \qquad (3.12)$$

Clearly, K is nonempty since it contains the end points of the subintervals of each S_n such as $0, \frac{1}{9}, \frac{2}{9}, \frac{1}{3}, \frac{2}{3}, \frac{7}{9}, \ldots$ Furthermore, K is closed since it is the intersection of closed sets; it is also bounded since it is a subset of the interval $[0, 1]$.

Ternary Representation of K

When a number has both terminating and nonterminating ternary representation, we agree to use the nonterminating form except when the terminating representation ends in 2. For example, $\frac{1}{3}$ has two ternary representations, .1 and $.0\bar{2}$, but we choose the nonterminating representation $.0\bar{2}$.

To show this, write

$$\frac{1}{3} = \frac{x_1}{3} + \frac{x_2}{3^2} + \frac{x_3}{3^3} + \cdots$$

The choice $x_1 = 1, x_i = 0$ for all $i > 1$ is discarded since this would lead to the terminating representation .1. Let $x_1 = 0$, then multiply both sides by 3^2 to obtain $3 = x_2 + \frac{x_3}{3} + \frac{x_4}{3^2} + \cdots$

Since x_2 takes only the values $0, 1$, or 2, we must take $x_2 = 2$. Subtracting 2 from both sides of the equation and multiplying by 3 leads to $x_3 = 2$. Hence, $\frac{1}{3} = .0\bar{2}$.

Similarly, one may show that $\frac{7}{9} = .20\bar{2}$.

We are going to show that $x \in K$ if and only if it has the ternary representation $.x_1x_2x_3\ldots$, where each x_i is either 0 or 2. To do this we write the intervals S_i in ternary representation. The set $S_1 = [0, .0\bar{2}] \cup [.2, .\bar{2}] = I_{11} \cup I_{12}$. If $x = .x_1 x_2 x_3 \ldots \in S_1$, then $x_1 = 0$ if $x \in I_{11}$, and $x_1 = 2$ if $x \in I_{12}$. We write the set S_2 in ternary representation as

$$S_2 = [0, .00\bar{2}] \cup [.02, .0\bar{2}] \cup [.2, .20\bar{2}] \cup [.22, .\bar{2}] = I_{21} \cup I_{22} \cup I_{23} \cup I_{24}.$$

Then,

1. $x \in I_{21}$ if and only if $x = .00x_3 x_4 \ldots$,

2. $x \in I_{22}$ if and only if $x = .02x_3 x_4 \ldots$,

3. $x \in I_{23}$ if and only if $x = .20x_3 x_4 \ldots$,

4. $x \in I_{24}$ if and only if $x = .22x_3 x_4 \ldots$

Continuing inductively, one may show that $x \in K$, if and only if $x_i \in \{0, 2\}$ for all $i = 1, 2, 3, \ldots$ (see Fig. 3.10).

We are now in a position to show that K is perfect. So let $x = .x_1 x_2 x_3 \ldots$ be a point in K where x_1 is either 0 or 2. Define a sequence $\{x^n\}$ in K, where $x^n = .x_1 x_2 \ldots x_n * * \ldots$ is a number in $[0, 1]$ that agrees with x in the first n ternary places. Then, $|x^n - x| \leq \sum_{i=n+1}^{\infty} \frac{2}{3^i} = \frac{1}{3^n} \to 0$ as $n \to \infty$. Hence, $x^n \to x$ as $n \to \infty$. Thus, x is a limit point of K.

Finally, K is totally disconnected since any interval in $[0, 1]$ must contain numbers with nonterminating 1 in their ternary representation. Therefore, K is a Cantor set.

Another interesting property of the Cantor middle-third set is that it provides us with an example of an **uncountable** set, which is totally disconnected (or a set of measure zero). This assertion may be concluded using a result from real analysis, which states that any nonempty perfect set of $\mathbb{R}$ must be uncountable [82]. However, we may prove this latter statement by using a rather simple and elegant arrangement. Define a map $h : [0, 1] \to K$ as follows. We write each point $x \in [0, 1]$ in a binary expansion $x = .x_1 x_2 x_3 \ldots$, where each x_i is either 0 or 1. We let $h(x) = y = .y_1 y_2 y_3 \ldots$, with $y_i = 2x_i$. Then, clearly h is one-to-one. Since the interval $[0,1]$ is uncountable, so is K.

Example 3.9
(Another Cantor Set). Consider the logistic map

$$F_\mu(x) = \mu x(1 - x) \text{ on } I = [0, 1],$$

where $\mu > 4$. Note that in this case $F_\mu(\frac{1}{2}) > 1$. Since $F_\mu(0) = 0$, it follows, by the intermediate value theorem, that there exists $\alpha_0 \in (0, \frac{1}{2})$ such that $F_\mu(\alpha_0) = 1$. Since F_μ is monotone on $[0, \frac{1}{2}]$, the interval $I_0 = [0, \alpha_0]$ consists of all points x to the left of $\frac{1}{2}$ where $F_\mu(x) \in I$. Similarly, there exists

$\alpha_1 > \frac{1}{2}$ with $F_\mu(\alpha_1) = 1$ and such that $F_\mu(x) \in I$ for all $x \in I_1 = [\alpha_1, 1]$ [see Fig. 3.11(a) and (b)].

Let $\quad A_1 = I_0 \cup I_1.$
Then $\quad A_1 = \{x \in I : F_\mu(x) \in I\}.$
Define $\quad A_2 = \{x \in I : F_\mu^2(x) \in I\} = \{x \in I : F_\mu(x) \in A_1\}.$

Then, A_2 consists of the four closed intervals $A_2 = I_{00} \cup I_{01} \cup I_{11} \cup I_{10}$, (see Fig. 3.12) where

$$I_{00} = \{x : x \in I_0 \quad \text{and} \quad F_\mu(x) \in I_0\},$$
$$I_{01} = \{x : x \in I_0 \quad \text{and} \quad F_\mu(x) \in I_1\},$$
$$I_{11} = \{x : x \in I_1 \quad \text{and} \quad F_\mu(x) \in I_1\},$$
$$I_{10} = \{x : x \in I_1 \quad \text{and} \quad F_\mu(x) \in I_0\}.$$

Continuing this process, we construct $A_n = \cup I_{s_0 s_1 \dots s_{n-1}}$, where s_i is either 0 or 1, and

$$I_{s_0 s_1 \dots s_j} = \{x \in I : x \in I_{s_0}, F_\mu(x) \in I_{s_1}, \dots, F_\mu^j(x) \in I_{s_j}\}$$

$$= \bigcap_{k=0}^{j} F_\mu^{-k}(I_{s_k})$$

$$= I_{s_0} \cap F_\mu^{-1}(I_{s_1 s_2 \dots s_j}). \quad \square \tag{3.13}$$

We first note that $A_n = \{x \in I : F_\mu^n(x) \in I\}$. Furthermore,

$$I_{s_0 s_1 \dots s_n} = I_{s_0 s_1 \dots s_{n-1}} \cap F_\mu^{-n}(I_{s_n}) \subset I_{s_0 s_1 \dots s_{n-1}}$$

Hence, $A_{n+1} \subset A_n$. Define the set

$$\Lambda = \bigcap_{n=1}^{\infty} A_n. \tag{3.14}$$

In the discussion that follows, we will show that Λ is a Cantor set.

We begin this task by computing points $\alpha_0 \in I_0$, and $\alpha_1 \in I_1$. This amounts to solving the equation $\mu x(1 - x) = 1$. Hence,

$$\alpha_0 = \frac{1}{2} - \frac{\sqrt{\mu^2 - 4\mu}}{2\mu},$$

and $\alpha_1 = \frac{1}{2} + \frac{\sqrt{\mu^2 - 4\mu}}{2\mu}$. To prove that Λ is a Cantor set we now assume that $\mu > 2 + \sqrt{5}$. Although this is true for $\mu > 4$, the proof becomes very much involved for $4 < \mu \le 2 + \sqrt{5}$ and we choose to leave it for more advanced texts.

We start the proof by using the following technical lemma.

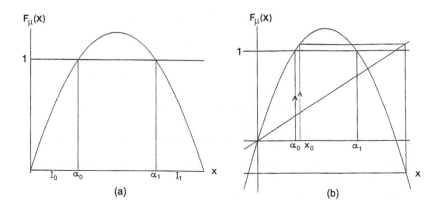

FIGURE 3.11
$A_1 = I_0 \cup I_1$. If $x_0 \notin A_1$, then $F_\mu(x_0) \notin I$.

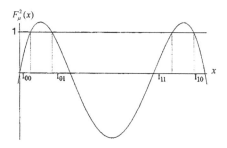

FIGURE 3.12
$A_2 = I_{00} \cup I_{01} \cup I_{11} \cup I_{10}$.

LEMMA 3.5

For $\mu > 2 + \sqrt{5}$, there exists $\varepsilon > 0$ such that $|F'_\mu(x)| > 1 + \varepsilon$, for all $x \in A_1$. Moreover, the length of each subinterval in A_n is less than $\frac{1}{(1+\varepsilon)^n}$.

PROOF   Let $\mu = 2 + \sqrt{5} + \delta$, for some $\delta > 0$. Then,

$$F'_\mu(\alpha_0) = (2 + \sqrt{5} + \delta) - 2(2 + \sqrt{5} + \delta) \left[\frac{1}{2} - \frac{\sqrt{1 + \delta^2 + 2\sqrt{5}\delta}}{2(2 + \sqrt{5} + \delta)} \right]$$

$$= \sqrt{1 + \delta^2 + 2\sqrt{5}\delta}$$

$$> 1 + \varepsilon, \quad \text{for some } \varepsilon > 0.$$

Since $F''_\mu = -2\mu < 0, F'_\mu$ is decreasing. So $F'_\mu(x) > F'_\mu(\alpha_0) > 1 + \varepsilon$ for all $x \in I_0$. By a similar argument, one may show the same result for all $x \in I_1$. This proves the first part of the lemma. The second part of the lemma is left as Problem 14.   ∎

Finally, we are ready to prove our main result.

THEOREM 3.7

The set Λ is a Cantor set.

PROOF   Since Λ is the intersection of closed sets, it is closed. Furthermore, Λ is bounded since it is contained in the interval $[0, 1]$ and nonempty since it contains at least the end points of all the intervals of the form $I_{s_0 s_1 \ldots s_n}, n \in \mathbb{Z}^+$, and $s_i = 0$ or 1. We leave it to you to show that Λ is perfect and totally disconnected (Problem 4).   ∎

3.7   Symbolic Dynamics

As we promised in the preceding section, we will now introduce the **sequence space** $\sum_2^+$ of all the one-sided sequences on the two symbols 0 and 1. This is defined as follows:

$$\sum_2^+ = \{x = \{x_n\}_{n=0}^\infty : x_n = 0 \ \text{ or } \ 1\}.$$

Thus, elements of $\sum_2^+$ are infinite strings of 0's and 1's such as

$$\{0 \ 0 \ 1 \ 0 \ 1 \ \ldots\}, \{0 \ 1 \ 0 \ 0 \ \ldots\}$$

etc. An overbar over a group of digits indicates that the group is repeated indefinitely. For example,

$$\{0\ 1\ 0\ 0\ \overline{1\ 0}\ \ldots\}$$

denotes the sequence

$$\{0\ 1\ 0\ 0\ 1\ 0\ 1\ 0\ 1\ 0\ \ldots\}.$$

On $\sum_2^+$, we define the shift map $\sigma : \sum_2^+ \to \sum_2^+$ by letting

$$\sigma\{x_0\ x_1\ x_2\ x_3\ \ldots\} = \{x_1\ x_2\ x_3\ \ldots\}.$$

For example, $\sigma\{1\ 0\ 1\ \bar{1}\ \ldots\} = \{0\ 1\ \bar{1}\ \ldots\}$.

The apparent simplicity of the map σ enables us to have a full understanding of its properties. There are two fixed points of σ: the constant sequences $\{0\ \bar{0}\ \ldots\}$ and $\{1\ \bar{1}\ \ldots\}$. Eventually fixed points are in abundance; they are of the forms $\{x_0\ x_1\ \ldots\ x_n\bar{1}\ \ldots\}$ and $\{x_0\ x_1\ \ldots\ x_n\bar{0}\ \ldots\}$ for all $n > 0$. The k-periodic points are of the form $\{\overline{x_0\ x_1\ \ldots\ x_{k-1}}\ \ldots\}$.

Since there are 2^k ways of arranging the block $x_0\ x_1\ \ldots\ x_{k-1}$, it follows that there are 2^k periodic points of period k. Eventually periodic points are also in abundance; they are of the form

$$\{x_0\ x_1\ \ldots\ x_n\ \overline{x_{n+1}\ x_{n+2}\ \cdots\ x_{n+k-1}}\ldots\}.$$

Next, we define a metric (or a distance function) on $\sum_2^+$ as follows: for $x = \{x_n\}$ and $y = \{y_n\}$ in $\sum_2^+$, we let

$$d(x,y) = \sum_{i=0}^{\infty} \frac{|x_i - y_i|}{2^i}. \tag{3.15}$$

We will now make a few observations about the distance function d.

REMARK 3.5

1. Since $|x_i - y_i|$ is either 0 or 1, we then have $d(x,y) \leq \sum_{i=0}^{\infty} \frac{1}{2^i} = 2$. Hence, this metric is bounded.

2. Suppose that $x_i = y_i$ for $i = 0, 1, 2, \ldots, n$. Then, $d(x,y) \leq \frac{1}{2^n}$. To show this, observe that

$$d(x,y) = \sum_{i=0}^{n} \frac{0}{2^i} + \sum_{i=n+1}^{\infty} \frac{|x_i - y_i|}{2^i}$$

$$\leq \sum_{i=n+1}^{\infty} \frac{1}{2^i}.$$

Thus,

$$d(x,y) \leq \frac{1}{2^n}.$$

3. If, on the other hand, $d(x, y) < 1/2^n$, then $x_i = y_i$ for $i = 0, 1, 2, \ldots, n$. To show this, assume that $x_j \neq y_j$ for some $j \leq n$.

Then, $d(x, y) = \sum_{i=0}^{\infty} \frac{|x_i - y_i|}{2^i} \geq \frac{1}{2^j} \geq \frac{1}{2^n}$, a contradiction. ∎

Based on the above remarks, the following conclusions can be made.

LEMMA 3.6
The function d defined by Equation (3.15) is a metric on Σ_2^+.

PROOF   This is left to the reader as Problem 4. ∎

Example 3.10
Find $d(x, y)$ if

1. $x = \{0\ 1\ \bar{1}\ \ldots\}$ and $y = \{1\ 0\ \bar{1}\ \ldots\}$.

2. $x = \{0\ 1\ 0\ 1\ 0\ 1\ \bar{0}\ \}$ and $y = \{0\ 1\ 0\ 1\ 0\ 1\ \bar{1}\ \ldots\}$.   ⬜

SOLUTION

1. $d(x, y) = \frac{|x_1 - y_1|}{2^0} + \frac{|x_2 - y_2|}{2^1} + \cdots = 1 + \frac{1}{2} + 0 + \cdots = \frac{3}{2}$

2. $d(x, y) = \sum_{i=6}^{\infty} \frac{1}{2^i} = \frac{\frac{1}{2^6}}{1 - \frac{1}{2}} = \frac{1}{2^5}$ ∎

The next step is to show that the shift map σ is chaotic on the sequence space. But before doing so, we need to establish its continuity.

LEMMA 3.7
The shift map $\sigma : \Sigma_2^+ \to \Sigma_2^+$ is continuous.

PROOF   Let $\varepsilon > 0$ be given and $x = \{x_0\ x_1\ x_2\ \ldots\} \in \Sigma_2^+$.
Then, for some $n \in \mathbb{Z}^+$, $\frac{1}{2^n} < \varepsilon$. Let $\delta = \frac{1}{2^{n+1}}$. If $y \in \Sigma_2^+$ with $d(x, y) < \delta$, then, from Remark 3.5(3) above, we conclude that $x_i = y_i$ for $i = 0, 1, \ldots, n+1$. Let $\sigma(x) = \tilde{x}$ and $\sigma(y) = \tilde{y}$. Then $\tilde{x}_i = \tilde{y}_i$ for $i = 0, 1, \ldots, n$. Remark 3.5(2) then gives $d(\tilde{x}, \tilde{y}) \leq \frac{1}{2^n} < \varepsilon$. Consequently, σ is continuous at x. Since x was arbitrarily chosen, σ is continuous on Σ_2^+. ∎

Now we are ready to establish the chaoticity of σ on Σ_2^+.

THEOREM 3.8

The shift map $\sigma : \sum_2^+ \rightarrow \sum_2^+$ is chaotic on $\sum_2^+$.

PROOF We first prove that the set of periodic points of σ is dense in $\sum_2^+$. So let $x = \{x_i\} = \{x_0, \ldots, x_n, \ldots\}$ be an arbitrary point in $\sum_2^+$. We will produce a sequence of periodic points that converges to x. This sequence is constructed as follows:

$$y_1 = \{\overline{x_0} \ldots\},$$
$$y_2 = \{\overline{x_0\ x_1} \ldots\},$$
$$\vdots$$
$$y_n = \{\overline{x_0\ x_2\ \ldots\ x_n} \ldots\}.$$

It is easy to prove that $y_n \rightarrow x$ as $n \rightarrow \infty$ (Why?).

Next, we exhibit a dense orbit in $\sum_2^+$. Consider the sequence

$$z = \left\{ \underbrace{01}_{1-\text{block}}, \underbrace{00\ 01\ 10\ 11}_{2-\text{blocks}}, \underbrace{000\ 001\ 010\ 100\ 011\ 101\ 110\ 111}_{3-\text{blocks}}, \ldots \right\}.$$

This sequence consists of all the blocks of 0's and 1's of length 1 (there are only 2), of length 2 (there are four 2-blocks), etc. We claim that $\overline{O(z)} = \sum_2^+$. To prove the claim, let $x \in \sum_2^+$ and $B_\delta(x)$ be a ball around x, for some $\delta > 0$. To show that $x \in \overline{O(z)}$, it suffices to show that $B_\delta(x) \cap O(z) \neq \emptyset$. Choose $n \in \mathbb{Z}^+$ such that $\frac{1}{2^n} < \delta$. Then, the first $n + 1$ terms $x_0\ x_1\ x_2\ \ldots\ x_n$ in the sequence x must appear in the n-blocks in the sequence z. Therefore, for some $k \in \mathbb{Z}^+, \sigma^k(z)$ agrees with x in the first n terms. Thus, $d(\sigma^k(z), x) \leq \frac{1}{2^n} < \delta$ and the proof of the claim is now complete. It follows from Theorem 3.5 that σ is chaotic on $\sum_2^+$.

Before ending this section, we want to show that $\sum_2^+$ is indeed a Cantor set. This will be accomplished with minimal effort. Define a map $h : K \rightarrow \sum_2^+$ as follows: for $x = 0.x_0\ x_1\ x_2\ \ldots$ with x_i is either 0 or 2, we let

$$h(x) = y = \{y_0, y_1, y_2, \ldots\}, \quad \text{where} \quad y_i = \frac{x_i}{2}. \tag{3.16}$$

It is not hard to verify that h is a homeomorphism; that is, h is one-to-one onto, continuous such that the inverse h^{-1} has the same properties. Thus, $\sum_2^+$ inherits all of the topological properties of K; thus, $\sum_2^+$ is a Cantor set (Problem 8). ■

Exercises - (3.6 and 3.7)

In Problems 1–3 we consider the map $h : \mathbb{R} \to \mathbb{R}$ defined by

$$h(x) = \begin{cases} 3x; & x \leq \frac{1}{2} \\ 3(1-x); & x > \frac{1}{2}. \end{cases}$$

1. Find all the fixed points of the map h and write down their ternary expansion.

2. Show that if $x \in (\frac{1}{9}, \frac{2}{9}) \cup (\frac{1}{3}, \frac{2}{3}) \cup (\frac{7}{9}, \frac{8}{9})$, then $h^n(x) \to -\infty$ as $n \to \infty$.

3. Define the set $E = \{x \in [0,1] : h^n(x) \in [0,1] \text{ for all } n \in \mathbb{Z}^+\}$.

    (a) Prove that E is the Cantor middle-third set K.

    (b) Use Part (a) to show that the map $f : E \to [0, \frac{1}{3}] \cap E$, defined by $f(x) = \frac{1}{3}x$, is a homeomorphism.

4. Prove that the set Λ defined by Formula (3.14) is perfect and totally disconnected.

    (Hint: Use Lemma 3.5.)

5. (The Cantor middle-fifth set). Let $\tilde{S}_0 = [0,1]$. Remove its open middle fifth $(\frac{2}{5}, \frac{3}{5})$ and denote the remaining set by $\tilde{S}_1$ so that $\tilde{S}_1 = [0, \frac{2}{5}] \cup [\frac{3}{5}, 1]$. Next, remove the open middle fifths of each of the two subintervals of $\tilde{S}_1$ to obtain $\tilde{S}_2 = [0, \frac{4}{25}] \cup [\frac{6}{25}, \frac{2}{5}] \cup [\frac{3}{5}, \frac{19}{25}] \cup [\frac{21}{25}, 1]$. We define $\tilde{S}_n$ inductively. Prove that the set $\tilde{K} = \bigcap_{n=0}^{\infty} \tilde{S}_n$ is a Cantor set.

6. A point x is a nonwandering point for a map g if, for any open interval J containing x, there exists $p \in J$ and $k \in \mathbb{Z}^+ \backslash \{0\}$ such that $g^n(p) \in J$. Let $\Omega(g)$ be the set of all nonwandering points of g.

    (a) Show that $\Omega(g)$ is a closed set.

    (b) Show that $\Omega(F_\mu) = \Lambda$, if $\mu > 2 + \sqrt{5}$.

7. Let E be the set of all sequences in Σ_2^+ that contain no consecutive 0's.

    (a) Show that E is invariant under the shift map σ.

    (b) How many periodic points of period k does σ have in E?

    (c) Is $\sigma : E \to E$ chaotic on E?

8. Use the map h in (3.16) above to show that Σ_2^+ is a Cantor set.

9. Let $\sum_N^+$ be the set of all sequences $x = x_0\, x_1\, x_2\, \ldots$ where x_i is one of the numbers in the set $\{0, 1, 2, \ldots, N-1\}$. Define a metric on $\sum_N^+$ by letting

$$d_N(x, y) = \sum_{i=0}^{\infty} \frac{|x_i - y_i|}{N^i}.$$

Show that d_N is a metric on $\sum_N^+$.

10. Consider the shift map $\sigma : \sum_N^+ \to \sum_N^+$, where $\sum_N^+$ is as defined in Problem 8. Prove that σ is chaotic on $\sum_N^+$.

11. Show that:

(a) the set of eventually periodic points of the shift map σ that are not periodic is dense in $\sum_2^+$.

(b) The set of points that are neither periodic nor eventually periodic, of the shift map σ, is dense in $\sum_2^+$.

12. For:

(a) $x = \{x_1\, x_2\, \ldots\} \in \sum_2^+$, define the stable set

$$W^s(x) = \left\{ y \in \sum_2^+ : d(\sigma^n(x), \sigma^n(y)) \to 0 \ \text{ as } \ n \to \infty \right\}.$$

Describe the set $W^s(x)$.

(b) Let $b = 10\overline{10}\cdots$. Determine $W^s(b)$.

13. Complete the proof of Lemma 3.5 by showing that the length of each interval in A_n is less than $\frac{1}{(1+\varepsilon)^n}$ for some $\varepsilon > 0$.

14. Show that the map h defined by (3.16) is a homeomorphism.

3.8 Conjugacy

So far, we have shown that the shift map σ, on the sequence space $\sum_2^+$ (Lemma 3.7) is chaotic. In Example 3.13, we will show that the double-angle map is chaotic on the unit circle S^1. In this section, we will show that if two maps are conjugate, they must then have identical topological properties. In particular, if a map is conjugate to either the shift map or the double-angle map, it is then chaotic.

Recall that if $f : J \to J$ is a one-to-one and onto map such that f and f^{-1} are continuous, then f is said to be a **homeomorphism**. For example, the

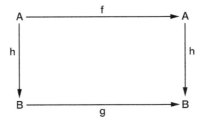

FIGURE 3.13
The map f is h-conjugate to the map g.

map $f(x) = x^3$ is a homeomorphism on $\mathbb{R}$, whereas the map $f(x) = x^2$ is not a homeomorphism on $\mathbb{R}$ but a homeomorphism on $\mathbb{R}^+$.

Next, we begin our exposition by introducing the notion of conjugacy.

DEFINITION 3.6   *Let $f : A \to A$ and $g : B \to B$ be given maps. Then f and g are said to be **conjugate**, denoted by $f \approx g$, if there exists a homeomorphism $h : A \to B$ such that $h \circ f = g \circ h$ (Fig. 3.13).*

We also say that f is h-conjugate to g to emphasize the importance of the homeomorphism h.

It is not difficult to show that the conjugacy relation $\approx$ is an equivalence relation (Problem 1). Furthermore, if $f \approx g$, then $f^k \approx g^k$ for all $k \in \mathbb{Z}^+$ (Problem 2).

Example 3.11
Consider the logistic map $F_\mu(x) = \mu x(1 - x)$, with $0 < \mu \le 4$ and the quadratic map $G(x) = ax^2 + bx + c$, where $a \ne 0$. Show that F_μ and G are conjugate via the homeomorphism

$$h(x) = -\frac{\mu}{a}x + \frac{\mu - b}{2a}. \quad \square \qquad (3.17)$$

SOLUTION   It is easy to verify that the map $h : [0, 1] \to \left[\frac{-\mu-b}{2a}, \frac{\mu-b}{2a}\right]$ is indeed a homeomorphism. Now, if $h(F_\mu(x)) = G(h(x))$, we must have

$$-\frac{\mu}{a}[\mu x(1 - x)] + \frac{\mu - b}{2a} = a\left[\frac{-\mu}{a}x + \frac{\mu - b}{2a}\right]^2 + b\left(\frac{-\mu}{a}x + \frac{\mu - b}{2a}\right) + c,$$

which gives

$$c = \frac{b^2 - \mu^2 + 2\mu - 2b}{4a}.$$

Therefore, for this value of c, F_μ is h-conjugate to G.

If the conjugacy map h between two maps f and g happens to be linear, as in the above example, we say that f is **linearly conjugate** to g. ∎

We illustrate this procedure in the next example.

Example 3.12
Show that $F_4(x) = 4x(1 - x)$ on $[0,1]$ is linearly conjugate to the map $f(x) = 2x^2 - 1$ on $[-1,1]$. ∎

SOLUTION   From the preceding example (Formula (3.17)), the map $h(x) = -2x + 1$ is the conjugation map which takes $[0,1]$ onto $[-1,1]$. Moreover,

$$h(F_4(x)) = 8x^2 - 8x + 1 = f(h(x)).$$

Since h is a homeomorphism, it conjugates F_4 and g. ∎

Next, we show that conjugacy preserves chaos.

THEOREM 3.9
Suppose that the map $f : A \to A$ is h-conjugate to the map $g : B \to B$. Then f is chaotic on A, if and only if g is chaotic on B.

PROOF   Suppose that f is chaotic on A. To show that g is chaotic, we first show that it is transitive. Let U and V be two open sets in B and suppose that g h-conjugates f. Then $h(U)$ and $h(V)$ are open sets in A. Since f is chaotic, there exists $k \in \mathbb{Z}^+$ such that $f^k(h(U)) \cap h(V) \neq \emptyset$. Hence, $h(g^k(U)) \cap h(V) \neq \emptyset$ (see Problem 2). Consequently, $g^k(U) \cap V \neq \emptyset$. Hence, g is transitive.

Next, we show that the set P of periodic points in B is dense in B. To this end, we let U be any open subset of B. Then, $h^{-1}(U)$ is an open subset of A and thus must contain a k-periodic point $x \in A$. Since $x = f^k(x)$, it follows that $h(x) = h(f^k(x)) = g^k(h(x))$. So $k(x)$ is a k-periodic point of g. Furthermore, $h(x) \in h(h^{-1}(U)) = U$, and consequently, the set P is dense in B. We now use Theorem 3.5 to conclude that the map g is chaotic on B. ∎

The first application of Theorem 3.9 is on the logistic map F_μ, which will be stated in the next result.

THEOREM 3.10
If $\mu > 2 + \sqrt{5}$, then the logistic map F_μ is chaotic on the space Λ.

PROOF   We prove the theorem by establishing a conjugacy between F_μ and the shift map σ. The conjugacy map $h : \Lambda \to \sum_2^+$ is defined as follows: For $x \in \Lambda$ we let

$$h(x) = \{a_0 \ a_1 \ a_2 \ \ldots\}, \quad \text{where} \quad a_n = \begin{cases} 0 \text{ if } F_\mu^n(x) \in I_0 \\ \\ 1 \text{ if } F_\mu^n(x) \in I_1. \end{cases} \tag{3.18}$$

In other words, $h(x) = \{a_0 \ a_1 \ a_2 \ \ldots\}$, if and only if $F_\mu^n(x) \in I_{a_n}$ for each $n \in \mathbb{Z}^+$. The sequence $h(x)$ is called the **itinerary** of x.

Next, we show that the map h is one-to-one and onto. To do so, we need to show that if $a = a_0 \ a_1 \ a_2 \ \ldots \ \in \sum_2^+$, then $h^{-1}(a)$ is exactly one point. Observe that if $x \in h^{-1}(a)$, then $x \in I_{a_0 \ a_1 \ \ldots \ a_n}$ for all $n \in \mathbb{Z}^+$; i.e., $h^{-1}(a) = \bigcap_{n=0}^{\infty} I_{a_0 \ a_1 \ \ldots \ a_n}$. Recall from Example 3.9 that

$$I_{a_0} \supset I_{a_0 \ a_1} \supset I_{a_0 \ a_1 \ a_2} \ \cdots \ \supset I_{a_0 \ a_1 \ \ldots \ a_n} \supset \cdots,$$

and from Lemma 3.5, the length of $I_{a_0 \ a_1 \ \ldots \ a_n}$ tends to 0 as $n \to \infty$. This implies by the nested interval theorem (Problem 14) that $h^{-1}(a)$ is indeed a single point in $\sum_2^+$.

The proof that h and h^{-1} are continuous is left to Problem 4. We also can show that F_μ is h-conjugate to σ (Problem 5).

Now, since σ is chaotic on $\sum_2^+$ (Theorem 3.8), it follows from Theorem 3.9 that F_μ is chaotic on Λ for $\mu > 2 + \sqrt{5}$. ∎

REMARK 3.6   Observe that in the proof of Theorem 3.9, we only need the fact that the conjugacy map h is onto, continuous, and open. Hence, we really did not need to assume that the conjugacy map h is one-to-one. This leads to the introduction of a weaker notion called **semiconjugacy**, which we will now explore. ∎

DEFINITION 3.7   *The map $f : A \to A$, and $g : B \to B$ are said to be* **semiconjugate** *if there exists a map $h : A \to B$, with $h \circ f = g \circ h$ and such that h is onto, continuous, and open.*

THEOREM 3.11
If $f : A \to A$ and $g : B \to B$ are semiconjugate, then f is chaotic, if and only if g is chaotic.

PROOF   Modify the proof of Theorem 3.9 (Problem 14). ∎

Before we illustrate the utility of conjugacy, we are going to revisit the double-angle map.

Example 3.13
(The Double Angle Map). Let $g : S^1 \to S^1$ be a map on the unit circle given by $g(\theta) = 2\theta$. Show that g is chaotic on S^1. ▯

SOLUTION   We observe first that a point θ in S^1 is k-periodic if and only if $g^k(\theta) = \theta$ or $2^k\theta = \theta$. This is true if and only if $2^k\theta = \theta + 2n\pi$, for some $n \in \mathbb{Z}^+$. Solving for θ, we obtain

$$\theta = \frac{2n\pi}{2^k - 1}. \tag{3.19}$$

Thus, we conclude that the point θ is k-periodic if and only if θ is of the form (3.19), for $n = 0, 1, 2, \ldots, 2^k - 2$. Hence, there are $(2^k - 1)$ periodic points of period k in this map, which may be written in the form[2]

$$x(n) = 2n\pi/(2^k - 1). \tag{3.20}$$

Let U be an open arc on S^1 defined as $U = \{\theta | \theta_1 < \theta < \theta_2\}$ (Fig. 3.14). Let $d = \frac{(\theta_2 - \theta_1)}{2}$.

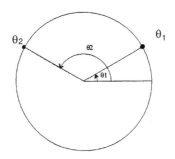

FIGURE 3.14
An open arc in $S^1 : U = \{\theta | \theta_1 < \theta < \theta_2\}$.

Now, if $\theta(n) = \frac{2n\pi}{2^k - 1}$ and $\theta(n+1) = \frac{2(n+1)\pi}{2^k - 1}$ are two consecutive angles in Formula (3.19), then $\theta(n+1) - \theta(n) = \frac{2\pi}{2^k - 1}$. Choose k sufficiently large such that $\frac{2\pi}{2^k - 1} < d$. Hence, there exist n, and $k \in \mathbb{Z}^+$ with $\theta(n) = 2n\pi/(2^k - 1) \in U$. Hence, the set of periodic points is dense in S^1. To prove that g is transitive, let U and V be two open arcs in S^1 such that the length of the

---

[2]The points $x(n)$ in Formula (3.19) are the roots of the equation $z^{(2^k - 1)} = 1$ in the complex domain. If we let $\omega = 2i\pi/(2^k - 1)$, then $x(0) = 1, x(1) = \omega, x(2) = \omega^2, \ldots, x(2^k - 2) = \omega^{2^k - 2}$.

arc U is δ. Then, $g(U)$ has length 2δ, $g^2(U)$ has length 4δ, and eventually $g^k(U) = S^1$ for some $k \in \mathbb{Z}^+$. Thus, $g^k(U) \cap V \neq \emptyset$.

By Theorem 3.6 (for metric spaces), g is chaotic. ∎

To illustrate the utility of the preceding theorem, we give the following example.

Example 3.14

1. Use the double-angle map $g : S^1 \to S^1$ given by $g(\theta) = 2\theta$ of Example 3.13 to show that the logistic map $F_4(x) = 4x(1-x)$ is chaotic on the interval $I = [0, 1]$.

2. Use the logistic map F_4 to show that the tent map T is chaotic on $I = [0, 1]$.

3. Show that the doubling map $D : [0, 1] \to [0, 1]$ and the double angle map $g : S^1 \to S^1$ are conjugate. Conclude that D is chaotic on $[0, 1]$.

◻

SOLUTION

1. Define the map $h : S^1 \to I$ by letting $h(\theta) = \sin^2(\frac{\theta}{2})$. This map is onto (not one-to-one), continuous, and open. Furthermore,

$$h(g(\theta)) = \sin^2(\theta) = F_4(h(\theta)).$$

Thus, h semiconjugates g with F_4. Since g is chaotic on S^1 it follows from Theorem 3.11 that F_4 is chaotic on I.

2. Define the map $h : I \to I$ by letting $h(x) = \sin^2(\frac{\pi}{2}x)$. Then, h semiconjugates T with F_4. To prove this, we will show that $F_4 \circ h = h \circ T$.

Now

$$F_4(h(x)) = 4\sin^2\left(\frac{\pi}{2}x\right)\left[1 - \sin^2\left(\frac{\pi}{2}x\right)\right]$$
$$= 4\sin^2\left(\frac{\pi}{2}x\right)\cos^2\left(\frac{\pi}{2}x\right)$$
$$= \sin^2(\pi x),$$

on the other hand

$$h(T(x)) = \begin{cases} h(2x) = \sin^2(\pi x) & \text{if } 0 \leq x \leq \frac{1}{2}, \\ h(2 - 2x) = \sin^2(\pi - \pi x) = \sin^2 \pi x & \text{if } \frac{1}{2} \leq x \leq 1. \end{cases}$$

Thus h semiconjugates T with F_4 and hence T is chaotic.

3. Use the conjugacy $h : [0,1] \to S^1$, defined by $h(x) = 2\pi x$. The details are left to the reader as Problem 9.

∎

Exercises - (3.8)

1. Show that the conjugacy relation $\approx$ between maps is an equivalence relation.

2. Show that if $f \approx g$, then $f^k \approx g^k$, for all $k \in \mathbb{Z}^+$.

In Problems 3–8 we consider the map $h : \Lambda \to \Sigma_2^+$ as defined by Equation (3.18), where $\mu > 2 + \sqrt{5}$.

3. For $\mu = 5$, find   (a) $h\left(\frac{\sqrt{5}-1}{2\sqrt{5}}\right)$,   (b) $h\left(\frac{1+\sqrt{5}}{2\sqrt{5}}\right)$.

4. Prove that the map h and h^{-1} are continuous.

5. Show that F_μ is h-conjugate to σ.

6. Show that the map $f(x) = x^2 - \frac{3}{4}$ on $[-\frac{3}{2}, \frac{3}{2}]$ and the logistic map $F_3(x) = 3x(1 - x)$ on $[0,1]$ are linearly conjugate.

7. Show that the map $G_\lambda(x) = 1 - \lambda x^2$ on $[-1,1]$, where $\lambda \in (0,2]$ is conjugate to the logistic map F_μ on the interval $\left[1 - \frac{\mu}{4}, \frac{\mu}{4}\right]$, where $\mu \in (2,4]$. Then, show that G_2 is chaotic on $[-1,1]$.

8. Suppose that f_1 is linearly conjugate to f_2, and f_2 is linearly conjugate to f_3. Prove that f_1 is linearly conjugate to f_3.

9. Show that the doubling map $D : [0,1] \to [0,1]$ is conjugate to the double angle map $g : S^1 \to S^1$. Then conclude that D is chaotic on $[0,1]$.

10. Consider the maps $f : S^1 \to S^1$ defined by $f(\theta) = 3\theta$ and $g : [-1,1] \to [-1,1]$ defined by $g(x) = 4x^3 - 3x$.

    (a) Show that f and g are semiconjugate.
    (b) Show that g is chaotic on $[-1,1]$.

In Problems 11–13 we consider another quadratic map defined by $Q_c(x) = x^2 + c$ on the interval $J = [-2,2]$.

11. Show that Q_{-2} is conjugate to the logistic map F_4 and then conclude that Q_{-2} is chaotic on the interval $[-2,2]$.

12. (a) Show that $c < -\dfrac{(5 + 2\sqrt{5})}{4}$ corresponds to the case $\mu > 2 + \sqrt{5}$ for the logistic map F_μ.

(b) Let $J_n = \{x \in J : Q_c^n(x) \in J\}$ and $\tilde{\Lambda} = \bigcap\limits_{n=1}^{\infty} J_n$. Show that if $c < -\dfrac{(5 + 2\sqrt{5})}{4}$, then $\tilde{\Lambda}$ is a Cantor set.

13. Prove that the map Q_c is chaotic on $\tilde{\Lambda}$ for $c < -\dfrac{(5 + 2\sqrt{5})}{4}$.

14. Prove Theorem 3.11.

15. Let $f : \mathbb{R} \to \mathbb{R}^1$ be a C^1-map and I_1, I_2 be two disjoint closed bounded intervals. Let $I = I_, \cup I_2$ and assume that $f(I_i) \supset I$, for $i = 1, 2$. Assume also that $|f'(x)| \geq \lambda > 1$ for all $x \in I \cap f^{-1}(I)$.

(a) Prove that $\Lambda = \bigcap\limits_{k=0}^{\infty} f^{-k}(I)$ is a Cantor set.

(b) Define $h : \Lambda \longrightarrow \Sigma_2^+$ as the itinerary map (3.18). Show that h is a conjugacy map.

(c) Show that f is chaotic on Λ.

3.9 Other Notions of Chaos

We are now in a position to present the three notions of chaos put forth by Li and Yorke [62], Block and Coppel [12], and Devaney [25]. In this presentation, we will follow the paper by Aulbach and Kieninger [3] and assume that the space X is compact, which in $\mathbb{R}^n$ means that X is closed and bounded. The assumption of compactness is not essential in the definition of chaos, but it makes the relationship among the various notions of chaos more transparent. We will use the definition of chaos in the sense of Devaney (Definition 3.4).

DEFINITION 3.8 *A continuous map $f : X \to X$ is L/Y-Chaotic if there exists an uncountable subset S of X such that*

1. $\limsup\limits_{n \to \infty} d(f^n(x), f^n(y)) > 0$ *for all $x, y \in S$, $x \neq y$,*

2. $\liminf\limits_{n \to \infty} d(f^n(x), f^n(y)) = 0$ *for all $x, y \in S$, $x \neq y$,*

3. $\limsup\limits_{n \to \infty} d(f^n(x), f^n(p)) > 0$ *for all $x \in S$, $p \in X$, p periodic.*

It may be shown [12] that condition (3) in L/Y-chaos is redundant, since conditions (1) and (2) imply condition (3). In their famous paper [62], T.Y. Li and J. Yorke had an additional condition for L/Y-chaos: "f has periodic points of all periods." This was the main reason behind the title of their paper ""Period 3 implies chaos." They proved that if a continuous map on $\mathbb{R}$ has period 3, then it must have points of all periods (see Theorem 2.9).

DEFINITION 3.9   *A continuous map $f : X \to X$ is B/C-chaotic if there exists a positive integer m and a compact f^m-invariant subset Y of X such that $f^m|_Y$ is semiconjugate to the shift map σ on $\sum_2^+$.*

THEOREM 3.12 [55]
If the continuous maps f and g are conjugates, then f is chaotic in any one of the three senses if and only if g is chaotic in the same sense.

The preceding theorem may be extended to the case when f and g are semiconjugates.

THEOREM 3.13 [3]
Let $f : I \to I$ be a continuous map on a closed and bounded interval I. Then the following hold:

(i) D-chaos $\Leftrightarrow$ f has a positive topological entropy,[3]

(ii) B/C-chaos $\Rightarrow$ L/Y-chaos $\nRightarrow$ D-chaos.

The following example illustrates the above result.

Example 3.15
(Truncated Tent Map). Consider the tent map T. Now, for each $\lambda \in (0, 1]$, we define $G_\lambda : [0, 1] \to [0, 1]$ by $G_\lambda(x) = \min\{\lambda, T(x)\}$.

---

[3]In the one-dimensional setting the topological entropy, which we denote by h, is a measure of the growth of the number of periodic cycles as a function of the length of the period

$$h(f) = \lim_{n \to \infty} \frac{\ln N_n}{n},$$

where N_n is the number of distinct periodic orbits of length n.
For general metric spaces, the topological entropy of f is defined by

$$h(f) = \lim_{\varepsilon \to 0} \left(\limsup_{n \to \infty} \frac{1}{n} \log N(n, \varepsilon) \right)$$

where, roughly speaking, $N(n, \varepsilon)$ represents the number of distinguishable orbit segments of length n, assuming we cannot distinguish points that are less than ε apart. The interested reader may consult Ott [75], or Block and Coppel [12] for more details on topological entropy.

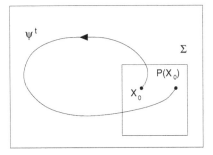

FIGURE 3.15

A 2-dimensional Poincaré map.

Define $\lambda_n := \min\{\lambda \in (0,1] : T$ has a 2^n-periodic point in $(0,\lambda]\}$. Then we may show that $\lambda^* = \lim_{n\to\infty} \lambda_n \approx 0.824908\dots$. The following observations may be shown [3]:

1. For each $\lambda \in (0,\lambda^*)$, G_λ is not chaotic in any of the three senses.

2. G_{λ^*} is L/Y-chaotic, but neither B/C-chaotic nor D-chaotic.

3. For each $\lambda \in (\lambda^*,1]$, G_λ is chaotic in the three senses.

☐

3.10 Rössler's Attractor

In Section 1.3.2 we defined a one-dimensional Poincaré map associated with a system of two differential equations. This definition may be extended to a system of three differential equations of the form

$$
\begin{aligned}
\dot{x} &= f(x,y,z) \\
\dot{y} &= g(x,y,z) \\
\dot{z} &= h(x,y,z).
\end{aligned}
\tag{3.21}
$$

The Poincaré map this time will be planar, that is a 2-dimensional map. Let Γ be a periodic orbit (limit cycle) of system (3.21) and let Σ be a plane intersecting Γ transversely (Σ is called a Poincaré section). For a point $X_0 \in \Sigma$, the point $P(X_0)$ is defined as the next intersection of the orbit through X_0 of system (3.21) with Σ, as illustrated in Fig. 3.15.

Let Y_0 be the initial point of the periodic orbit Γ. Then Y_0 is a fixed point of the Poincaré map P. Hence P reduces the study of stability of a periodic orbit of Equation (3.21) to the study of the stability of the fixed point Y_0 under the Poincaré map P.

We now apply this procedure to studying a famous attractor, named after its creator Otto Rössler.

In 1977, O.E. Rössler [87] was able to extract simpler, asymmetric attracting structures from the Lorenz attractor. He proposed the following system of three differential equations.

$$\dot{x} = -y - z$$
$$\dot{y} = x + ay \qquad\qquad (3.22)$$
$$\dot{z} = b + z(x - c).$$

Notice that the only nonlinear term zx appears in the third equation.

At this stage we are not assuming any background in differential equations. But we will use simple but important observations:

(a) $\dot{x}(t) > 0$ implies $x(t)$ is increasing,

(b) $\dot{x}(t) < 0$ implies $x(t)$ is decreasing,

(c) $\dot{x}(t) = 0$ implies $x(t)$ is constant.

If one assumes that z is small enough to be negligible, then we have

$$\dot{x} = -y$$
$$\dot{y} = x + ay$$

which can be written as

$$\ddot{x} - a\dot{x} + x = 0. \qquad\qquad (3.23)$$

To solve this second order differential equation, we let $x(t) = e^{\lambda t}$ (compare with difference equations: $x(n) = \lambda^n$). Then substituting $\dot{x} = \lambda e^{\lambda t}$, $\ddot{x} = \lambda^2 e^{\lambda t}$, we get $\lambda^2 e^{\lambda t} - a\lambda e^{\lambda t} + e^{\lambda t} = 0$. Dividing by $e^{\lambda t}$ we obtain the characteristic equation

$$\lambda^2 - a\lambda + 1 = 0$$
$$\lambda_{1,2} = \frac{a \pm \sqrt{a^2 - 4}}{2}.$$

Clearly for $0 < a < 2$, $\lambda_{1,2} = (a \pm i\sqrt{a^2 - 4})/2$. Moreover $x(t) = e^{\frac{a}{2}t}(c_1 \cos \beta t + c_2 \sin \beta t)$, where $\beta = \sqrt{4 - a^2}/2$ which is an unstable spiral (focus).

Now in the full system (3.23), orbits (trajectories) near the $x - y$ plane spiral outward from the origin. This produces spreading of adjacent trajectories, which is a key ingredient in the mixing action of chaos. Note that this spreading is achieved with only linear terms. If the original system is fully linear, then the spreading would merely continue and all trajectories (orbits) diverge far away from the origin. The nonlinear term will dramatically change this

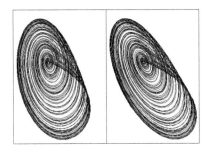

FIGURE 3.16

Steroscopic view of the Rössler attractor for parametric values $a = 0.432$, $b = 2$, $c = 4$.

scenario as the constant c will play the role of a control. In the third equation if x is less than c, the z subsystem is asymptotically stable (Why?) and tends to $z = -b/(x-c)$ (this is obtained by letting $\dot{z} = 0$ and thus $0 = b + z(x-c)$). However, if $x > c$, the z-subsystem diverges (unstable). Choosing $b > 0$ ensures that this divergence will be toward positive z.

Figure 3.15 shows an orbit spiraling outwards while appearing to remain in a plane near to and parallel to the $x-y$ plane. When x becomes large enough, the z subsystem switches on and the orbit leaps upwards. Once a becomes large, the z tem in the first equation comes into play, and $\dot{x}$ becomes large and negative, throwing the orbit back toward smaller x. Eventually x decreases below c, the z variable becomes self restoring, and the orbit lands near the $x - y$ plane again. Through the feedback of z to the $\dot{x}$ equation, orbits are folded back and reinserted closer to the origin, where they begin an outward spiral once more. Rössler named this chaotic attractor "spiral chaos."

Next we will construct a Poincaré section, which would appear on this scale to be a line segment. Hence the point where an orbit on the attractor crosses this half-plane can be identified by giving the distance from the center, that is, the x-coordiante of the point alone. This key observation allows one to study the dynamics of the attractor via a one-dimensional map. To find this map, we consider an orbit on the attractor, and letting $x(n+1)$ be the x-coordiante of the $(n+1)$th crossing of the Poincaré section as a function of $x(n)$. Plotting $x(n + 1)$ versus $x(n)$ produces Fig. 3.17 which is reminiscent of the logistic map $F_\mu(x) = \mu x(1 - x)$.

For fixed values of $a = b = 0.2$ and for different values of c, Fig. 3.17 shows that the differential system (3.22) undergoes period-doubling in its route to chaos.

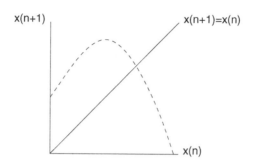

FIGURE 3.17
One-dimensional map constructed from an orbit of the differential system (3.22).

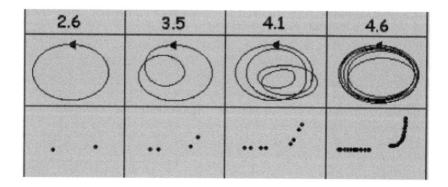

FIGURE 3.18
The Poincaré section is intersected twice for each periodic orbit. For $a = b = 0.2$ and (i) $c = 2.6$, (ii) $c = 3.5$, (iii) $c = 4.1$, (iv) $c = 4.6$. At the value $c = 4.6$, the sytem is in full-blown chaos.

FIGURE 3.19
Saturn with its rings.

FIGURE 3.20
Mimas, largest moon of Saturn.

3.11 Saturn's Rings

The rings of Saturn [35] are almost circular and nearly planar in nature, (Fig. 3.19). They are 250,000 Km across, but no more than 1.5 Km thick. The rings are composed mainly of water ice ranging in size from 1 cm to several meters. A nonlinear map which produces a ring pattern qualitatively resembling the rings of Saturn has been developed by Fröyland [41]. This map is three dimensional in essence but may be written in the following form.

$$\theta(n+1) = \theta(n) + 2\pi \left(\frac{\sigma}{r(n)} \right)^{\frac{3}{2}}$$

$$r(n+1) = 2r(n) - r(n-1) - a\frac{\cos\theta(n)}{(r(n) - \sigma)^2}$$

where $r(n)$ is the radial distance of a ring particle from the center of Saturn after the nth revolution, $\theta(n)$ the angular position of a ringle particle with respect to Mimas after n revolutions, and σ is the radial distance of Mimas from Saturn's center. Recall that Mimas is the is the largest moon out of 18 moons of Saturn (Fig. 3.20).

The above equation may be written as a 3-dimensional system of first order difference equation by letting $\theta(n) = \theta(n)$, $r(n) = r(n)$, $z(n) = r(n-1)$. This yields

$$\theta(n+1) = \theta(n) + 2\pi \left(\frac{\sigma}{r(n)}\right)^{\frac{3}{2}}$$

$$r(n+1) = 2r(n) - z(n) - a\frac{\cos\theta(n)}{(r(n)-\sigma)^2} \qquad (3.24)$$

$$z_{n+1} = r(n)$$

or

$$F\begin{pmatrix}\theta\\r\\z\end{pmatrix} = \begin{pmatrix}\theta + 2\pi\left(\frac{\sigma}{r}\right)^{\frac{3}{2}}\\2r - z - a\frac{\cos\theta(n)}{(r(n)-\sigma)^2}\\r\end{pmatrix}. \qquad (3.24')$$

In this model, there are two important forces acting on the ring particles, the dominant effect of Saturn's attracting gravitational force and the perturbing influence of Mimas. The effect of Saturn may be explained as follows. Each time Mimas completes an orbit of radius σ with a period T_σ, it undergoes an angular change of 2π radians. If T_n is the period for any other satellite object in its nth revolutions, the angle θ that the object makes with respect to Mimas on the $n+1$st revolution will be given by

$$\theta(n+1) = \theta(n) + 2\pi\left(\frac{T_\sigma}{T_n}\right). \qquad (3.25)$$

Now Kepler's third law for planetary orbits states that the period T of an object orbiting a planet of mass M_p in a circular orbit of radius r is given by

$$T^2 = \frac{4\pi^2}{GM_p}r^3 \qquad (3.26)$$

where $G = 6.67 \times 10^{-11}\ N - m^2/Kg^2$ is the gravitational constant. Letting $r(n)$ be the distance of a ring particle from Saturn's center after n revolutions, then

$$\frac{T_\sigma}{T_n} = \left(\frac{\sigma}{r(n)}\right)^{\frac{3}{2}}. \qquad (3.27)$$

Substituting (3.27) into (3.25) yields the first equation in (3.24).

The effect of Mimas is to perturb the radial distance r of a ring particle, causing the distance to change from one orbit to the next. By Newton's second law, a particle's radial acceleration is given by

$$m\frac{d^2r}{dt^2} = F_r \qquad (3.28)$$

where F_r is the radial component of the gravitational force between Mimas and the particle of mass m. Using Euler's method (see Section 1.3.1) we replace $\frac{d^2r}{dt^2}$ by $(r(n+1) - 2r(n) + r(n-1))/(\Delta t)^2$, where we set $\Delta t = T_\sigma$. This gives

$$r(n+1) = 2r(n) - r(n-1) + f(r(n), \theta(n))$$

where $f(r(n), \theta(n)) = \frac{(T_\sigma)^2 F_r(r(n), \theta(n))}{m}$. To find f we apply Newton's law of gravitation to the interaction between Mimas (map M_σ) and a particle. Hence

$$f = -a\frac{g(\theta(n))}{(r(n) - \sigma)^2}$$

with $a = GM_\sigma(T_\sigma)^2 = 4\pi^2\sigma^3\frac{M_\sigma}{M_s}$, M_s being the mass of Saturn, and the angular-dependent map $g(\theta(n))$ is given by $\cos(\theta(n))$. The parameter a is approximately equal to 17, and $M_s = 5.68 \times 10^{26} Kg$.

Conservative versus nonconservative maps

Maps that conserve area or volume are called <u>conservative</u> maps. Consider a two dimensinal map

$$F\begin{pmatrix} x \\ y \end{pmatrix} = \begin{pmatrix} f(x, y) \\ g(x, y) \end{pmatrix}.$$

Under this map an infinitesimal area $dx\,dy$ maps into the new area

$$d\bar{x}\,d\bar{y} = |det\ J(x, y)|\,dxdy$$

where

$$J = \begin{pmatrix} \frac{\partial f}{\partial x} & \frac{\partial f}{\partial y} \\ \frac{\partial g}{\partial x} & \frac{\partial g}{\partial y} \end{pmatrix}$$

is the Jacobian matrix. Clearly, a map is conservative if $|det\ J| = 1$. For three-dimensional system of the form

$$F\begin{pmatrix} x \\ y \\ z \end{pmatrix} = \begin{pmatrix} f(x, y, z) \\ g(x, y, z) \\ h(x, y.z) \end{pmatrix},$$

the Jacobian is defined as

$$J = \begin{pmatrix} \frac{\partial f}{\partial x} & \frac{\partial f}{\partial y} & \frac{\partial f}{\partial z} \\ \frac{\partial g}{\partial x} & \frac{\partial g}{\partial y} & \frac{\partial g}{\partial z} \\ \frac{\partial h}{\partial x} & \frac{\partial h}{\partial y} & \frac{\partial h}{\partial z} \end{pmatrix}.$$

For our map $F\begin{pmatrix} \theta \\ r \\ z \end{pmatrix}$ in (3.24)$'$, we have

$$J = \begin{pmatrix} 1 & \frac{-3\sigma^{\frac{3}{2}}}{r^{\frac{5}{2}}} & 0 \\ a\frac{\sin\theta}{(r-\sigma)^2} & 2 + \frac{2a\cos\theta}{(r-\sigma)^3} & -1 \\ 0 & 1 & 0 \end{pmatrix}.$$

Notice that *det J* $= 1$, and hence our map is conservative, i.e., it preserves volumes.

4

Stability of Two-Dimensional Maps

Is evolution a matter of survival of the fittest or survival of the most stable?

A. M. Waldrop

4.1 Linear Maps vs. Linear Systems

Recall from linear algebra that a map $L : \mathbb{R}^2 \to \mathbb{R}^2$ is called a linear transformation if

1. $L(U_1 + U_2) = L(U_1) + L(U_2)$ for $U_1, U_2 \in \mathbb{R}^2$

2. $L(\alpha U) = \alpha L(U)$ for $U \in \mathbb{R}^2$ and $\alpha \in \mathbb{R}$.

Moreover, it is always possible to represent f (with a given basis for $\mathbb{R}^2$) by a matrix A. A typical example is

$$L\begin{pmatrix} x \\ y \end{pmatrix} = \begin{pmatrix} ax + by \\ cx + dy \end{pmatrix}$$

which may be written in the form

$$L\begin{pmatrix} x \\ y \end{pmatrix} = \begin{pmatrix} a & b \\ c & d \end{pmatrix}\begin{pmatrix} x \\ y \end{pmatrix}$$

or

$$L(U) = AU, \tag{4.1}$$

where $U = \begin{pmatrix} x \\ y \end{pmatrix}$ and $A = \begin{pmatrix} a & b \\ c & d \end{pmatrix}$.

By iterating L, we conclude that $L^n(U) = A^n U$. Hence, the orbit of U under f is given by

$$\{U, AU, A^2U, \dots, A^n U, \dots\} \tag{4.2}$$

Thus, to compute the orbit of U, it suffices to compute $A^n U$ for $n \in \mathbb{Z}^+$.

Another way of looking at the same problem is by considering the following two-dimensional system of difference equations

$$\begin{aligned} x(n+1) &= ax(n) + by(n) \\ y(n+1) &= cx(n) + dy(n), \end{aligned} \qquad (4.3)$$

or

$$U(n+1) = AU(n). \qquad (4.4)$$

By iteration, one may show that the solution of Equation (4.4) is given by

$$U(n) = A^n U(0). \qquad (4.5)$$

So, if we let $U_0 = U(0)$, then $L^n(U_0) = U(n)$.

The form of Equation (4.3) is more convenient when we are considering applications in biology, engineering, economics, and so forth. For example, $x(n)$ and $y(n)$ may represent the population sizes at time period n of two competitive cooperative species, or preys and predators.

In the next section, we will develop the necessary machinery to compute A^n for any matrix of order two. The general theory may be found in [32, 33, 60].

4.2 Computing A^n

Consider a matrix $A = (a_{ij})$ of order 2×2. Then, $p(\lambda) = \det(A - \lambda I)$ is called the **characteristic polynomial** of A and its zeros are called the **eigenvalues** of A. Associated with each eigenvalue λ of A a nonzero eigenvector $V \in \mathbb{R}^2$ with $AV = \lambda V$.

Example 4.1

Find the eigenvalues and the eigenvectors of the matrix

$$A = \begin{pmatrix} 2 & 3 \\ 1 & 4 \end{pmatrix}. \quad \Box$$

SOLUTION First we find the eigenvalues of A by solving the characteristic equation $\det(A - \lambda I) = 0$ or

$$\begin{vmatrix} 2-\lambda & 3 \\ 1 & 4-\lambda \end{vmatrix} = 0$$

which is

$$\lambda^2 - 6\lambda + 5 = 0.$$

Hence, $\lambda_1 = 1$ and $\lambda_2 = 5$. To find the corresponding eigenvector V_1, we solve the vector equation $AV_1 = \lambda V_1$ or $(A - \lambda_1 I)V_1 = 0$.
  For $\lambda_1 = 1$, we have

$$\begin{pmatrix} 1 & 3 \\ 1 & 3 \end{pmatrix} \begin{pmatrix} v_{11} \\ v_{21} \end{pmatrix} = \begin{pmatrix} 0 \\ 0 \end{pmatrix}.$$

Hence, $v_{11} + 3v_{21} = 0$. Thus, $v_{11} = -3v_{21}$. So, if we let $v_{21} = 1$, then $v_{11} = -3$. It follows that the eigenvector V_1 corresponding to λ_1 is given by $V_1 = \begin{pmatrix} -3 \\ 1 \end{pmatrix}$.
  For $\lambda_2 = 5$, the corresponding eigenvector may be found by solving the equation $(A - \lambda_2 I)V_2 = 0$. This yields

$$\begin{pmatrix} -3 & 3 \\ 1 & -1 \end{pmatrix} \begin{pmatrix} v_{12} \\ v_{22} \end{pmatrix} = \begin{pmatrix} 0 \\ 0 \end{pmatrix}.$$

Thus, $-3v_{12} + 3v_{22} = 0$ or $v_{12} = v_{22}$. It is then appropriate to let $v_{12} = v_{22} = 1$ and hence $V_2 = \begin{pmatrix} 1 \\ 1 \end{pmatrix}$. ∎

  To find the general form for A^n for a general matrix A is a formidable task even for a 2×2 matrix such as in Example 4.1. Fortunately, however, we may be able to transform a matrix A to another simpler matrix B whose nth power B^n can easily be computed. The essence of this process is captured in the following definition.

DEFINITION 4.1   *The matrices A and B are said to be similar if there exists a nonsingular[1] matrix P such that*

$$P^{-1}AP = B.$$

We note here that the relation "similarity" between matrices is an equivalence relation, i.e.,

1. A is similar to A.

2. If A is similar to B then B is similar to A.

3. If A is similar to B and B is similar to C, then A is similar to C.

The most important feature of similar matrices, however, is that they possess the same eigenvalues.

---

[1] A matrix P is said to be nonsingular if its inverse P^{-1} exists. This is equivalent to saying that det $P \neq 0$, where det denotes determinant.

THEOREM 4.1

Let A and B be two similar matrices. Then A and B have the same eigenvalues.

PROOF   Suppose that $P^{-1}AP = B$ or $A = PBP^{-1}$. Let λ be an eigenvalue of A and V be the corresponding eigenvector. Then, $\lambda V = AV = PBP^{-1}V$. Hence, $B(P^{-1}V) = \lambda(P^{-1}V)$. Consequently, λ is an eigenvalue of B with $P^{-1}V$ as the corresponding eigenvector.   ∎

The notion of similarity between matrices corresponds to linear conjugacy, which we have encountered in Chapter 3. In other words, two linear maps are conjugate if their corresponding matrix representations are similar. Thus, the linear maps L_1, L_2 on $\mathbb{R}^2$ are linearly conjugate if there exists an invertible map h such that

$$L_1 \circ h = h \circ L_2$$

or

$$h^{-1} \circ L_1 \circ h = L_2.$$

The next theorem tells us that there are three simple "canonical" forms for 2×2 matrices.

THEOREM 4.2

Let A be a 2×2 real matrix. Then A is similar to one of the following matrices:

1. $\begin{pmatrix} \lambda_1 & 0 \\ 0 & \lambda_2 \end{pmatrix}$

2. $\begin{pmatrix} \lambda & 1 \\ 0 & \lambda \end{pmatrix}$

3. $\begin{pmatrix} \alpha & \beta \\ -\beta & \alpha \end{pmatrix}$

PROOF   Suppose that the eigenvalues λ_1 and λ_2 are real. Then, we have two cases to consider. The first case is where $\lambda_1 \neq \lambda_2$. In this case, we may easily show that the corresponding eigenvectors V_1 and V_2 are linearly independent (Problem 10). Hence, the matrix $P = (V_1, V_2)$, i.e., the matrix P whose columns are these eigenvectors, is nonsingular. Let $P^{-1}AP = J = \begin{pmatrix} e & f \\ g & h \end{pmatrix}$. Then,

$$AP = PJ. \tag{4.6}$$

Comparing both sides of Equation (4.6), we obtain

$$AV_1 = eV_1 + gV_2.$$

Hence,

$$\lambda_1 V_1 = e V_1 + g V_2.$$

Thus, $e = \lambda_1$ and $g = 0$.

Similarly, one may show that $f = 0$ and $h = \lambda_2$. Consequently, J is a diagonal matrix of the form (a).

The second case is where $\lambda_1 = \lambda_2 = \lambda$. There are two subcases to consider here. The first subcase occurs if we are able to find two linearly independent eigenvectors V_1 and V_2 corresponding to the eigenvalue λ. This subcase is then reduced to the preceding case. We note here that this scenario happens when $(A - \lambda I)V = 0$ for all $V \in \mathbb{R}^2$. In particular, one may let $V_1 = \begin{pmatrix} 1 \\ 0 \end{pmatrix}$ and $V_2 = \begin{pmatrix} 0 \\ 1 \end{pmatrix}$, which are clearly linearly independent.

The second subcase occurs when there exists a nonzero vector $V_2 \in \mathbb{R}^2$ such that $(A - \lambda I)V_2 \neq 0$. Equivalently, we are able to find only one eigenvector (not counting multiples) V_1 with $(A - \lambda I)V_1 = 0$. In practice, we find V_2 by solving the equation

$$(A - \lambda I)V_2 = V_1.$$

The vector V_2 is called a generalized eigenvector of A. Note that $AV_1 = \lambda V_1$ and $AV_2 = \lambda V_2 + V_1$. Now, we let $P = (V_1, V_2)$ and $P^{-1}AP = J$. Then,

$$AP = PJ. \tag{4.7}$$

Comparing both sides of Equation (4.7) yields

$$J = \begin{pmatrix} \lambda & 1 \\ 0 & \lambda \end{pmatrix}. \tag{4.8}$$

The matrix J is in a **Jordan form**.

Next, we assume that A has a complex eigenvalue $\lambda_1 = \alpha + i\beta$. Since A is assumed to be real, it follows that the second eigenvalue λ_2 is a conjugate of λ_1, that is, $\lambda_2 = \alpha - i\beta$. Let $V = V_1 + iV_2$ be the eigenvector corresponding to λ_1. Then,

$$AV = \lambda_1 V$$
$$A(V_1 + iV_2) = (\alpha + i\beta)(V_1 + iV_2).$$

Hence,

$$AV_1 = \alpha V_1 - \beta V_2$$
$$AV_2 = \beta V_1 + \alpha V_2,$$

letting $P = (V_1, V_2)$ we get $P^{-1}AP = J$. Hence,

$$AP = PJ. \tag{4.9}$$

Comparison of both sides of Equation (4.9) yields

$$J = \begin{pmatrix} \alpha & \beta \\ -\beta & \alpha \end{pmatrix}. \tag{4.10}$$

Theorem 4.2 gives us a simple method of computing the general form of A^n for any 2×2 real matrix. In the first case, when $P^{-1}AP = D = \begin{pmatrix} \lambda_1 & 0 \\ 0 & \lambda_2 \end{pmatrix}$, we have

$$\begin{aligned} A^n &= (PDP^{-1})^n \\ &= PD^nP^{-1} \\ &= P\begin{pmatrix} \lambda_1^n & 0 \\ 0 & \lambda_2^n \end{pmatrix}P^{-1}. \end{aligned} \tag{4.11}$$

In the second case, when $P^{-1}AP = J = \begin{pmatrix} \lambda & 1 \\ 0 & \lambda \end{pmatrix}$, then

$$\begin{aligned} A^n &= PJ^nP^{-1} \\ &= P\begin{pmatrix} \lambda^n & n\lambda^{n-1} \\ 0 & \lambda^n \end{pmatrix}P^{-1}. \end{aligned} \tag{4.12}$$

Equation (4.12) may be easily proved by mathematical induction (Problem 11).

In the third case, we have $P^{-1}AP = J = \begin{pmatrix} \alpha & \beta \\ -\beta & \alpha \end{pmatrix}$. Let $\omega = \arctan(\beta/\alpha)$. Then $\cos\omega = \alpha/|\lambda_1|$, $\sin\omega = \beta/|\lambda_1|$. Now, we write the matrix J in the form

$$J = |\lambda_1|\begin{pmatrix} \alpha/|\lambda_1| & \beta/|\lambda_1| \\ -\beta/|\lambda_1| & \alpha/|\lambda_1| \end{pmatrix} = |\lambda_1|\begin{pmatrix} \cos\omega & \sin\omega \\ -\sin\omega & \cos\omega \end{pmatrix}.$$

By mathematical induction one may show that (Problem 11)

$$J^n = |\lambda_1|^n\begin{pmatrix} \cos n\omega & \sin n\omega \\ -\sin n\omega & \cos n\omega \end{pmatrix}. \tag{4.13}$$

and thus

$$A^n = |\lambda_1|^n P\begin{pmatrix} \cos n\omega & \sin n\omega \\ -\sin n\omega & \cos n\omega \end{pmatrix}P^{-1}. \tag{4.14}$$

Example 4.2
Solve the system of difference equations

$$X(n+1) = AX(n) \tag{4.15}$$

where

$$A = \begin{pmatrix} -4 & 9 \\ -4 & 8 \end{pmatrix}, \quad X(0) = \begin{pmatrix} 1 \\ 0 \end{pmatrix}. \quad \Box$$

SOLUTION   The eigenvalues of A are repeated: $\lambda_1 = \lambda_2 = 2$. The only eigenvector that we are able to find is $V_1 = \begin{pmatrix} 3 \\ 2 \end{pmatrix}$. To construct P we need to find a generalized eigenvector V_2. This is accomplished by solving the equation $(A - 2I)V_2 = V_1$. Then, V_2 may be taken as any vector $\begin{pmatrix} x \\ y \end{pmatrix}$, with $3y - 2x = 1$. We take $V_2 = \begin{pmatrix} 1 \\ 1 \end{pmatrix}$. Now if we put $P = \begin{pmatrix} 3 & 1 \\ 2 & 1 \end{pmatrix}$, then $P^{-1}AP = J = \begin{pmatrix} 2 & 1 \\ 0 & 2 \end{pmatrix}$. Thus, the solution of Equation (4.15) is given by

$$X(n) = PJ^n P^{-1} x(0)$$
$$= \begin{pmatrix} 3 & 1 \\ 2 & 1 \end{pmatrix} \begin{pmatrix} 2^n & n2^{n-1} \\ 0 & 2^n \end{pmatrix} \begin{pmatrix} 1 & -1 \\ -2 & 3 \end{pmatrix} \begin{pmatrix} 1 \\ 0 \end{pmatrix}$$
$$= 2^n \begin{pmatrix} 1 - 3n \\ -2n \end{pmatrix}. \quad \blacksquare$$

REMARK 4.1   If a map $f : \mathbb{R}^2 \to \mathbb{R}^2$ is given by $f(X_0) = AX_0$, then $f^n(X_0) = A^n X_0 = PJ^n P^{-1} X_0$. In particular, if $X_0 = \begin{pmatrix} 1 \\ 0 \end{pmatrix}$, then $f^n(X_0) = 2^n \begin{pmatrix} 1 - 3n \\ -2n \end{pmatrix}$ for all $n \in \mathbb{Z}^+$. $\blacksquare$

Exercises - (4.1 and 4.2)

In Problems 1–5, find the eigenvalues and eigenvectors of the matrix A and compute A^n.

1. $A = \begin{pmatrix} -4.5 & 5 \\ -7.5 & 8 \end{pmatrix}$

2. $A = \begin{pmatrix} 4.5 & -1 \\ 2.25 & 1.5 \end{pmatrix}$

3. $A = \begin{pmatrix} 8/3 & 1/3 \\ -4/3 & 4/3 \end{pmatrix}$

4. $A = \begin{pmatrix} 2 & 3 \\ -3 & 2 \end{pmatrix}$

5. $A = \begin{pmatrix} -2 & -3 \\ 1 & 1 \end{pmatrix}$

6. Let $L : \mathbb{R}^2 \to \mathbb{R}^2$ be defined by $L(X) = AX$, where A is as in Problem 1. Find $L^n \begin{pmatrix} 0 \\ 1 \end{pmatrix}$.

7. Solve the difference equation $X(n+1) = AX(n)$, where A is as in Problem 3, and $X(0) = \begin{pmatrix} 1 \\ 1 \end{pmatrix}$.

8. Solve the difference equation $X(n+1) = AX(n)$, where A is as in Problem 4, and $X(0) = X_0$.

9. Let $f : \mathbb{R}^2 \to \mathbb{R}^2$ be defined by $f(X) = AX$, with A as in Problem 5. Find $f^n \begin{pmatrix} 0 \\ 1 \end{pmatrix}$.

10. Let A be a 2×2 matrix with distinct real eigenvalues. Show that the corresponding eigenvectors of A are linearly independent.

11. (a) If $J = \begin{pmatrix} \lambda & 1 \\ 0 & \lambda \end{pmatrix}$, show that $J^n = \begin{pmatrix} \lambda^n & n\lambda^{n-1} \\ 0 & \lambda^n \end{pmatrix}$.

    (b) If $J = \begin{pmatrix} \alpha & \beta \\ -\beta & \alpha \end{pmatrix}$, show that $J^n = |\lambda|^n \begin{pmatrix} \cos n\omega & \sin n\omega \\ -\sin n\omega & \cos n\omega \end{pmatrix}$, where $|\lambda| = \sqrt{\alpha^2 + \beta^2}$, $\omega = \arctan\left(\frac{\beta}{\alpha}\right)$.

12. Let a matrix A be in the form

$$A = \begin{pmatrix} 0 & 1 \\ -p_2 & -p_1 \end{pmatrix}.$$

   (a) Show that if A has distinct eigenvalues λ_1 and λ_2, then

$$P^{-1}AP = \begin{pmatrix} \lambda_1 & 0 \\ 0 & \lambda_2 \end{pmatrix},$$

   where $P = \begin{pmatrix} 1 & 1 \\ \lambda_1 & \lambda_2 \end{pmatrix}$.

   (b) Show that if A has a repeated eigenvalue λ, then

$$P^{-1}AP = \begin{pmatrix} \lambda & 1 \\ 0 & \lambda \end{pmatrix},$$

   where $P = \begin{pmatrix} 1 & 0 \\ \lambda & 1 \end{pmatrix}$.

   (c) Show that if A has complex eigenvalues $\lambda_1 = \alpha + i\beta$ and $\lambda_2 = \alpha - i\beta$, then

$$P^{-1}AP = \begin{pmatrix} \alpha & \beta \\ -\beta & \alpha \end{pmatrix},$$

   where $P = \begin{pmatrix} 1 & 0 \\ \alpha & \beta \end{pmatrix}$.

4.3   Fundamental Set of Solutions

Consider the linear system

$$X(n+1) = AX(n),$$ (4.16)

where A is a 2×2 matrix. Then, two solutions $X_1(n)$ and $X_2(n)$ of Equation (4.16) are said to be linearly independent if $X_2(n)$ is not a scaler multiple of $X_1(n)$ for all $n \in \mathbb{Z}^+$. In other words, if $c_1 X_1(n) + c_2 X_2(n) = 0$ for all $n \in \mathbb{Z}^+$, then $c_1 = c_2 = 0$. A set of two linearly independent solutions $\{X_1(n), X_2(n)\}$ is called a fundamental set of solutions of Equation (4.16).

DEFINITION 4.2   *Let $\{X_1(n), X_2(n)\}$ be a fundamental set of solutions of Equation (4.16). Then*

$$X(n) = k_1 X_1(n) + k_2 X_2(n), \quad k_1, k_2 \in \mathbb{R}$$ (4.17)

is called a general solution of Equation (4.16).

Finding $X_1(n)$ and $X_2(n)$ is generally an easy task. We now give an explicit derivation.

In the sequel λ_1, λ_2 denote the eigenvalues of A; V_1, V_2 are the corresponding eigenvectors of A.

We have three cases to consider.

Case (i)

Suppose that $P^{-1}AP = J = \begin{pmatrix} \lambda_1 & 0 \\ 0 & \lambda_2 \end{pmatrix}$. Then a general solution may be given by

$$X(n) = A^n X(0) = PJ^n P^{-1} X(0)$$
$$= (V_1, \ V_2) \begin{pmatrix} \lambda_1^n & 0 \\ 0 & \lambda_2^n \end{pmatrix} \begin{pmatrix} k_1 \\ k_2 \end{pmatrix}$$

where $\begin{pmatrix} k_1 \\ k_2 \end{pmatrix} = P^{-1} X(0)$. Then,

$$X(n) = k_1 \lambda_1^n V_1 + k_2 \lambda_2^n V_2.$$ (4.18)

Here, $X_1(n) = \lambda_1^n V_1$ and $X_2(n) = \lambda_2^n V_2$ constitute a fundamental set of solutions since in this case V_1 and V_2 are linearly independent eigenvectors. Note that one may check directly that $\lambda_1^n V_1$ and $\lambda_2^n V_2$ are indeed solutions of Equation (4.16) (Problem 13a).

Case (ii)

Suppose that $P^{-1}AP = J = \begin{pmatrix} \lambda & 1 \\ 0 & \lambda \end{pmatrix}$. Then, a general solution may be given by

$$
\begin{aligned}
X(n) &= PJ^n P^{-1} X(0) \\
&= (V_1, \ V_2) \begin{pmatrix} \lambda^n & n\lambda^{n-1} \\ 0 & \lambda^n \end{pmatrix} \begin{pmatrix} k_1 \\ k_2 \end{pmatrix} \\
&= k_1 \lambda^n V_1 + k_2(n\lambda^{n-1}V_1 + \lambda^n V_2) \qquad (4.19)
\end{aligned}
$$

Hence, $X_1(n) = \lambda^n V_1$ and $X_2(n) = \lambda^n V_2 + n\lambda^{n-1}V_1$ constitute a fundamental set of solutions of Equation (4.16) (Problem 13b).

Case (iii)

Suppose that $P^{-1}AP = J = \begin{pmatrix} \alpha & \beta \\ -\beta & \alpha \end{pmatrix}$. If $\omega = \arctan(\beta/\alpha)$, then the general solution may be given by

$$
\begin{aligned}
X(n) &= PJ^n P^{-1} X(0) \\
&= (V_1 V_2)|\lambda_1|^n \begin{pmatrix} \cos n\omega & \sin n\omega \\ -\sin n\omega & \cos n\omega \end{pmatrix} \begin{pmatrix} k_1 \\ k_2 \end{pmatrix} \\
&= |\lambda_1|^n [k_1 \cos n\omega + k_2 \sin n\omega)V_1 \\
&\quad + (-k_1 \sin n\omega + k_2 \cos n\omega)V_2]. \qquad (4.20)
\end{aligned}
$$

Hence, $X_1(n) = |\lambda_1|^n[(k_1 \cos n\omega)V_1 - (k_1 \sin(n\omega))V_2]$ and $X_2(n) = (|\lambda_1|^n[(k_2 \sin(n\omega))V_1 + (k_2 \cos(n\omega)]V_2$ constitute a fundamental set of solutions (Problem 13c).

Example 4.3
Solve the system of difference equations

$$
X(n+1) = AX(n), \ X(0) = \begin{pmatrix} 1 \\ 2 \end{pmatrix},
$$

where

$$
A = \begin{pmatrix} -2 & -3 \\ 3 & -2 \end{pmatrix}. \quad \square
$$

SOLUTION The eigenvalues of A are $\lambda_1 = -2+3i$ and $\lambda_2 = -2-3i$. The corresponding eigenvectors are $V = \begin{pmatrix} -1 \\ i \end{pmatrix}$ and $\overline{V} = \begin{pmatrix} -1 \\ -i \end{pmatrix}$, respectively.

This time, we take a short cut and use Equation (4.20). The vectors V_1 and V_2 referred to in this formula are the real part of V, $V_1 = \begin{pmatrix} -1 \\ 0 \end{pmatrix}$, and the imaginary part of V, $V_2 = \begin{pmatrix} 0 \\ 1 \end{pmatrix}$. Now, $|\lambda_1| = \sqrt{13}$, $\omega = \arctan(\frac{-3}{2}) \approx$ 123.69°. Thus,

$$X(n) = (13)^{n/2} \left[(k_1 \cos n\omega + k_2 \sin n\omega) \begin{pmatrix} -1 \\ 0 \end{pmatrix} \right.$$

$$\left. + (-k_1 \sin n\omega + k_2 \cos n\omega) \begin{pmatrix} 0 \\ 1 \end{pmatrix} \right]$$

$$X(0) = \begin{pmatrix} 1 \\ 2 \end{pmatrix} = k_1 \begin{pmatrix} -1 \\ 0 \end{pmatrix} + k_2 \begin{pmatrix} 0 \\ 1 \end{pmatrix}.$$

Hence, $k_1 = 1, k_2 = 2$. Thus,

$$X(n) = (13)^{n/2} \left[(\cos n\omega + 2 \sin n\omega) \begin{pmatrix} 1 \\ 0 \end{pmatrix} + (-\sin n\omega + 2 \cos n\omega) \begin{pmatrix} 0 \\ 1 \end{pmatrix} \right]$$

$$= (13)^{n/2} \begin{pmatrix} -\cos n\omega - 2 \sin n\omega \\ -\sin n\omega + 2 \cos n\omega \end{pmatrix}.$$

4.4 Second-Order Difference Equations

A second-order difference equation with constant coefficients is a scalar equation of the form

$$u(n + 2) + p_1 u(n + 1) + p_2 u(n) = 0 \qquad (4.21)$$

Although one may solve this equation directly, it is sometimes beneficial to convert it to a two-dimensional system. The trick is to let $u(n) = x_1(n)$ and $u(n + 1) = x_2(n)$.

Then we have

$$x_1(n + 1) = x_2(n)$$
$$x_2(n + 1) = -p_2 x_1(n) - p_1 x_2(n)$$

which is of the form

$$X(n + 1) = AX(n) \qquad (4.22)$$

where

$$X(n) = \begin{pmatrix} x_1(n) \\ x_2(n) \end{pmatrix}, \quad \text{and} \quad A = \begin{pmatrix} 0 & 1 \\ -p_2 & -p_1 \end{pmatrix}.$$

The characteristic equation of A is given by

$$\lambda^2 + p_1 \lambda + p_2 = 0. \qquad (4.23)$$

Observe that we may obtain the characteristic Equation (4.23) by letting $u(n) = \lambda^n$ in Equation (4.21). Thus, if λ_1 and λ_2 are the roots of Equation (4.23), then $u_1(n) = \lambda_1^n$ and $u_2(n) = \lambda_2^n$ are solutions of Equation (4.21). Using Eqs. (4.18), (4.19), and (4.20), we can make the following conclusions:

1. If $\lambda_1 \neq \lambda_2$ and both are real, then the general solution of Equation (4.21) is given by

$$u(n) = c_1\lambda_1^n + c_2\lambda_2^n, \qquad (4.24)$$

2. If $\lambda_1 = \lambda_2 = \lambda$, then the general solution of Equation (4.21) is given by

$$u(n) = c_1\lambda^n + c_2 n\lambda^n, \qquad (4.25)$$

3. If $\lambda_1 = \alpha + i\beta$, $\lambda_2 = \alpha - i\beta$, then the general solution of Equation (4.21) is given by

$$u(n) = |\lambda_1|^n(c_1 \cos n\omega + c_2 \sin n\omega), \qquad (4.26)$$

where $\omega = \arctan(\beta/\alpha)$.

Example 4.4
Solve the second-order difference equation

$$x(n + 2) + 6x(n + 1) + 9x(n) = 0, \; x(0) = 1, \; x(1) = 0. \quad \square$$

SOLUTION   The characteristic equation associated with the equation is given by $\lambda^2 + 6\lambda + 9 = 0$.

Hence, the characteristic roots are $\lambda_1 = \lambda_2 = -3$. The general solution is given by

$$x(n) = 9(-3)^n + c_2 n(-3)^n$$
$$x(0) = 1 = c_1$$
$$x(1) = 0 = -3c_1 - 3c_2.$$

Thus, $c_2 = -1$ and, consequently,

$$x(n) = (-3)^n - n(-3)^n$$
$$= (-3)^n(1 - n) \quad \blacksquare$$

Exercises - (4.3 and 4.4)

1. Solve the system

$$x_1(n + 1) = -x_1(n) + x_2(n)$$
$$x_2(n + 1) = 2x_2(n)$$
$$\text{with } x_1(0) = 1, \; x_2(0) = 2.$$

2. Find the general solution of the system

$$X(n+1) = AX(n), \quad \text{where} \quad A = \begin{pmatrix} 1 & -5 \\ 1 & -1 \end{pmatrix}.$$

3. Solve the problem $X(n+1) = AX(n)$, where $A = \begin{pmatrix} 1 & 1 \\ -2 & 4 \end{pmatrix}$.

4. Solve the system

$$X(n+1) = AX(n), \quad \text{with} \quad A = \begin{pmatrix} 2 & -1 \\ 0 & 4 \end{pmatrix}, \quad X(0) = \begin{pmatrix} 1 \\ 2 \end{pmatrix}.$$

5. Solve the system $x(n+1) = Ax(n)$, with $A = \begin{pmatrix} 3 & 1 \\ 0 & 3 \end{pmatrix}$.

6. Solve the difference equation

$$x(n+2) - 5x(n+1)6x(n) = 0, \quad x(0) = 2$$

   (a) By converting it to a system,

   (b) Directly as it is.

7. Solve the difference equation

$$F(n+2) = F(n+1) + F(n), \quad F(1) = 1, \quad F(2) = 1.$$

   (This is called the Fibonacci sequence.)

8. The Chebyshev polynomials of the first and second kind are defined as follows:

$$T_n(x) = \cos(n \cos^{-1}(x)),$$
$$U_n(x) = \frac{1}{\sqrt{1-x^2}} \sin[(n+1)\cos^{-1}(x)], \quad \text{for } |x| < 1.$$

   (a) Show that $T_n(x)$ satisfies the difference equation

$$T_{n+2}(x) - 2xT_{n+1}(x) + T_n(x) = 0, \quad T_0(x) = 1, \quad T_1(x) = x.$$

   (b) Solve for $T_n(x)$.

   (c) Show that

$$U_{n+2}(x) - 2xU_{n+1}(x) + U_n(x) = 0, \quad U_0(x) = 1, \quad U_1(x) = 2x.$$

   (d) Write down the first four terms of $T_n(x)$ and $U_n(x)$.

9. Solve the equation $x(n+2) + 16x(n) = 0$.

10. Let A be a 2×2 real matrix with distinct eigenvalues λ_1 and λ_2. Prove that the corresponding eigenvectors V_1 and V_2 are linearly independent.

11. Let A be a 2×2 real matrix with a repeated eigenvalue λ. Let V_1 be an eigenvector corresponding to λ and let V_2 be a generalized eigenvector. Show that V_1 and V_2 are linearly independent.

12. Let A be a 2×2 real matrix with complex eigenvalues $\lambda_1 = \alpha + i\beta$ and $\lambda_2 = \alpha - i\beta$. Suppose that $V = V_1 + iV_2$ is the eigenvector corresponding to λ_1. Prove that the matrix $P = (V_1, V_2)$ is nonsingular. (*Hint: It suffices to show that V_1 and V_2 are linearly independent.*)

13. (a) Show that $X_1(n)$ and $X_2(n)$, obtained from Equation (4.18), are solutions of Equation (4.16).

    (b) Show that $X_1(n)$ and $X_2(n)$, obtained from Equation (4.19), are solutions of Equation (4.16).

    (c) Show that $X_1(n)$ and $X_2(n)$, obtained from Equation (4.20), are solutions of Equation (4.16).

    In Problems 14 and 15, consider the nonhomogeneous equation

    $$Y(n+1) = AY(n) + g(n) \tag{4.27}$$

    where A is a 2×2 matrix and g is a function defined on $\mathbb{Z}^+$.

14. Show that

    $$Y(n) = A^n Y(0) + \sum_{k=0}^{n-1} A^{n-k-1} g(k). \tag{4.28}$$

    (This is called the variation of constants formula.)

15. Use Formula (4.28) to find the solution of Equation (4.27) with

    $$A = \begin{pmatrix} 2 & 1 \\ 0 & 2 \end{pmatrix}, \quad g(n) = \begin{pmatrix} n \\ 1 \end{pmatrix}, \quad Y(0) = \begin{pmatrix} 1 \\ 0 \end{pmatrix}.$$

16. Solve the equation $y(n+2) - 5y(n+1) + 4y(n) = 4^n$.

4.5 Phase Space Diagrams

One of the best graphical methods to illustrate the various notions of stability is the phase portrait or the phase space diagram. Let $f : \mathbb{R}^2 \to \mathbb{R}^2$ be a given map. Then, starting from an initial point $X_0 = \begin{pmatrix} x_1(0) \\ x_2(0) \end{pmatrix}$, we plot the sequence of point $X_0, f(X_0), f^2(X_0), f^3(X_0), \ldots$ and then connect the points by straight lines. An arrow is placed on these connecting lines to indicate the direction of the motion on the orbit. In many instances, we need to be prudent in choosing our initial points in order to get a better phase portrait.

In this section, we consider linear systems for which $f(X) = AX$, where A is a 2×2 matrix. Observe that if $A - I$ is nonsingular, i.e., $\det(A-I) \neq 0$, then the origin $\begin{pmatrix} 0 \\ 0 \end{pmatrix}$ is the only fixed point of the map f. Equivently, $X^* = \begin{pmatrix} 0 \\ 0 \end{pmatrix}$ is the only fixed point of the system

$$X(n+1) = AX(n) \tag{4.29}$$

As stipulated in Theorem 4.2, there exists a nonsingular matrix P such that $P^{-1}AP = J$ where J is one of the forms (1), (2), or (3) in Theorem 4.2. If we let

$$X(n) = PY(n) \tag{4.30}$$

in Equation (4.29), we obtain

$$Y(n+1) = JY(n). \tag{4.31}$$

Our plan here is to draw the phase portrait of Equation (4.31), then use the transformation (4.30) to obtain the phase portrait of the original system (4.29).

(I). We begin our discussion by assuming that J is in the diagonal form $J = \begin{pmatrix} \lambda_1 & 0 \\ 0 & \lambda_2 \end{pmatrix}$, where λ_1 and λ_2 are not necessarily distinct. Here we have two linearly independent solutions:

$$Y_1(n) = \lambda_1^n V_1, \quad \text{and} \quad Y_2(n) = \lambda_2^n V_2, \quad \text{where} \quad V_1 = \begin{pmatrix} 1 \\ 0 \end{pmatrix} \text{ and } V_2 = \begin{pmatrix} 0 \\ 1 \end{pmatrix}$$

are the eigenvectors of A corresponding to the eigenvalues λ_1 and λ_2, respectively.

Observe that $Y_1(n)$ is a multiple of V_1, and thus must stay on the line emanating from the origin in the direction of V_1; in this case, the x axis. Similarly, $Y_2(n)$ must stay on the line passing through the origin and in the direction of V_2; in this case, the y axis. These two solutions are called straight line solutions. The general solution is given by

$$Y(n) = k_1\lambda_1^n \begin{pmatrix} 1 \\ 0 \end{pmatrix} + k_2\lambda_2^n \begin{pmatrix} 0 \\ 1 \end{pmatrix}, \ Y(0) = \begin{pmatrix} k_1 \\ k_2 \end{pmatrix}. \tag{4.32}$$

We have the following cases to consider:

1. If $|\lambda_1| < 1$ and $|\lambda_2| < 1$, then all solutions tend to the origin as $n \to \infty$. Observe that if $|\lambda_1| < |\lambda_2| < 1$, then, $|\lambda_1^n|$ goes to zero faster than $|\lambda_2^n|$.

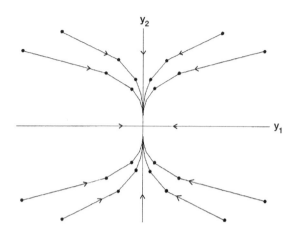

FIGURE 4.1a
(a) A sink: $0 < \lambda_1 < \lambda_2$.

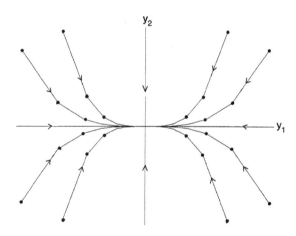

FIGURE 4.1b
(b) A sink: $0 < \lambda_2 < \lambda_1 < 1$.

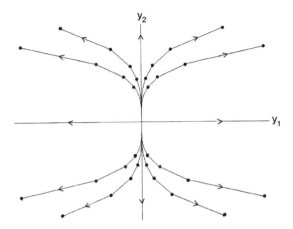

FIGURE 4.2a
(a) A source: $\lambda_1 > \lambda_2 > 1$.

And consequently, any solution $Y(n)$ in the form (4.31) is asymptotic to the straight line solution $Y_2(n) = \lambda_2^n \begin{pmatrix} 0 \\ 1 \end{pmatrix}$ (see Fig. 4.1a).

On the other hand, if $|\lambda_1| > |\lambda_2|$, then $Y(n)$ is asymptotic to $Y_1(n) = \lambda_1^n \begin{pmatrix} 1 \\ 0 \end{pmatrix}$ (see Fig. 4.1b).

Phase portraits 4.1a and 4.1b are called <u>sinks</u>.

2. If $|\lambda_1| > 1$, and $|\lambda_2| > 1$, then we obtain an source as illustrated in Figs. 4.2a and 4.2b.

Note that if $|\lambda_1| > |\lambda_2| > 1$, then $Y(n)$ is asymptotic to $Y_2(n) = \lambda_2^n \begin{pmatrix} 0 \\ 1 \end{pmatrix}$ (the y axis) when $n \to -\infty$ and is dominated by $Y_1(n) = \lambda_1^n \begin{pmatrix} 1 \\ 0 \end{pmatrix}$ when $n \to \infty$.

3. If $|\lambda_1| < 1$ and $|\lambda_2| > 1$, then we obtain a saddle (Fig. 4.3). In this case, when $n \to \infty$ $Y(n)$, is asymptotic to $Y_2(n) = \lambda_2^n \begin{pmatrix} 0 \\ 1 \end{pmatrix}$ as $n \to \infty$ and is asymptotic to $Y_1(n) = \lambda_1^n \begin{pmatrix} 1 \\ 0 \end{pmatrix}$ as $n \to -\infty$. Similar analysis is readily available for the case $|\lambda_1| > 1$ and $|\lambda_2| < 1$.

4. If $\lambda_1 = \lambda_2$, then

$$Y(n) = k_1 \lambda^n \begin{pmatrix} 1 \\ 0 \end{pmatrix} + k_2 \lambda^n \begin{pmatrix} 0 \\ 1 \end{pmatrix} = \lambda^n \begin{pmatrix} k_1 \\ k_2 \end{pmatrix}.$$

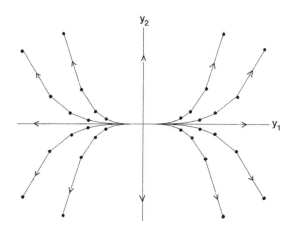

FIGURE 4.2b
(b) A source: $\lambda_2 > \lambda_1 > 1$.

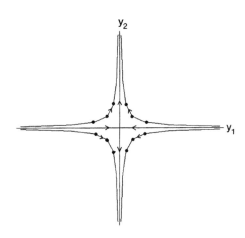

FIGURE 4.3
A saddle: $0 < \lambda_1 < 1, \lambda_2 > 1$.

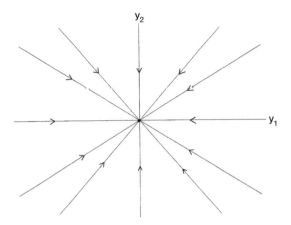

FIGURE 4.4
A sink: $0 < \lambda_1 < \lambda_2 < 1$.

Hence, every solution $Y(n)$ lies on a line passing through the origin with a slope k_2/k_1 (see Fig. 4.4).

Observe that in each of the four subcases, the presence of a negative eigenvalue will cause the solution $Y(n)$ to oscillate around the origin and the phase portrait will not look as nice as in Figs. 4.1a–4.4.

(II). Suppose that J is in the form

$$J = \begin{pmatrix} \lambda & 1 \\ 0 & \lambda \end{pmatrix}.$$

Then, we only have one straight line solution, $Y_1(n) = \lambda^n V_1 = \lambda^n \begin{pmatrix} 1 \\ 0 \end{pmatrix}$. The general solution is given by

$$Y(n) = k_1 \lambda^n \begin{pmatrix} 1 \\ 0 \end{pmatrix} + k_2 \left(n\lambda^{n-1} \begin{pmatrix} 1 \\ 0 \end{pmatrix} + \lambda^n \begin{pmatrix} 0 \\ 1 \end{pmatrix} \right)$$

$$= (k_1 \lambda + k_2 n)\lambda^{n-1} \begin{pmatrix} 1 \\ 0 \end{pmatrix} + k_2 \lambda^n \begin{pmatrix} 0 \\ 1 \end{pmatrix}.$$

Now, if $|\lambda| < 1$, then, $Y(n) \to 0$ as $n \to \infty$, since $\lim\limits_{n\to\infty} n\lambda^{n-1} = 0$ (by L'Hopital Rule). Since the term $k_1 \lambda^n \begin{pmatrix} 0 \\ 1 \end{pmatrix}$ tends to the origin, as $n \to \infty$, faster than the term $(k_1 \lambda + k_2 n)\lambda^n \begin{pmatrix} 1 \\ 0 \end{pmatrix}$, our solution $Y(n)$ tends to the origin asymptotic to the x axis (see Fig. 4.5a). In this case, the origin is called a

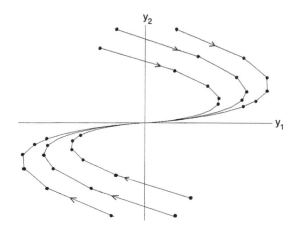

FIGURE 4.5a

(a) A degenerate sink: $\lambda_1 = \lambda_2 = \lambda, \ 0 < \lambda < 1$.

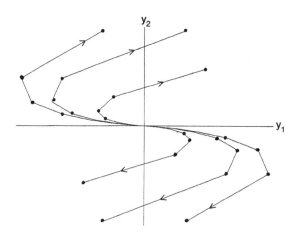

FIGURE 4.5b

(b) A degenerate source: $\lambda_1 = \lambda_2 = \lambda, \ \lambda > 1$.

degenerate sink. Figure 4.5b depicts the case when $|\lambda| > 1$ and in this case, the origin is called a degenerate source.

(III). Suppose that J is in the form

$$J = \begin{pmatrix} \alpha & \beta \\ -\beta & \alpha \end{pmatrix}.$$

In this case, we have no straight line solutions due to the presence of $\cos n\beta$ and $\sin n\beta$ in the solutions $Y_1(n) = |\lambda_1|^n (k_1 \cos n\beta + k_2 \sin n\beta) \begin{pmatrix} 1 \\ 0 \end{pmatrix}$ and $Y_2(n) = |\lambda_1|^n (-k_1 \sin n\beta + k_2 \cos n\beta) \begin{pmatrix} 0 \\ 1 \end{pmatrix}$. The general solution is given by

$$Y(n) = |\lambda_1|^n \begin{pmatrix} k_1 \cos n\beta + k_2 \sin n\beta \\ -k_1 \sin n\beta + k_2 \cos n\beta \end{pmatrix}$$

with $Y(0) = \begin{pmatrix} k_1 \\ k_2 \end{pmatrix}$. Define an angle γ by setting $\cos \gamma = k_1/r_0$, and $\sin \gamma = k_2/r_0$, where $r_0 = \sqrt{k_1^2 + k_2^2}$. Then

$$y_1(n) = |\lambda_1|^n r_0 \cos(n\omega - \gamma)$$
$$y_2(n) = -|\lambda_1|^n r_0 \sin(n\omega - \gamma).$$

Thus, the solution in polar coordinates is given by

$$r(n) = \sqrt{y_1^2(n) + y_2^2(n)}$$
$$= r_0 |\lambda_1|^n \qquad (4.33)$$
$$\theta(n) = \arctan\left(\frac{y_2(n)}{y_1(n)}\right)$$
$$= -(n\omega - \gamma) \qquad (4.34)$$

It follows from Eqs. (4.33) and (4.34) that

1. If $|\lambda_1| < 1$, then we have a stable focus where each orbit spirals toward the origin [Fig. 4.6(a)]. On the other hand, if $|\lambda_1| > 1$, then we have an unstable focus, where each orbit spirals away from the origin [Fig. 4.6(b)].

2. If $|\lambda_1| = 1$, then we have a center, where the orbits follow a circular path [Fig. 4.6(c)]. This is due to the fact that $y_1^2(n) + y_2^2(n) = r_0^2$.

To this end, we have obtained the phase portraits of Equation (4.31), which may be called "canonical" phase portraits. To obtain the phase portraits of the original system (4.29), we apply (4.30), i.e., we apply P to the orbits of Equation (4.31). Since $P \begin{pmatrix} 1 \\ 0 \end{pmatrix} = (V_1, V_2) \begin{pmatrix} 0 \\ 1 \end{pmatrix} = V_1$, and $P \begin{pmatrix} 0 \\ 1 \end{pmatrix} =$

$(V_1, \ V_2) \begin{pmatrix} 0 \\ 1 \end{pmatrix} = V_2$, applying P to the orbits $Y(n)$ amounts to rotating the coordinates; the x axis to V_1 and the y axis to V_2. In other words, the straight line solutions are now along the eigenvectors V_1 and V_2. Using this observation, one may opt to sketch the phase portrait of Equation (4.29) directly and without going through the canonical forms.

The set of points on the line emanating from the origin along V_1 is called the <u>stable</u> subspace W^s; the set of points on the line passing through the origin in the direction of V_2 is called the <u>unstable</u> subspace W^u. Hence,

$$W^s = \{X \in \mathbb{R}^2 : A^n X \to 0 \text{ as } n \to \infty\}, \tag{4.35}$$
$$W^u = \{X \in \mathbb{R}^2 : A^{-n} X \to 0 \text{ as } n \to \infty\} \tag{4.36}$$

The following example illustrates the above-described direct method to sketch the phase portrait.

Example 4.5
Sketch the phase portrait of the system $X(n+1) = AX(n)$, where

$$A = \begin{pmatrix} 1 & 1 \\ 0.25 & 1 \end{pmatrix} \quad \Box$$

SOLUTION   The eigenvalues of A are $\lambda_1 = \frac{3}{2}$, and $\lambda_2 = \frac{1}{2}$; the corresponding eigenvectors are $V_1 = \begin{pmatrix} 2 \\ 1 \end{pmatrix}$ and $V_2 = \begin{pmatrix} 2 \\ -1 \end{pmatrix}$, respectively. Hence, we have two straight line solutions, $X_1(n) = (1.5)^n \begin{pmatrix} 2 \\ 1 \end{pmatrix}$ and $X_2(n) = (0.5)^n \begin{pmatrix} 2 \\ -1 \end{pmatrix}$. The general solution is given by $X(n) = k_1(1.5)^n \begin{pmatrix} 2 \\ 1 \end{pmatrix} + k_2(0.5)^n \begin{pmatrix} 2 \\ -1 \end{pmatrix}$. Note that $x(n)$ is asymptotic to the line through $V_1 = \begin{pmatrix} 2 \\ 1 \end{pmatrix}$ (see Fig. 4.7).   ∎

4.6   Stability Notions

Consider the map $f : \mathbb{R}^2 \to \mathbb{R}^2$ and let $X^* = \begin{pmatrix} x_1^* \\ x_2^* \end{pmatrix}$ be a fixed point of f; i.e., $f(X^*) = X^*$.

Our main objective in this section is to introduce the main stability notions pertaining to the fixed point x^*. Observe that these notions were previously

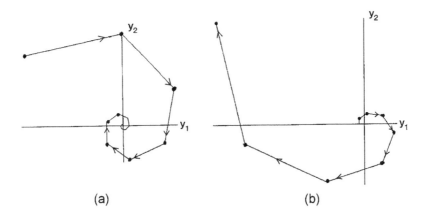

(a)   (b)

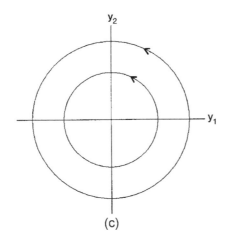

(c)

FIGURE 4.6
(a) Stable focus: $|\lambda_1| < 1$. (b) Source: $|\lambda_1| > 1$. (c) Center: $\lambda_{1,2} = \alpha \pm i\beta$, $|\lambda_{1,2}| = 1$.

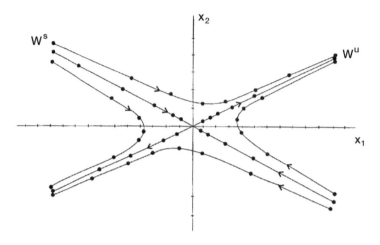

FIGURE 4.7
Saddle: $\lambda_1 > 1$, $0 < \lambda_2 < 1$. Stable and unstable subspaces: W^s, W^u.

introduced in Chapter 1. The only difference in $\mathbb{R}^2$ is that we replace the absolute value by a convenient "norm" on $\mathbb{R}^2$. Roughly speaking, a norm of a vector (point) in $\mathbb{R}^2$ is a measure of its magnitude. A formal definition follows.

DEFINITION 4.3 *A real valued function on a vector space V is said to be a norm on V, denoted by $|\ |$, if the following properties hold:*

1. *$|X| \geq 0$ and $|X| = 0$ if and only if $X = 0$, for $X \in V$.*

2. *$|\alpha X| = |\alpha||X|$ for $X \in V$ and any scalar α.*

3. *$|X + Y| \leq |X| + |Y|$ for $X, Y \in V$ (the triangle inequality).*

In the sequel, we choose the ℓ_1 norm on $\mathbb{R}^2$ defined for $X = \begin{pmatrix} x_1 \\ x_2 \end{pmatrix}$ as

$$|X| = |x_1| + |x_2| \qquad (4.37)$$

For each vector norm on $\mathbb{R}^2$ there corresponds a norm $||\ ||$ on all 2×2 matrices $A = (a_{ij})$ defined as follows

$$||A|| = \sup\{|AX| : |X| \leq 1\}. \qquad (4.38)$$

It may be easily shown that for $X \in \mathbb{R}^2$,

$$|AX| \leq ||A||\ |X| \qquad (4.39)$$

Let $\rho(A)$ be the **spectral radius** of A defined as $\rho(A) = \max\{|\lambda_1|, |\lambda_2| :$ λ_1, λ_2 are the eigenvalues of $A\}$. Then it may be shown that for our selected vector norm

$$||A||_1 = \max\{(|a_{11}| + |a_{21}|), (|a_{12}| + |a_{22}|)\}. \qquad (4.40)$$

(For a proof see [49].)

For example, if $X = \begin{pmatrix} 1 \\ 2 \end{pmatrix}$, $|X| = 3$. And for the matrix $A = \begin{pmatrix} 1 & 3 \\ -2 & -4 \end{pmatrix}$, $||A||_1 = \max\{3, 7\} = 7$. The eigenvalues of A are $\lambda_1 = -2$, $\lambda_2 = -1$. Note that $\rho(A) = \max\{|-2|, |-1|\} = 2$, and thus $\rho(A) \leq ||A||_1$.

It is left to the reader to prove, in general, that $\rho(A) \leq ||A||_1$ for any matrix A (Problem 14).

Without any further delay, we now give the required stability definitions.

DEFINITION 4.4   *A fixed point X^* of a map $f : \mathbb{R}^2 \to \mathbb{R}^2$ is said to be*

1. **stable** *if given $\varepsilon > 0$ there exists $\delta > 0$ such $|X - X^*| < \delta$ implies $|f^n(X) - X^*| < \varepsilon$ for all $n \in \mathbb{Z}^+$ (see Fig. 4.8a).*

2. **attracting** *(sink) if there exists $\nu > 0$ such that $|X - X^*| < \nu$ implies $\lim_{n \to \infty} f^n(X) = X^*$. It is **globally attracting** if $\nu = \infty$ (see Fig. 4.9).*

3. **asymptotically stable** *if it is both stable and attracting. It is **globally asymptotically stable** if it is both stable and globally attracting, (see Fig. 4.12(a))*

4. **unstable** *if it is not stable (see Fig. 4.8b).*

REMARK 4.2   In [91], Sedaghat showed that a globally attracting fixed point of a continuous one-dimensional map must be stable. Kenneth Palmer pointed out to me that this result may be found in the book of Block and Coppel [12]. Moreover, the proof in Block and Coppel requires only local attraction (see Appendix for a proof). The situation changes dramatically in two- or higher dimensional continuous maps, for there are continuous maps that possess a globally attracting unstable fixed point. We are going to present one of these maps and put several others as problems for you to investigate.

Example 4.6

Consider the two-dimensional map in polar coordinates

$$g \begin{pmatrix} r \\ \theta \end{pmatrix} = \begin{pmatrix} \sqrt{r} \\ \sqrt{2\pi\theta} \end{pmatrix}, \quad r > 0, 0 \leq \theta \leq 2\pi.$$

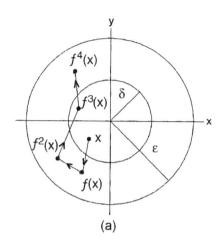

(a)

FIGURE 4.8a
(a) The fixed point $X^* = 0$ is stable.

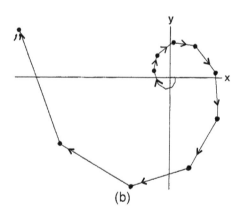

(b)

FIGURE 4.8b
(b) $X^* = 0$ is an unstable fixed point.

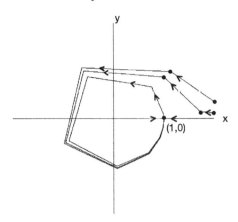

FIGURE 4.9

$x^* = \begin{pmatrix} 1 \\ 0 \end{pmatrix}$ is an unstable globally attracting fixed point.

Then,

$$g^n \begin{pmatrix} r \\ \theta \end{pmatrix} = \begin{pmatrix} r^{2^{-n}} \\ (2\pi)^{(1-2^{-n})}\theta^{2^{-n}} \end{pmatrix}.$$

Clearly $\lim_{n\to\infty} g^n \begin{pmatrix} r \\ \theta \end{pmatrix} = \begin{pmatrix} 1 \\ 2\pi \end{pmatrix} \equiv \begin{pmatrix} 1 \\ 0 \end{pmatrix}$. Thus, each orbit is attracted to the

fixed point $\begin{pmatrix} 1 \\ 0 \end{pmatrix}$. However, if $\theta = \delta\pi$, $0 < \delta < 1$, then the orbit of $\begin{pmatrix} r \\ \theta \end{pmatrix}$ will

spiral clockwise around the fixed point $\begin{pmatrix} 1 \\ 0 \end{pmatrix}$ before converging to it. Hence,

$\begin{pmatrix} 1 \\ 0 \end{pmatrix}$ is globally attracting but not asymptotically stable (see Fig. 4.9). ☐

4.7   Stability of Linear Systems

In this section, we focus our attention on linear maps where $f(X) = AX$, and A is a 2×2 matrix. Equivalently, we are interested in the difference equation

$$X(n+1) = AX(n). \tag{4.41}$$

For such linear maps, we can provide complete information about the stability of the fixed point $X^* = \begin{pmatrix} 0 \\ 0 \end{pmatrix}$. The main result now follows.

THEOREM 4.3

The following statements hold for Equation (4.41):

(a) If $\rho(A) < 1$, then the origin is asymptotically stable.

(b) If $\rho(A) > 1$, then the origin is unstable.

(c) If $\rho(A) = 1$, then the origin is unstable if the Jordan form is of the form $\begin{pmatrix} \lambda & 1 \\ 0 & \lambda \end{pmatrix}$, and stable otherwise.

PROOF   Suppose that $\rho(A) < 1$. Then it follows from Eqs. (4.18), (4.19), (4.20), that $\lim_{n\to\infty} X(n) = 0$. Thus, the origin is (globally) attracting. To prove stability, we consider three cases.

(i) Suppose that the solution $X(n)$ is given by Equation (4.18). This is the case when the eigenvalues of the matrix A are real and there are two linearly independent eigenvectors.

$$X(n) = P \begin{pmatrix} \lambda_1^n & 0 \\ 0 & \lambda_2^n \end{pmatrix} P^{-1} X(0).$$

Hence,

$$|X(n)| \leq ||P|| \, ||P^{-1}|| \rho(A) |X(0)|$$
$$\leq M |X(0)|$$

where $M = ||P|| \, ||P^{-1}|| \, \rho(A)$.

Now, given $\varepsilon > 0$, let $\delta = \varepsilon/M$. Then $|X(0)| < \delta$ implies that $|X(n)| < M\delta = \varepsilon$. This shows that the origin is stable.

(ii) Suppose that the solution $X(n)$ is given by Equation (4.19). This case occurs if the matrix A has a repeated eigenvalue λ and only one eigenvector.

$$X(n) = P \begin{pmatrix} \lambda^n & n\lambda^{n-1} \\ 0 & \lambda^n \end{pmatrix} P^{-1} X(0)$$
$$|X(n)| \leq ||P|| \, ||P^{-1}|| \, (n|\lambda|^{n-1} + |\lambda|^n)|X(0)|.$$

Since $n|\lambda|^n \to 0$ as $n \to \infty$, there exists $N \in \mathbb{Z}^+$, such that the term $(n|\lambda|^{n-1} + |\lambda|^n)$ is bounded by a positive number L. Hence,

$$|X(n)| \leq M |X(0)|$$

where $M = L||P|| \, ||P^{-1}||$.

The proof of the stability of the origin may be completed by seting $\delta = \varepsilon/M$ as in part (a).

(iii) If A is not in the form $\begin{pmatrix} \lambda & 1 \\ 0 & \lambda \end{pmatrix}$, then it is either diagonalizable to $J = \begin{pmatrix} \lambda_1 & 0 \\ 0 & \lambda_2 \end{pmatrix}$, where $|\lambda_1| < 1$ and $|\lambda_2| = 1$ or $|\lambda_2| < 1$ and $|\lambda_1| = 1$. In either case, J^n is bounded and hence the origin is stable. ∎

The proofs of parts (b) and (c) are left to you as Problem 11.

Exercises - $(4.5 - 4.7)$

In Problems 1–9, sketch the phase portrait of the system $X(n+1) = AX(n)$, when A is the given matrix. Determine the stability of the origin.

1. $\begin{pmatrix} 1/2 & 0 \\ 0 & 1/2 \end{pmatrix}$

2. $\begin{pmatrix} 1/5 & 0 \\ 0 & 2 \end{pmatrix}$

3. $\begin{pmatrix} 2 & 1 \\ 0 & 2 \end{pmatrix}$

4. $\begin{pmatrix} -1/2 & 1 \\ 0 & -1/2 \end{pmatrix}$

5. $\begin{pmatrix} 0.5 & 0.25 \\ -0.25 & 0.5 \end{pmatrix}$

6. $\begin{pmatrix} 1 & 1 \\ -1 & 3 \end{pmatrix}$

7. $\begin{pmatrix} -3 & -4 \\ 7.5 & 8 \end{pmatrix}$

8. $\begin{pmatrix} -2 & 1 \\ -7 & 3 \end{pmatrix}$

9. $\begin{pmatrix} 1 & 2 \\ -1 & -1 \end{pmatrix}$

10. $\begin{pmatrix} 1 & 3 \\ -1 & 1 \end{pmatrix}$

11. $\begin{pmatrix} 1 & 1 \\ \frac{1}{4} & 1 \end{pmatrix}$

12. Show that if X^* is a fixed point of a linear map f on $\mathbb{R}^2$ and is asymptotically stable, then it must be globally asymptotically stable.

13. Complete the proof of Theorem 4.3.

14. Prove that for any 2×2 matrix $A, \rho(A) < ||A||$.

4.8   The Trace-Determinant Plane

4.8.1   Stability Analysis

Table (4.1) provides a partial summary of everything we have done so far. In this section we provide another important way of presenting these results, namely, trace-determinant plane, where we employ pictures, rather than a table. This turns out to be a better scheme when one is interested in studying bifurcation in two-dimensional systems.

The following two results provide the framework for using the trace-determinant plane. Recall that for matrix $A = \begin{pmatrix} a_{11} & a_{12} \\ a_{21} & a_{22} \end{pmatrix}$, $tr\ A = a_{11} + a_{22}$, and $det\ A = a_{11}a_{22} + a_{12}a_{21}$.

THEOREM 4.4
Let $A = (a_{ij})$ be a 2×2 matrix. Then $\rho(A) < 1$ if and only if

$$|tr\ A| - 1 < det\ A < 1. \tag{4.42}$$

PROOF

(i) Assume that $\rho(A) < 1$. If λ_1, λ_2 are the eigenvalues of A, then $|\lambda_1| < 1$ and $|\lambda_2| < 1$. The characteristic equation of the matrix A is given by $det\ (A - \lambda I) = \lambda^2 - (a_{11} + a_{22})\lambda - (a_{11}a_{22} - a_{12}a_{21}) = 0$, or $\lambda^2 - (tr\ A)\lambda + det\ A = 0$. Hence the eigenvalues are

$$\lambda_1 = \frac{1}{2}\left[tr\ A + \sqrt{(tr\ A)^2 - 4\ det\ A}\right],$$
$$\lambda_2 = \frac{1}{2}\left[tr\ A - \sqrt{(tr\ A)^2 - 4\ det\ A}\right].$$

Case (a) λ_1 and λ_2 are real roots, i.e., $(tr\ A)^2 - 4\ det\ A \geq 0$. Now $-1 < \lambda_1$, $\lambda_2 < 1$ implies that

$$-2 < tr\ A \pm \sqrt{(tr\ A)^2 - 4\ det\ A} < 2$$

TABLE 4.1
Partial summary of the stability of linear systems.

Type	Eigenvalue	Phase Portrait
Saddle	$0 < \lambda_1 < 1 < \lambda_2$	
Sink	$0 < \lambda_2 < \lambda_1 < 1$	
Source	$\lambda_2 > \lambda_1 > 1$	
Spiral Sink	$\lambda = \alpha \pm i\beta,\ |\lambda| < 1,\ \beta \neq 0$	
Spiral Source	$\lambda = \alpha \pm i\beta,\ |\lambda| > 1,\ \beta \neq 0$	
Center	$\lambda = \alpha \pm i\beta,\ |\lambda| = 1,\ \beta \neq 0$	
"Oscillatory" Saddle	$-1 < \lambda_1 < 0,\ \lambda_2 < -1$	
"Oscillatory" Source	$\lambda_1 > 1,\ \lambda_2 > -1$	

or

$$-2 - tr\ A < \sqrt{(tr\ A)^2 - 4\ det\ A} < 2 - tr\ A \qquad (4.43)$$

$$-2 - tr\ A < -\sqrt{(tr\ A)^2 - 4\ det\ A} < 2 - tr\ A. \qquad (4.44)$$

Squaring the second inequality (4.43) yields

$$1 - tr\ A + det\ A > 0. \qquad (4.45)$$

Similarly, if we square the first inequality in (4.44) we obtain

$$1 + tr\ A + det\ A > 0. \qquad (4.46)$$

Now from the second inequality (4.43) and the first inequality in (4.44) we obtain $2 + tr\ A > 0$ and $2 - tr\ A > 0$ or $|tr\ A| < 2$. Since $(tr\ A)^2 - 4\ det\ A \geq 0$, it follows that

$$det\ A \leq (tr\ A)^2/4 < 1. \qquad (4.47)$$

Combining (4.45), (4.46) and (4.47) yields (4.42).

Case (b) λ_1 and λ_2 are complex conjugates, i.e.,

$$(tr\ A)^2 - 4\ det\ A < 0. \qquad (4.48)$$

In this case we have $\lambda_{1,2} = \frac{1}{2}\left[tr\ A \pm i\sqrt{4\ det\ A - (tr\ A)^2}\right]$ and

$$|\lambda_1|^2 = |\lambda_2|^2 = \frac{(tr\ A)^2}{4} + \frac{4\ det\ A}{4} - \frac{(tr\ A)^2}{4} = det\ A.$$

Hence $0 < det\ A < 1$. To show that inequalities (4.45) and (4.46) hold, note that since $det\ A > 0$ either (4.45) (if $tr\ A > 0$) or (4.46) (if $tr\ A < 0$) holds. Without loss of generality, assume that $tr\ A > 0$. Then (4.45) holds. From (4.48), $tr\ A < 2\sqrt{det\ A}$. If we let $det\ A = x$, then $x \in (0,1)$ and $f(x) = 1 + x - 2\sqrt{x} < 1 + det\ A - tr\ A$. Note that $f(0) = 1$ and $f'(0) = 1 - \frac{1}{\sqrt{x}}$ indicate that f is decreasing on $(0,1)$ with range $(0,1)$. This implies that $1 + det\ A - tr\ A > f(x) > 0$ and this completes the proof.

(ii) Conversely, assume that (4.42) holds. Then we have two cases to consider.

Case (a) $(tr\ A)^2 - 4\ det\ A \geq 0$. Then

$$\begin{aligned}
|\lambda_{1,2}| &= \frac{1}{2}\left|tr\ A \pm \sqrt{(tr\ A)^2 - 4\ det\ A}\right| \\
&< \frac{1}{2}\left|tr\ A \pm \sqrt{(det\ A + 1)^2 - 4\ det\ A}\right| \\
&< \frac{1}{2}\left(det\ A + 1 + \sqrt{(det\ A - 1)^2}\right) \\
&= \frac{1}{2}\left(det\ A + 1 - (det\ A - 1)\right) = 1.
\end{aligned}$$

Case (b) $(tr\ A)^2 - 4\ det\ A < 0$. Then

$$|\lambda_{1,2}| = \frac{1}{2}\left|tr\ A \pm i\sqrt{4\ det\ A - (tr\ A)^2}\right|$$
$$= \frac{1}{2}\sqrt{(tr\ A)^2 + 4\ det\ A - (tr\ A)^2}$$
$$= \sqrt{det\ A} < 1.$$

∎

As a by-product of the preceding result, we obtain the following criterion for asymptotic stability.

COROLLARY 4.1
The origin in Equation (4.41) is asymptotically stable if and only if condition (4.42) holds true.

Note that condition (4.42) may be spelled out in the following three inequalities:

$$1 + tr\ A + det\ A > 0, \quad \text{or}$$

$$\boxed{det\ A > -tr\ A - 1}(4.45)' \tag{4.49}$$

$$1 - tr\ A + det\ A > 0, \quad \text{or}$$

$$\boxed{det\ A > tr\ A - 1}(4.46)' \tag{4.50}$$

$$\boxed{det\ A < 1.}(4.47)' \tag{4.51}$$

Viewing $det\ A$ as a function of $tr\ A$, the above three inequalities give us the stability region as the interior of the triangle bounded by the lines $det\ A = -tr\ A - 1$, $det\ A = tr\ A - 1$, and $det\ A = 1$, as shown in Fig. 4.10.

Next we delve a little deeper into finding the exact values of the eigenvalues of the matrix A along the boundaries of the triangle enclosing the stability region. The following result provides us with the needed answers. Let $\lambda_1 = \frac{1}{2}\left(tr\ A + \sqrt{(tr\ A)^2 - 4\ det\ A}\right)$, $\lambda_2 = \frac{1}{2}\left(tr\ A - \sqrt{(tr\ A)^2 - 4\ det\ A}\right)$ be the eigenvalues of A.

THEOREM 4.5
The following statements hold for any 2×2 matrix A.

(i) *If $|tr\ A| - 1 = det\ A$, then we have*

    (a) *the eigenvalues of A are $\lambda_1 = 1$ and $\lambda_2 = det\ A$ if $tr\ A > 0$,*

    (b) *the eigenvalues of A are $\lambda_2 = -1$ and $\lambda_1 = -det\ A$ if $tr\ A < 0$,.*

(ii) If $|tr\ A| - 1 < det\ A$, and $det\ A = 1$, then the eigenvalues of A are $e^{\pm i\theta}$, where $\theta = \cos^{-1}(tr\ A/2)$.

PROOF

(i) Let $|tr\ A| - 1 = det\ A$. Then $(tr\ A)^2 - 4\ det\ A = (det\ A + 1)^2 \geq 0$. This implies that the eigenvalues are real numbers. Moreover, $\lambda_{1,2} = \frac{1}{2}\left(tr\ A \pm \sqrt{(tr\ A)^2 - 4\ det\ A}\right) = \frac{1}{2}(tr\ A \pm (det\ A - 1))$.

   (a) If $tr\ A > 0$, then $tr\ A = 1 + det\ A$, and consequently,

$$\lambda_{1,2} = \begin{cases} 1 \\ det\ A \end{cases}.$$

   (b) If $tr\ A < 0$, then $tr\ A = -1 - det\ A$, and consequently,

$$\lambda_{1,2} = \begin{cases} 1 \\ det\ A \end{cases}.$$

(ii) Let $|tr\ A| - 1 < det\ A$, and $det\ A = 1$. Then $(tr\ A)^2 - 4\ det A < (det\ A + 1)^2 - 4 = 0$. Hence, the eigenvalues are complex conjugates. Moreover,

$$\lambda_{1,2} = \frac{1}{2}\left((tr\ A)^2/4 \pm \sqrt{4\ det\ A - (tr\ A)^2}\right)$$
$$= \frac{1}{2}tr\ A \pm \sqrt{1 - (tr\ A)^2}.$$

Thus $|\lambda_{1,2}| = \sqrt{(tr\ A)^2/4 + 1 - (tr\ A)^2/4} = 1$. Furthermore, $\theta = \arctan(\lambda_{1,2})$ $= \tan^{-1}\left(\frac{\pm\sqrt{1-(tr\ A)^2/4}}{(tr\ A)^2/4}\right) = \cos^{-1}(tr\ A/2)$, which give $\lambda_{1,2} = e^{\pm i\theta} = \cos\theta \pm$ $i\sin\theta$. ∎

4.8.2 Navigating the Trace-Determinant Plane

The trace-determinant plane is effective in the study of linear systems with parameters. It provides us a chart of those locations where we can expect dramatic changes in the phase portrait. There are three critical loci. Let T denotes the trace and D denote the determinant. Then there are three critical lines: $D = tr\ A - 1$, $D = -tr\ A - 1$, and $D = 1$; they enclose the stability region in the trace-determinant planes.

   We now illustrate our analysis by the following example.

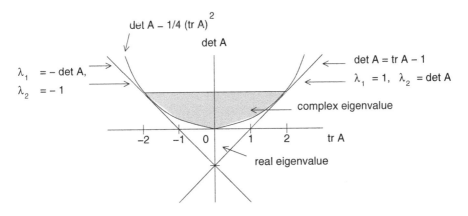

FIGURE 4.10a
(a) The stability region for Equation (4.41) is the shaded triangle.

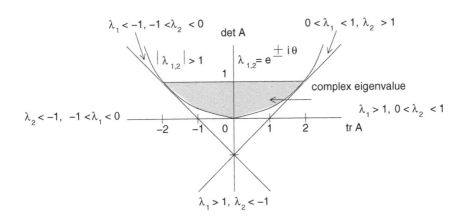

FIGURE 4.10b
(b) The determination of eigenvalues in different regions.

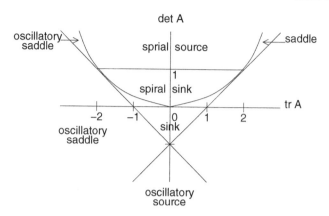

FIGURE 4.10c
(c) Description of the dynamics of Equation (4.41) in all the regions in the det-trace plane.

Example 4.7

Consider the one-parameter family of linear systems $X(n+1) = AX(n)$, where

$$A = \begin{pmatrix} -1 & a \\ -2 & 1 \end{pmatrix}$$

which depends on the parameter a. As a varies, the determinant of the matrix, $det\ A$, is always $2a - 1$, while the trace of the matrix, $tr\ A$, is always 0. As we vary the parameter a from negative to positive values, the corresponding point (T, D) moves vertically along the line $T = 0$. Now if $D < -1$, which occurs if $2a - 1 < -1$ or $a < 0$, we have a degenerate case, $\lambda_1 = 1$ and $\lambda_2 = -1$ with corresponding eigenvectors $\begin{pmatrix} 0 \\ 1 \end{pmatrix}$ and $\begin{pmatrix} 1 \\ 1 \end{pmatrix}$. Thus every point on the y-axis is a fixed point and every other point in the plane is periodic of period 2. For $0 < a \le \frac{1}{2}$, we have a sink, and for $\frac{1}{2} < a < 1$ we have a spiral sink. At exactly $a = 1$ we have a center, and if $a > 1$ we have a spiral source (see Fig 4.10b).

The values of a where critical dynamical changes occur are called bifurcation values. In this example, the bifurcation values of a are $0, \frac{1}{2}, 1$. ☐

Exercises - 4.8

In problems 1-6, consider the one-parameter families of linear systems $X(n+1) = AX(n)$, where A depends on a parameter α. In a brief essay, discuss different types of dynamical behavior exhibited by the systems as α increases along the real line, modeled after Example 4.7.

1. $A = \begin{pmatrix} 2 & 3 \\ 1+\alpha & 4 \end{pmatrix}$

2. $A = \begin{pmatrix} 3+\alpha & 2 \\ -2 & 3 \end{pmatrix}$

3. $A = \begin{pmatrix} 2+\alpha & 1 \\ 0 & 2 \end{pmatrix}$

4. $A = \begin{pmatrix} \alpha+1 & 1 \\ \alpha & \alpha+1 \end{pmatrix}$

5. $A = \begin{pmatrix} \alpha+1 & \alpha^2+\alpha \\ 1 & \alpha+1 \end{pmatrix}$

6. $A = \begin{pmatrix} \alpha+1 & \sqrt{1-\alpha^2} \\ 1 & 1 \end{pmatrix}$, $-1 \le \alpha \le 1$

In problems 7-9, we consider two-parameter families of the linear system $X(n+1) = AX(n)$, where A depends on two parameters α, β.

In the ab-plane, identify all regions where the system possesses a saddle, a sink, a spiral sink, and so on.

7.* $A = \begin{pmatrix} \alpha+1 & 1 \\ \beta & 2 \end{pmatrix}$

8.* $A = \begin{pmatrix} \alpha+1 & \beta \\ \beta & \alpha+1 \end{pmatrix}$

In the trace-determinant plane (Fig. 4.10d).

1. Show that we have a saddle in regions in ③ and ⑤.

2. Show that we have an oscillatory saddle in regions ⑦ and ⑧.

3. (a) Show that we have an oscillatory source in region ⑥.

   (b) Show that we have a spiral source in region ③.

4.9 Liapunov Functions for Nonlinear Maps

In 1892, the Russian mathematician A. M. Liapunov (sometimes transliterated as Lyapunov) introduced a new method to investigate the stability of

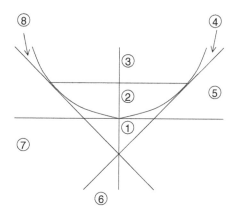

FIGURE 4.10d

nonlinear differential equations. This method, now known as Liapunov's second method, allows one to determine the stability of solutions to a differential equation without actually solving it.

In this section, we will adapt Liapunov's second method to two-dimensional maps/difference equations. The adaptation process is more or less straightforward and follows closely to LaSalle [60] and Elaydi [32].

Consider the difference equation

$$X(n+1) = f(X(n)) \tag{4.52}$$

where $f : G \rightarrow \mathbb{R}^2, G \subset \mathbb{R}^2$, is continuous. Let X^* be a fixed point of f, that is, $f(X^*) = X^*$. For $V : \mathbb{R}^2 \rightarrow \mathbb{R}$, we define the variation ΔV of V relative to Equation (4.52) as

$$\Delta V(X) = V(f(X)) - V(X).$$

Hence,

$$\Delta V(X(n)) = V(X(n+1)) - V(X(n)).$$

So, if $\Delta V \leq 0$, then V is nonincreasing along the orbits of f.

DEFINITION 4.5   *A real valued function $V : G \rightarrow \mathbb{R}, G \subset \mathbb{R}^2$, is said to be a Liapunov function on G if*

   1. V is continuous on G

   2. $\Delta V(X) \leq 0$, whenever X and $f(X) \in G$.

Let $B(X, \gamma) = \{Y \in \mathbb{R}^2 : |Y - X| < \gamma\}$ denote the open ball around X. Then, we say that the Liapunov function is positive definite at X^* if $V(X) > 0$

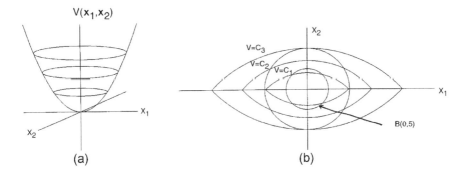

FIGURE 4.11
A Liapunov function V and its level curves.

for all $X \in B(X^*, \delta)$, for some $\delta > 0$, $X \neq X^*$, and $V(X^*) = 0$. The function V is said to be negative definite if $-V$ is positive definite.

We now present the reader with an informal geometrical discussion on the first Liapunov stability theorem. Without loss of generality, we focus our attention on the stability of the fixed point $X^* = 0$. Suppose that there exists a positive definite Liapunov function V defined on $B(0, \eta), \eta > 0$. Figure 4.11(a) illustrates the graph of V in a three-dimensional coordinate system, while Fig. 4.11(b) depicts the level curves $V \begin{pmatrix} x_1 \\ x_2 \end{pmatrix} = c$ in the plane. Assume that for some $\varepsilon > 0$, $B(0, \varepsilon)$ contains the level curve $V \begin{pmatrix} x_1 \\ x_2 \end{pmatrix} = \tilde{c}$ and this level curve, in turn, contains the ball $B(0, \delta), 0 < \delta \leq \varepsilon$.

Now, if $X \in B(0, \delta)$, then $V(X) \leq \tilde{c}$. Since $\Delta V \leq 0$, it follows that $V(f^n(X)) \leq V(X) \leq \tilde{c}$, for all $n \in \mathbb{Z}^+$. Consequently, the orbit of X stays indefinitely in $B(0, \varepsilon)$, and hence 0 is a stable fixed point. On the other hand, if $\Delta V < 0$, then $V(f^n(X)) < V(X) < \tilde{c}$ for all $n \in \mathbb{Z}^+$, which intuitively leads to the conclusion that $f^n(X) \to 0$ as $n \to \infty$. This is the essence of the proof of the next theorem. A more rigorous proof follows.

THEOREM 4.6

Suppose that V is a positive definite Liapunov function defined on an open ball $G = B(X^*, \gamma)$ around a fixed point X^* of a continuous map f on $\mathbb{R}^2$. Then, X^* is stable. If, in addition, $\Delta V(X) < 0$, whenever X and $f(X) \in G, X \neq X^*$, then X^* is asymptotically stable on G. Moreover, if $G = \mathbb{R}^2$ and $V(X) \to \infty$ as $|X| \to \infty$, then X^* is globally asymptotically stable.

PROOF Choose $\alpha_1 > 0$ such that $B(X^*, \alpha_1) \subset G \cap H$. Since f is continuous, there is $\alpha_2 > 0$ such that if $X \in B(X^*, \alpha_2)$ then $f(X) \in B(X^*, \alpha_1)$.

Let $0 < \varepsilon \leq \alpha_2$ be given. Define $\psi(\varepsilon) = \min\{V(X)|\varepsilon \leq |X - X^*| \leq \alpha_1\}$. By the intermediate value theorem, there exists $0 < \delta < \varepsilon$ such that $V(X) < \psi(\varepsilon)$ whenever $|X - X^*| < \delta$.

Realize now that if $X_0 \in B(X^*, \delta)$, then $X(n) \in B(X^*, \varepsilon)$ for all $n \geq 0$. This claim is true because, if not, there exists $X_0 \in B(X^*, \delta)$ and a positive integer m such that $X(r) \in B(X^*, \varepsilon)$ for $1 \leq r \leq m$ and $X(m+1) \notin B(X^*, \varepsilon)$. Since $X(m) \in B(X^*, \varepsilon) \subset B(X^*, \alpha_2)$, it follows that $X(m+1) \in B(X^*, \alpha_1)$. Consequently, $V(X(m+1)) \geq \psi(\varepsilon)$. However, $V(X(m+1)) \leq, \ldots, \leq V(X_0) < \psi(\varepsilon)$ and we thus have a contradiction. This establishes stability.

To prove asymptotic stability, assume that $X_0 \in B(X^*, \delta)$. Then $X(n) \in B(X^*, \varepsilon)$ holds true for all $n \geq 0$. If $\{X(n)\}$ does not converge to X^*, then it has a subsequence $\{X(n_i)\}$ that converges to $Y \in R^k$. Let $E \subset B(X^*, \alpha_1)$ be an open neighborhood of Y with $X^* \notin E$. Having already defined on E the function $h(x) = V(f(X))/V(X)$, we may consider h as well defined, continuous, and $h(X) < 1$ for all $X \in E$. Now if $\eta \in (h(Y), 1)$ then there exists $\delta > 0$ such that $X \in B(Y, \delta)$ implies $h(X) \leq \eta$. Thus for sufficiently large n_i,

$$V\big(f(X(n_i))\big) \leq \eta V\big(X(n_i - 1)\big) \leq \eta^2 V\big(X(n_i - 2)\big), \ldots, \leq \eta^{n_i} V(X_0).$$

Hence,

$$\lim_{n_i \to \infty} V\big(X(n_i)\big) = 0.$$

But, since $\lim_{n_i \to \infty} V\big(X(n_i)\big) = V(Y)$, this statement implies that $V(Y) = 0$, and consequently $Y = X^*$.

To prove the global asymptotic stability, it suffices to show that all solutions are bounded and then repeat the above argument. Begin by assuming there exists an unbounded solution $X(n)$, and then some subsequence $\{X(n_i)\} \to \infty$ as $n_i \to \infty$. Since $V(X) \to \infty$, as $|X| \to \infty$, this assumption implies that $V\big(X(n_i)\big) \to \infty$ as $n_i \to \infty$, which is a contradiction since $V(X_0) > V\big(X(n_i)\big)$ for all i. This concludes the proof.   ∎

Example 4.8

Consider the two-dimensional system

$$x(n+1) = \frac{ay(n)}{1 + x^2(n)}$$
$$y(n+1) = \frac{bx(n)}{1 + y^2(n)} \tag{4.53}$$

or the map

$$F\begin{pmatrix} x \\ y \end{pmatrix} = \begin{pmatrix} ay/(1 + x^2) \\ bx/(1 + y^2) \end{pmatrix}.$$

Discuss the stability of the zero solution of Equation (4.53).   ⬜

SOLUTION   Take $V(x, y) = x^2 + y^2$. Then V is positive definite. Moreover,

$$\Delta V \begin{pmatrix} x \\ y \end{pmatrix} = V \left(F \begin{pmatrix} x \\ y \end{pmatrix} \right) - V \begin{pmatrix} x \\ y \end{pmatrix}$$

$$= \frac{a^2 y^2}{(1 + x^2)^2} + \frac{b^2 x^2}{(1 + y^2)^2} - x^2 - y^2 \qquad (4.54)$$

$$= \left(\frac{b^2}{(1 + y^2)^2} - 1 \right) x^2 + \left(\frac{a^2}{(1 + x^2)^2} - 1 \right) y^2$$

$$\leq (b^2 - 1)x^2 + (a^2 - 1)y^2.$$

Now, we have three cases to consider.

The first case is if $a^2 < 1$ and $b^2 < 1$, then $\Delta V < 0$ and thus we conclude from Theorem 4.3 that the origin is asymptotically stable. Furthermore, since $V \begin{pmatrix} x \\ y \end{pmatrix} \to \infty$ as $\left| \begin{pmatrix} x \\ y \end{pmatrix} \right| \to \infty$, the origin is globally asymptotically stable (see Fig. 4.12(a)).

However, in the second case, if $a^2 \leq 1$ and $b^2 \leq 1$, then $\Delta V \leq 0$ and we can only conclude from Theorem 4.3 that the origin is stable.

In the final case, when $a^2 > 1$ and $b^2 > 1$, Theorem 4.3 fails to provide us with information about the stability (or lack thereof) of the origin. ∎

It is now evident that finer analysis is needed to fully understand the stability in the last two cases. Subsequently, we are led to an important result due to LaSalle [60], which is commonly known as LaSalle's invariance principle [32]. To prepare for such important results, we should become familiar with certain terminology, some old and some new.

Recall that a set H is (positively) invariant under a map $f : \mathbb{R}^2 \to \mathbb{R}^2$ if $f(H) \subset H$. The (positive) limit set $\Omega(x)$ of $x \in \mathbb{R}^2$ is defined to be the set of all limit points of its positive orbit $O(x)$. It may be shown (Problem 2) that

$$\Omega(x) = \bigcap_{i=0}^{\infty} \overline{\bigcup_{n=i}^{\infty} f^n(x)}. \qquad (4.55)$$

Furthermore, $\Omega(x)$ is closed and (positively) invariant (Problem 3).

A closed set H is said to be invariantly connected if it is not the union of two nonempty disjoint closed invariant sets.

The nagging question still persists as to whether or not $\Omega(X)$ is nonempty for a given $X \in \mathbb{R}^2$. The next lemma settles this question.

LEMMA 4.1
If $O(x)$ is bounded, then $\Omega(x)$ is nonempty, compact, and invariant.

PROOF   The proof is left to the reader as Problem 2. ∎

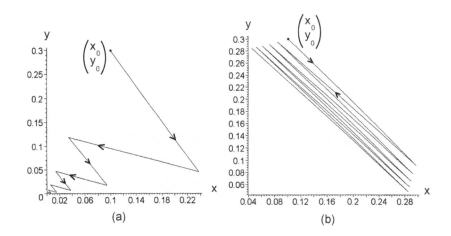

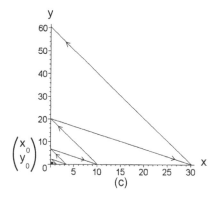

FIGURE 4.12

(a) $a^2 < 1$, $b^2 < 1$, the origin is globally asymptotically stable. (b) $a^2 = 1$, $b^2 = 1$, an orbit approaching a 4-cycle. (c) $a^2 > 1$, $b^2 > 1$, the origin is unstable.

Now, let V be a Liapunov function on a subset $G \subset \mathbb{R}^2$. Define

$$E = \{X \in \overline{G} : \Delta V(X) = 0\}.$$

Let M be the maximal invariant subset of E, and for $c \in \mathbb{R}^+$, $V^{-1}(c) = \{X : V(X) = c, \ X \in \mathbb{R}^2\}$.

THEOREM 4.7 (LaSalle's Invariance Principle)

Suppose that V is a positive definite Liapunov function defined on an open ball $G = B(X^*, \gamma)$ around a fixed point X^* of a two-dimensional map f. If for $X \in G$, $O(X)$ is bounded and contained in G, then for some $c \in \mathbb{R}^+$, $f^n(X) \to M \cap V^{-1}(c)$ as $n \to \infty$.

PROOF   Let $X \in \mathbb{R}^2$ such that $O(x)$ is bounded. Then, by Lemma 4.1, $\Omega(X) \neq \phi$. Now, since $V(f^n(X))$ is nonincreasing and bounded below, $\lim_{n \to \infty} V(f^n(X)) = c$, for some $c \in \mathbb{R}^+$. For $Y \in \Omega(X)$, $f^{n_i}(X) \to Y$ as $n_i \to \infty$, for some subsequence of positive integers n_i. By the continuity of V, it follows that

$$\lim_{n_i \to \infty} V(f^{n_i}(X)) = V(Y) = c.$$

Hence, $\Omega(X) \subset V^{-1}(c)$. Furthermore, since $\Omega(X)$ is invariant, $V(f(Y)) = V(Y)$ and consequently, $\Delta V(Y) = 0$. Hence $\Omega(X) \subset E$. But, $\Omega(X)$ is invariant, which implies that it must be contained in M. Therefore, $f^n(X) \to M \cap V^{-1}(c)$.   ∎

A remark is now in order. Note that if $M = \{X^*\}$ is a singleton, then the preceding theorem tells us that X^* is definitely an asymptotically stable fixed point. This observation leads to the complete analysis of part *(b)* of Example 4.8.

Example 4.9

(Example 4.8 Revisited). Let us reexamine case *(b)* in Example 4.8 in light of LaSalle's invariance principle. Here are the two subcases to consider.

Case one is where $a^2 \leq 1$, $b^2 \leq 1$, and $a^2 + b^2 < 2$. Without loss of generality, we may assume that $a^2 < 1$ and $b^2 = 1$. Then, $\Delta V \leq (a^2 - 1)y^2$ which is zero when $y = 0$. Thus E is the x axis. To find the largest invariant subset M of E, note that for $\begin{pmatrix} x \\ 0 \end{pmatrix} \in E$, $F\begin{pmatrix} x \\ 0 \end{pmatrix} = \begin{pmatrix} 0 \\ bx \end{pmatrix}$. Hence, $M = \left\{ \begin{pmatrix} 0 \\ 0 \end{pmatrix} \right\}$. Consequently, the origin is asymptotically stable.

In case two, $a^2 = 1$ and $b^2 = 1$. It follows from Equation (4.54) that $\Delta V = 0$ if $x = 0$ or $y = 0$. Thus $E = M =$ the union of the two coordinate axes. LaSalle's invariance now tells us that there exists $c > 0$ such that each orbit $O^+(u)$ approaches the set $\left\{ \begin{pmatrix} \pm c \\ 0 \end{pmatrix}, \begin{pmatrix} 0 \\ \pm c \end{pmatrix} \right\}$.

Now, if $c \neq 0$, $F^4 \begin{pmatrix} c \\ 0 \end{pmatrix} = \begin{pmatrix} c \\ 0 \end{pmatrix}$, and that $\begin{pmatrix} c \\ 0 \end{pmatrix}$ is a point of period 4 with the cycle $\left\{ \begin{pmatrix} c \\ 0 \end{pmatrix}, \begin{pmatrix} 0 \\ bc \end{pmatrix}, \begin{pmatrix} abc \\ 0 \end{pmatrix}, \begin{pmatrix} 0 \\ ac \end{pmatrix} \right\}$. Similarly, we may show that $\left\{ \begin{pmatrix} 0 \\ c \end{pmatrix}, \begin{pmatrix} ac \\ 0 \end{pmatrix}, \begin{pmatrix} 0 \\ abc \end{pmatrix}, \begin{pmatrix} bc \\ 0 \end{pmatrix} \right\}$ is also a 4-cycle. Since $\Omega(u)$ is invariantly connected, each orbit must approach only one of these 4-cycles [Fig. 4.12(b)]. Finally, we observe that if $ab = 1$, then we have 2-cycles instead. □

We now end this section by giving a result about instability. This will enable us to treat the remaining case of Example 4.8.

THEOREM 4.8

Let $V : G \subset \mathbb{R}^2 \to \mathbb{R}$ be a continuous function such that relative to Equation (4.52), ΔV positive definite (negative definite) on a neighborhood of a fixed point x^*. If there exists a sequence $x_i \to x^*$ as $i \to \infty$ with $V(x_i) > 0$ ($V(x_i) < 0$), then x^* is unstable.

PROOF   Assume that $\Delta V(x) > 0$ for $x \in B(\eta), x \neq 0, \Delta V(0) = 0$. We will prove Theorem 4.8 by contradiction, first assuming that the zero solution is stable. Then for $\varepsilon < \eta$, there will exist $\delta < \varepsilon$ such that $\|x_0\| < \delta$ implies $\|x(n, 0, x_0)\| < \varepsilon, n \in Z^+$.

Since $a_i \to 0$, pick $x_0 = a_j$ for some j with $V(x_0) > 0$, and $\|x_0\| < \delta$. Hence $\overline{0(x_0)} \subset \overline{B(\varepsilon)} \subset B(\eta)$ is closed and bounded (compact). Since its domain is compact, $V(x(n))$ is also compact and therefore bounded above. Since $V(x(n))$ is also increasing, it follows that $V(x(n)) \to c$. Following the proof of LaSalle's invariance principal, it is easy to see that $\lim_{n \to \infty} x(n) = 0$. Therefore, we would be led to believe that $0 < V(x_0) < \lim_{n \to \infty} V(x(n)) = 0$. This statement is unfeasible—so the zero solution cannot be stable, as we first assumed. The zero solution of Equation (4.37) is thus unstable.

The conclusion of the theorem also holds if ΔV is negative definite and $V(a_i) < 0$.   ∎

We are now in a position to tackle the remaining case in Example 4.8. Assume now that $a^2 > 1$ and $b^2 > 1$. Let δ be sufficiently small such that $b^2 > (1 + \delta^2)^2$ and $a^2 > (1 + \delta^2)^2$. Define a function $V \begin{pmatrix} x \\ y \end{pmatrix} = x^2 + y^2$ on the open disc $B(0, \delta)$ centered at the origin and with radius δ. Then V is clearly positive definite. Moreover, for $\begin{pmatrix} x \\ y \end{pmatrix} \in B(0, \delta)$ we have

$$\Delta V \begin{pmatrix} x \\ y \end{pmatrix} = \left(\frac{b^2}{(1+y^2)^2} - 1 \right) x^2 + \left(\frac{a^2}{(1+x^2)^2} - 1 \right) y^2$$

$$\geq \left(\frac{b^2}{1+\delta^2)^2} - 1 \right) x^2 + \left(\frac{a^2}{(1+\delta^2)^2} - 1 \right) y^2$$

$$> 0 \ \text{if} \ \begin{pmatrix} x \\ y \end{pmatrix} \neq \begin{pmatrix} 0 \\ 0 \end{pmatrix}.$$

Hence by Theorem 4.3, the origin is unstable (see Fig. 4.12(c)).

4.10 Linear Systems Revisited

In Sec. 4.3, we have settled most of the questions concerning the stability of second-order linear systems:

$$X(n+1) = AX(n). \tag{4.56}$$

In this section, we are going to construct suitable Liapunov functions for System (4.56). This is important for our program since by modifying such Liapunov functions we can find appropriate Liapunov functions for a large class of nonlinear equations of the form

$$Y(n+1) = AY(n) + g(Y(n))$$

which we will study in the next section. But, before embarking on our task we need to introduce a few preliminaries about definite matrices.

Let $X \in \mathbb{R}^2$, $B = (b_{ij})$ a real symmetric 2×2 matrix, and consider the quadratic form $V : \mathbb{R}^2 \to \mathbb{R}$ defined by

$$V(X) = X^T B X = \sum_{i=1}^{2} \sum_{j=1}^{2} b_{ij} X_i X_j \tag{4.57}$$

where X^T denotes the transpose of vector X.

A matrix B is said to be positive (negative) definite if the corresponding $V(X)$ as defined in Equation (4.57) is positive (negative) for all $0 \neq X \in \mathbb{R}^2$. If, however, $V(X) \geq 0$ $(V(X) \leq 0)$ for all $X \in \mathbb{R}^2$, then B is positive (negative) semidefinite. The following result, due to Sylvester [74], gives a complete characterization of the notion of definiteness of V.

THEOREM 4.9
Let V be a quadratic form as defined in Equation (4.57). Then the following statements hold true:

1. V is positive definite if and only if all principle minors of B are positive, i.e., if and only if

$$b_{11} > 0 \quad and \quad \det B > 0.$$

2. V is negative definite if and only if

$$b_{11} < 0 \quad and \quad \det B > 0.$$

3. V is positive (negative) definite if and only if all eigenvalues of B are nonzero and positive (negative), respectively.

4. If λ_1, λ_2 are the eigenvalues of B, $\lambda_m = \min_i |\lambda_i|$, $\lambda_M = \max_i |\lambda_i|$, $i = 1, 2$, then

$$\lambda_m |X|^2 \leq V(X) \leq \lambda_M |X|^2. \tag{4.58}$$

5. V is semidefinite (positive or negative) if and only if the nonzero eigenvalues of B have the same sign.

Let $B = \begin{pmatrix} 3 & 2 \\ 2 & 5 \end{pmatrix}$. Then, the principal minors are 3 and $\begin{pmatrix} 3 & 2 \\ 2 & 5 \end{pmatrix}$ where both have positive determinants. Hence, B is positive definite by Theorem 4.9. Notice that the eigenvalues of B are $\lambda_1 = 4\sqrt{5}$, and $\lambda_2 = 4 - \sqrt{5}$, which are also positive. Moreover, if we let $V(X) = X^T B X$, then

$$V(X) = (x_1 \ x_2) \begin{pmatrix} 3 & 2 \\ 2 & 5 \end{pmatrix} \begin{pmatrix} x_1 \\ x_2 \end{pmatrix}$$
$$= 3x_1^2 + 4x_1 x_2 + 5x_2^2.$$

Let us now go back to Equation (4.56) and consider the function $V(X) = X^T B X$, where B is positive definite as a prospective candidate for a Liapunov function. Then,

$$\Delta V(X(n)) = X^T(n+1)BX(n+1) - X^T(n)BX(n)$$
$$= X^T(n)A^T BAX(n) - X^T(n)BX(n)$$
$$= X^T(n)(A^T BA - B)X(n).$$

Thus, $\Delta V(X(n)) < 0$ if and only if

$$A^T BA - B = -C \tag{4.59}$$

for some positive definite matrix C. Equation (4.59) is called the Liapunov equation of System (4.56). The preceding argument established part of the following result whose complete proof is omitted.

THEOREM 4.10

$\rho(A) < 1$ if and only if for every symmetric and positive definite matrix C, Equation (4.59) has a unique solution B, which is also symmetric and positive definite.

PROOF   See [74].   ∎

An immediate corollary of Theorem 4.10, which will be useful in the next section, now follows.

COROLLARY 4.2

If $\rho(A) > 1$, then there exists a real symmetric matrix B that is not positive semidefinite such that Equation (4.59) holds for some symmetric positive definite matrix C.

Exercises - (4.9) and (4.10)

1. Consider the system

$$x_1(n+1) = x_2(n) - x_2(n)[x_1^2(n) + x_2^2(n)]$$
$$x_2(n+1) = x_1(n) - x_1(n)[x_1^2(n) + x_2^2(n)].$$

   Prove that the zero solution is asymptotically stable.

2. Prove Lemma 4.1.

3. Consider the system

$$x_1(n+1) = g_1(x_1(n), x_2(n))$$
$$x_2(n+1) = g_2(x_1(n), x_2(n))$$

   with $g_1(0,0) = g_2(0,0) = 0$, and for $x = (x_1, x_2)$ in a neighborhood of the origin

$$g_1(x_1, x_2)\, g_2(x_1, x_2) > x_1 x_2.$$

   Show that the origin is unstable.

   (*Hint: Let $V = x_1 x_2$.*)

4. Determine the stability of the equilibrium points of the system

$$x_1(n+1) = x_1^2(n) - x_2^2(n)$$
$$x_2(n+1) = 2x_1(n)x_2(n)$$

   by converting it to a polar system $(x_1(n) = r(n)\cos\theta(n), x_2(n) = r(n)\sin\theta(n))$. Draw the phase portrait of the system and show that the unit circle is a limit cycle.

5. Determine the stability of the origin for the map

$$f\begin{pmatrix} x \\ y \end{pmatrix} = \begin{pmatrix} 2y - 2yx^2 \\ \frac{1}{2}x + xy^2 \end{pmatrix}.$$

Draw the phase portrait of the system.

6. Verify Formula (4.55).

7. Determine the stability of the origin for the map

$$F\begin{pmatrix} x \\ y \end{pmatrix} = \begin{pmatrix} y \\ \alpha x/(1 + \beta y^2) \end{pmatrix}, \quad \beta > 0.$$

8. Suppose that $V : \mathbb{R}^2 \to \mathbb{R}$ is a continuous function with $\Delta^2 V(f^n(X)) > 0$ for $f^n(X) \neq 0$, where f is a two-dimensional continuous map. Prove that either $O(X)$ is unbounded or it tends to the origin. [Here $\Delta^2 = \Delta \circ \Delta$; $\Delta^2 V(f^n(X)) = V(f^{n+2}(X)) - 2V(f^{n+1}(X)) + V(f^n(X))$.]

9. Suppose that V is a Liapunov function for a continuous map f on $\mathbb{R}$ such that

   (a) $G_\lambda = \{X : V(X) < \lambda\}$ is bounded for each $\lambda \in \mathbb{R}^2$,

   (b) the set M is compact (M is the maximal invariant subset of E).

   Prove that M is a global attractor.

10. Wade through Problem 8 again, after replacing the condition $\Delta^2 V(f^n(X)) > 0$ by $\Delta^2 V(f^n(X)) < 0$.

11. Consider the system

$$x(n + 1) = y(n)$$
$$y(n + 1) = x(n) + f(x(n))$$

such that $\Delta[y(n)f(x(n))] > 0$ for all $n \in \mathbb{Z}^+$. Show that the solutions are either unbounded or tend to the origin.

(*Hint: Use $V(x, y) = xy$ and then use Problem 8.*)

12.* Show that the origin of the system

$$x(n + 1) = x(n) + \frac{x^2(n)(y(n) - x(n)) + y^5(n)}{r^2(n) + r^6(n)}$$
$$y(n + 1) = y(n) + \frac{y^2(n)(y(n) - 2x(n))}{r^2(n) + r^6(n)}$$

is globally attracting, but unstable.

13. Let B be a positive definite symmetric matrix with eigenvalues $\lambda_1 < \lambda_2$. Let $V(X) = X^T B X$. Show that

$$\lambda_1 |X|^2 \le V(X) \le \lambda_2 |X|^2.$$

14. Suppose that $\rho(A) < 1$, where A is a real 2×2 matrix. Show that the matrix $B = \sum_{r=0}^{\infty} (A^T)^r C A^r$ is symmetric and positive definite whenever C is. Then prove that this matrix B is a solution of Equation (4.59).

15. Prove that if a matrix B is a positive definite matrix, all of its eigenvalues are positive.

16. Prove Theorem 4.10.

17. Prove Corollary 4.2.

4.11 Stability via Linearization

In Chapter 1, we saw how the values of the derivative $f'(o^*)$ of a nonlinear, one-dimensional map f at a hyperbolic fixed point o^* determines completely the stability of o^*. For if $|f'(o^*)| < 1$, then o^* is asymptotically stable or a sink if $|f'(o^*)| > 1$, then o^* is unstable or a repeller. In essence, what we are saying is that the behavior of the linear difference equation

$$Y(n + 1) = f'(X^*)Y(n) \qquad (4.60)$$

determines the behavior of the original equation

$$X(n + 1) = f(X(n)) \qquad (4.61)$$

near the fixed points.

In the language of maps, this amounts to saying that the behavior of the linear map $g(X) = f'(X^*)X$ determines the behavior of the nonlinear map $f(X)$ near the fixed point X^*. Such a process is commonly called a linearization of the nonlinear map f or the difference equation (4.61). Now, consider a two-dimensional map $f : G \subset \mathbb{R}^2 \to \mathbb{R}^2$, where G is an open subset of $\mathbb{R}^2$. Then f is said to be continuously differentiable, (or a C^1 map) if its partial derivatives $\frac{\partial f}{\partial x_1}$ and $\frac{\partial f}{\partial x_2}$ exist and are continuous. If there is a 2×2 matrix A such that

$$\lim_{X \to Q} \frac{|f(X) - f(Q) - A(X - Q)|}{|X - Q|} = 0,$$

then the derivative $Df(Q)$ of f at Q is defined as $Df(Q) = A$. Hence, if $f = \begin{pmatrix} f_1 \\ f_2 \end{pmatrix}$, then

$$Df(Q) = A = \begin{pmatrix} \frac{\partial f_1}{\partial x_1} & \frac{\partial f_1}{\partial x_2} \\ \frac{\partial f_2}{\partial x_1} & \frac{\partial f_2}{\partial x_2} \end{pmatrix} \Bigg|_Q .$$

The matrix A is often called the Jacobian matrix of f. By the mean value theorem, we have

$$f(X) = f(Q) + A(X - Q) + g(X, P) \tag{4.62}$$

where

$$\lim_{X \to Q} \frac{g(X, Q)}{|X - Q|} = 0. \tag{4.63}$$

Statement (4.63) may be expressed in the little "o" language as $g(X, Q) = o(|X - Q|)$ as X tends to Q.

Suppose now that $Q = X^*$ is a fixed point of f, that is, $f(X^*) = X^*$. Then Equation (4.62) yields

$$f(X) - X^* = A(X - X^*) + g(X, X^*). \tag{4.64}$$

It is clear that $g(X^*, X^*) = 0$.

To simplify our exposition, we make the change of variables $Y = X - X^*$. Equation (4.64) becomes

$$f(Y + X^*) - X^* = AY + g(Y). \tag{4.65}$$

If we now let $h(Y) = f(Y + X^*) - X^*$ in Equation (4.65), we get

$$h(Y) = AY + g(Y) \tag{4.66}$$

with $g(Y) = o(|Y|)$ as Y tends to 0.

We now make two important observations concerning the relationship between the maps f and h. First, since $h(0) = f(X^*) - X^* = 0$, it follows that 0 is a fixed point of h if and only if X^* is a fixed point of f. Second, note that $h^n(Y) \to 0$ as $n \to \infty$ if and only if $f^n(X) = f^n(Y + X^*) \to X^*$ as $n \to \infty$. Hence, 0 is stable (asymptotically stable) under h if and only if X^* is stable (asymptotically stable) under f. A similar statement can be made about instability. Hence, without loss of generality, we may work directly with the map h in Equation (4.66). In other words, it suffices to consider the nonhomogeneous system

$$Y(n+1) = AY(n) + g(Y(n)) \tag{4.67}$$

with $A = Df(X^*)$, $g(Y) = o(|Y|)$, and $g(0) = 0$.

The linear part of Equation (4.67) is the homogeneous equation

$$X(n+1) = AX(n). \tag{4.68}$$

The main result of this section now follows.

THEOREM 4.11

Let $f : G \subset \mathbb{R}^2 \to \mathbb{R}^2$ be a C^1 map, where G is an open subset of $\mathbb{R}^2$, X^* is a fixed point of f, and $A = Df(X^*)$. Then the following statements hold true:

1. If $\rho(A) < 1$, then X^* is asymptotically stable.

2. If $\rho(A) > 1$, then X^* is unstable.

3. If $\rho(A) = 1$, then X^* may or may not be stable.

PROOF

1. Assume that $\rho(A) < 1$. Then by virtue of Theorem 4.10 , there exists a real symmetric and positive definite matrix B such that $A^T BA - B = -C$, where C is positive definite. Now, consider the Liapunov function $V(Y) = Y^T By$. Then the variation of V relative to Equation (4.67) is given by

$$\Delta VY = -Y^T CY + 2Y^T A^T Bg(Y) + V(g(Y)). \tag{4.69}$$

   Now, Equation (4.31) allows us to pick a $\gamma > 0$ such that $Y^T CY \geq 4\gamma|Y|^2$ for all $Y \in \mathbb{R}^2$. There exists $\delta > 0$ such that if $|Y| < \delta$, then $|A^T Bg(Y)| \leq \gamma|Y|$ and $V(g(Y)) \leq \gamma|Y|$. Hence, it follows from Equation (4.69) that $\Delta V(Y(n)) \leq -\gamma|Y(n)|^2$ which implies by Theorem 4.3 that the zero solution of Equation (4.67) is asymptotically stable.

2. Assume that $\rho(A) > 1$. Then, we use Corollary 4.2 to choose a real, symmetric 2×2 matrix B such that $B^T AB - B = -C$ is negative definite, where B is not positive semidefinite. Thus, the function $V(Y) = Y^T BY$ is negative at points arbitrarily close to the origin. Now, as in part 1, $\Delta V(Y(n)) \leq -\gamma|Y(n)|^2$. Thus, by Theorem 4.10, the zero solution of Equation (4.67) is unstable.

3. We prove this part by using the following example.

Example 4.10

1. Consider the system

$$x_1(n+1) = x_1(n) + x_2^2(n) + x_1^2(n)$$
$$x_2(n+1) = x_2(n).$$

The linear part has the matrix

$$A = \begin{pmatrix} 1 & 0 \\ 0 & 1 \end{pmatrix}$$

with $\rho(A) = 1$. To determine the stability of the origin we use the Liapunov function $V(X) = x_1 + x_2$. Then, V is not positive and

$$\Delta V(X(n)) = x_1^2(n) + x_2^2(n) > 0 \quad \text{if} \quad (x_1, x_2) \neq (0, 0).$$

Hence, by Theorem 4.10, the origin is unstable.

2. Let us now consider the system

$$x_1(n+1) = x_1(n) - x_1^3(n)x_2^2(n)$$
$$x_2(n+1) = x_2(n)$$

with the linear part as in (a). This time we use the Liapunov function $V(X) = x_1^2 + x_2^2$. Then

$$\Delta V(X(n)) = x_1^4(n)x_2^2(n)[-2 + x_1^2(n)x_2^2(n)].$$

Hence, $\Delta V \leq 0$ if $x_1^2 x_2^2 < 2$. Thus, the origin is stable by Theorem 4.3.
□

∎

Example 4.11
(Pielou Logistic Delay Equation [77]). One of the most popular continuous models for the growth of a population is the well-known Verhulst-Pearl differential equation given by

$$x'(t) = x(t)[a - bx(t)], \quad a, b > 0$$

where $x(t)$ is the size of the population at time t, $x'(t) = \frac{dx}{dt}$, a is the rate of growth of the population if the resources were unlimited and the individuals did not affect one another, and $-bx^2(t)$ represents the negative effect on the growth of the population due to crowdedness and limited resources.   □

The solution to this equation may be obtained by separation of variables x and t, and then integrating both sides. Hence,

$$x(t) = \frac{a/b}{1 + (e^{-at}/cb)}.$$

This implies that

$$x(t+1) = \frac{a/b}{1 + \left(e^{-a(t+1)}/cb\right)}$$

$$= \frac{(a/b)\,e^a}{1 + (e^{-at}/cb) + (e^a - 1)}.$$

Dividing by the quantity $[1 + (e^{-at}/cb)]$ both the numerator and the denominator on the right-hand side we obtain

$$x(t+1) = \frac{e^a x(t)}{1 + \frac{b}{a}(e^a - 1)x(t)}$$

or

$$x(t+1) = \frac{\alpha x(t)}{1 + \beta x(t)}.$$

This equation is called the Pielou logistic equation.

Now, if we assume that there is a delay of time period 1 in the response of the growth rate per individual to density change, then we obtain the difference equation (replace t by n)

$$x(n+1) = \frac{\alpha x(n)}{1 + \beta x(n-1)}. \qquad (4.70)$$

As an example of a population that can be modeled by Equation (4.70) is the blowfly (*Lucilia cuprina*) [54]. We now write Equation (4.70) in system form. Let $x_1(n) = x(n-1)$, and $x_2(n) = x(n)$. Then,

$$\begin{pmatrix} x_1(n+1) \\ x_2(n+1) \end{pmatrix} = \begin{pmatrix} x_2(n) \\ \frac{\alpha x_2(n)}{1+\beta x_1(n)} \end{pmatrix} = \begin{pmatrix} f_1(x_1, x_2) \\ f_2(x_1, x_2) \end{pmatrix}. \qquad (4.71)$$

There are two fixed points $\begin{pmatrix} 0 \\ 0 \end{pmatrix}$ and $\begin{pmatrix} (\alpha-1)/\beta \\ (\alpha-1)/\beta \end{pmatrix}$.

1. The fixed point $Z_1^* = \begin{pmatrix} 0 \\ 0 \end{pmatrix}$. Here,

$$A = Df(0) = \begin{pmatrix} 0 & 1 \\ 0 & \alpha \end{pmatrix}$$

with eigenvalues 0 and α. Since $\alpha > 1$, the origin is unstable by Theorem 4.11.

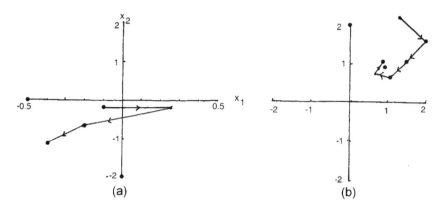

(a)   (b)

FIGURE 4.13
For the Pielou Logistic equation, (a) the trivial solution is unstable. (b) The
equilibrium $Z_2^* = ((\alpha - 1)/\beta, \ (\alpha - 1)\beta)$ is asymptotically stable.

2. The fixed point $Z_2^* = \begin{pmatrix} (\alpha - 1)/\beta \\ (\alpha - 1)/\beta \end{pmatrix}$. In this case,

$$A = Df(z_2^*) = \begin{pmatrix} 0 & 1 \\ \frac{1-\alpha}{\alpha} & 1 \end{pmatrix}.$$

By Theorem 4.11, $\rho(A) < 1$ if and only if

$$|tr\ A| < 1 + \det A < 2$$
if and only if
$$1 < 1 + \frac{\alpha - 1}{\alpha} < 2$$
if and only if
$$0 < \frac{\alpha - 1}{\alpha} < 1$$

Clearly this is satisfied if $\alpha > 1$. Hence, by Theorem 4.11, z_2^* is asymptotically stable (see Fig. 4.13).

4.11.1   The Hartman-Grobman Theorem

Similar to the one-dimensional case, we say a fixed point x^* of a planar map f
is hyperbolic if $|\lambda| \neq 1$, for all eigenvalues of $A = Df(x^*)$. Theorem 4.11 has
given us almost complete information about the stability of hyperbolic fixed
points.

An extension of this result to periodic points is straightforward and will
be left to the reader. In proving Theorem 4.11, we have relied heavily on
Liapunov function techniques. One may arrive at the same conclusion by
looking at matters through conjugacy, a topic treated in Chapter 3. This is

the essence of the so-called Hartman-Grobman theorem, which roughly states that near a hyperbolic fixed point, a map is conjugate to the linear map induced by its derivative at the fixed point. But this comes with a price, we require the map f to be C^1 diffeomorphism, that is a homeomorphism such that f, f', and their inverses are continuously differentiable.

THEOREM 4.12
(Hartman-Grobman Theorem). *Let $f : \mathbb{R}^2 \to \mathbb{R}^2$ be a C^1 diffeomorphism with a hyperbolic fixed point X^*. Then there exist open neighborhoods G of X^* and H of the origin and a homeomorphism $h : H \to G$ such that $f(h(X)) = h(AX)$ for all $X \in H$, where $A = Df(X^*)$.*

In fact, Hartman [85, cf.] showed that the conjugacy map h is C^1 if f is C^2. As a corollary of the Hartman-Grobman theorem, one may easily establish Theorem 4.11.

4.11.2 The Stable Manifold Theorem

Finally, our discussion of the stability of nonlinear maps will not be complete without the stable manifold theorem. Roughly speaking, this theorem states that if the origin is a saddle under the linear map induced by the derivative of a planar map f, then under f the origin exhibits a saddle-like behavior.

An accurate statement of the theorem now follows. However, its proof, is omitted and the interested reader is referred to [25, 85]. Here we assume that the eigenvalues of $A = Df(X^*)$ are $|\lambda_1| < 1$ and $|\lambda_2| > 1$ with corresponding eigenvectors V_1 and V_2.

THEOREM 4.13
(The Stable Manifold Theorem). *Let X^* be a hyperbolic fixed point of a C^1 map $f : G \subset \mathbb{R}^2 \to \mathbb{R}^2$. Then there exists $\varepsilon > 0$ and C^1 curves $\gamma_1 : (-\varepsilon, \varepsilon) \to \mathbb{R}^2$ and $\gamma_2 : (-\varepsilon, \varepsilon) \to \mathbb{R}^2$ such that*

1. $\gamma_1(0) = \gamma_2(0) = X^*$.

2. $\gamma_1'(0) = V_1$, and $\gamma_2'(0) = V_2$.

3. If $Q = \gamma_1(t)$, then $f^n(Q) \to X^*$ as $n \to \infty$.

4. If $Q \in \gamma_2(t)$, then $f^{-n}(Q) \to X^*$ as $n \to \infty$.

The curve γ_1 is called the stable manifold and is usually denoted by $W^s(X^*)$; likewise the curve γ_2 is called the unstable manifold and is denoted by $W^u(X^*)$. Note that for the linear map $A = Df(X^*)$, the stable manifold is the line in the direction of the eigenvector V_1, whereas the unstable manifold is the line in the direction of the eigenvector V_2. Moreover, at the origin, the stable

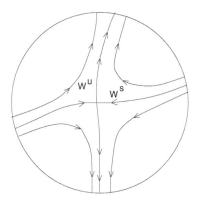

FIGURE 4.14
The stable manifold W^s and the unstable manifold W^u.

(unstable) manifold of the map f is tangent to the eigenvector V_1 (V_2) (see Fig. 4.14).

Exercises - (4.10 and 4.11)

1. Show that the zero solution of the equation

$$x(n+2) - \frac{1}{2}x(n+1) + 2x(n+1)x(n) + \frac{13}{16}x(n) = 0$$

   is asymptotically stable.

2. Linearize the map $f\begin{pmatrix} x \\ y \end{pmatrix} = \begin{pmatrix} \sin y - \frac{1}{2}x \\ \frac{y}{(0.6+x)} \end{pmatrix}$ around the origin and determine the stability of the origin. Draw the phase space diagram for both f and $Df(0)$.

3. Find conditions on α such that the origin is asymptotically stable under the map

$$f\begin{pmatrix} x \\ y \end{pmatrix} = \begin{pmatrix} y \\ \frac{\alpha x}{(1+\beta y^2)} \end{pmatrix}, \qquad \beta > 0.$$

4. Determine the values of a for which the origin is

   (a) attracting
   (b) repelling
   (c) a saddle point

for the map

$$f\begin{pmatrix} x \\ y \end{pmatrix} = \begin{pmatrix} y \\ a\sin x - y \end{pmatrix}.$$

Then draw the phase space diagram of f for $a = -.2, 1, 3$.

5. Determine the conditions for the asymptotic stability of the zero solution of the system

$$x_1(n+1) = \frac{ax_1(n)}{[1 + x_2(n)]}$$
$$x_2(n+1) = [\beta x_2(n) - x_1(n)][1 + x_1(n)].$$

6. Determine the stability of all the fixed points of the map

$$f\begin{pmatrix} x \\ y \end{pmatrix} = \begin{pmatrix} y^2 - \frac{1}{2}x \\ \frac{1}{4}x + \frac{1}{2}y \end{pmatrix},$$

and draw the phase space diagram.

7. Determine the conditions on a and b so that the origin is asymptotically stable under the map

$$f\begin{pmatrix} x \\ y \end{pmatrix} = \begin{pmatrix} y \\ a\sin x + by \end{pmatrix}.$$

8. The cat map C is defined by

$$C\begin{pmatrix} x \\ y \end{pmatrix} = \begin{pmatrix} (x+y)\bmod 1 \\ (x+2y)\bmod 1 \end{pmatrix}.$$

 (a) Show that the origin is a saddle.
 (b) Find the two 2-cycles of C and determine their stability.
 (c) Show that if $A = DC(0)$, then

$$A^n = \begin{pmatrix} c_{2n-1} & c_{2n} \\ c_{2n} & c_{2n+1} \end{pmatrix}, \qquad n \geq 2$$

where $c_1 = 1 = c_2$ and $c_{n+1} = c_n + c_{n-1}$ (c_n is called the Fibonacci sequence).

9. Determine the stability of all fixed points of

$$f\begin{pmatrix} x \\ y \end{pmatrix} = \begin{pmatrix} x^2 + \frac{1}{4} \\ 4x - y^2 \end{pmatrix}$$

and draw the phase space diagram of f.

10. Suppose that the zero solution of the system $x(n+1) = Ax(n)$ is asymptotically stable. Prove that the zero solution of the system $y(n + 1) = (A + B(n))y(n)$, where $B(n) = (b_{ij}(n))$ is a matrix function defined on $\mathbb{Z}^+$, is asymptotically stable if $\sum_{n=0}^{\infty} ||B(n)|| < \infty$.

11. Verify Equation (4.69).

12. Consider the map

$$E_\lambda \begin{pmatrix} x \\ y \end{pmatrix} = \begin{pmatrix} e^x - \lambda \\ -\frac{\lambda}{2} \arctan y \end{pmatrix}.$$

(a) Find all fixed points of E_λ and determine their stability.

(b) Identify the stable and unstable manifolds (if any) and sketch them.

13.* Let X_1^* and X_2^* be saddle points for a homeomorphism $f : \mathbb{R}^2 \to \mathbb{R}^2$. A point p is called **heteroclinic** if $\lim_{n\to\infty} f^n(p) = X_1^*$ and $\lim_{n\to\infty} f^{-n}(p) = X_2^*$. If $X_1^* = X_2^*$, then p is called a **homoclinic** point of f. Prove that conjugacy preserves homoclinic and heteroclinic points.

14.* Let $f : \mathbb{R}^2 \to \mathbb{R}^2$ be given by

$$f \begin{pmatrix} x \\ y \end{pmatrix} = \begin{pmatrix} \frac{1}{2}x \\ 2y - \frac{15}{8}x^3 \end{pmatrix}.$$

(a) Show that the origin is a saddle point of $Df(0)$.

(b) Find the stable and unstable manifolds $W^s(0)$ and $W^u(0)$.

(c) Find the conjugacy homeomorphism between f and $Df(0)$.

(d) Identify the heteroclinic points of f.

4.12 Applications

4.12.1 The Kicked Rotator and the Hénon Map

Consider a damped rotator that is periodically kicked by an external force F (see Fig. 4.15).

Then, its equation of motion is given by

$$\ddot{\theta} + \Gamma\dot{\theta} = F \tag{4.72}$$

where

$$F = Kf(\theta) \sum_{n=0}^{\infty} \delta(t - nT), \quad n \in \mathbb{Z}^+ \tag{4.73}$$

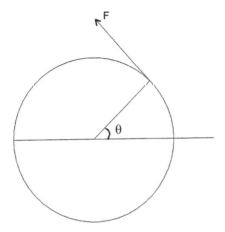

FIGURE 4.15
A damped rotator kicked by an external force F.

and $\dot{\theta} = \frac{d\theta}{dt}$ is the angular velocity, and where, Γ denotes the damping constant and T the time period between two kicks. In Expression (4.73), the function δ is the Dirac delta function[2] which is defined as

$$\delta(t - t_0) = 0 \text{ if } t \neq t_0$$

$$\int_{-\infty}^{\infty} \delta(t - t_0)dt = 1.$$

It is frequently convenient to represent impulse forcing functions by means of the delta function since it is amenable to the techniques of Laplace transform.

Analogous to difference equations, the second-order differential equation (4.72) may be put as a system of first-order equations as follows. Letting $x = \theta$ and $y = \dot{\theta}$ yields

$$\dot{x} = y$$

$$\dot{y} = -\Gamma y + K f(x) \sum_{n=0}^{\infty} \delta(t - nT). \qquad (4.74)$$

Next, we discretize System (4.74) by letting $x(n) = \lim_{\varepsilon \to 0} x(nT - \varepsilon)$ and $y(n) = \lim_{\varepsilon \to 0} y(nT - \varepsilon)$. By integrating the second part of Equation (4.74) for $nT - \varepsilon < t < (n+1)T - \varepsilon$, we obtain

$$y(t) = y(n)e^{-\Gamma(t-nT)} + K \sum_{m=0}^{\infty} f(x(m)) \int_{nT-\varepsilon}^{t} e^{\Gamma(s-t)}\delta(s - mT)ds$$

---

[2]Paul A. M. Dirac (1902) was awarded the Nobel Prize (with Erwin Schrödinger) in 1933 for his fundamental work in quantum mechanics.

or

$$y(n) = y(t)e^{\Gamma(t-nT)} - Ke^{\Gamma(t-n)} \sum_{m=0}^{\infty} f(x(m)) \int_{nT-\varepsilon}^{t} e^{\Gamma(s-t)}\delta(s-mT)ds$$

and

$$y(n+1) = e^{-\Gamma T}[\{y(t)e^{\Gamma(t-nT)}$$

$$- Ke^{\Gamma(t-n)} \sum_{m=0}^{\infty} f(x(m)) \int_{nT-\varepsilon}^{t} e^{\Gamma(s-t)}\delta(s-mT)dt\}$$

$$+ ke^{\Gamma(t-n)} \sum_{m=0}^{\infty} f(x(m)) \int_{nT-\varepsilon}^{(n+1)T-\varepsilon} e^{\Gamma(s-t)}\delta(s-mT)dt].$$

Thus,

$$y(n+1) = e^{-\Gamma T}[y(n) + Kf(x(n))]. \tag{4.75}$$

Similarly, by integrating the first equation of System (4.74) and substituting for y we have

$$x(n+1) = x(n) + \frac{1 - e^{-\Gamma T}}{\Gamma}(y(n) + Kf(x(n))) \tag{4.76}$$

In the next section, we will introduce a concrete realization of this model.

4.12.2   The Hénon Map

In 1976, the French Astronomer Michael Hénon [48] suggested a simplified model for the dynamics of the Lorenz system, which we alluded to in Chapter 3. The Hénon map (and the Hénon attractor) has become in recent years one of the icons of chaos theory and no treatment of dynamical systems is complete without it. In this section, our aim is limited to the discussion of stability and attractors of the Hénon map, given by

$$\begin{aligned} x(n+1) &= 1 - ax^2(n) + y(n) \\ y(n+1) &= bx(n) \end{aligned} \tag{4.77}$$

where a and b are real parameters with $|b| < 1$. The Hénon map may be derived from Eqs. (4.75) and (4.76) by simple manipulations and by letting $T = 1$. Let us write Equation (4.76) as

$$x(n+1) = x(n) + \frac{e^{\Gamma} - 1}{\Gamma}y(n+1).$$

Then,

$$y(n+1) = (x(n+1) - x(n))\frac{\Gamma}{(e^{\Gamma} - 1)}.$$

Substituting for $y(n)$ and $y(n+1)$ in Equation (4.75) yields

$$x(n+1) + e^{-\Gamma}x(n-1) = (1 + e^{-\Gamma})x(n) + \frac{1 - e^{-\Gamma}}{\Gamma} Kf(x(n)) \qquad (4.78)$$

If we now put $b = -e^{-\Gamma}$ and

$$\frac{1 - e^{-\Gamma}}{\Gamma} Kf(x(n)) + (1 + e^{-\Gamma})x(n) = 1 - ax^2(n),$$

then Equation (4.78) becomes

$$x(n+1) = 1 - ax^2(n) + bx(n-1).$$

The last second-order difference equation is equivalent to the Hénon map. We now make a few important observations.

First, the Hénon map contracts areas for $|b| < 1$. [If $b = 0$, we get the quadratic map $x(n+1) = 1 - ax^2(n)$.] To see this, we find the determinant of the Jacobian matrix of H. If $|\det DH| < 1$ for all (x, y), the map is area contracting. Now, from vector calculus we know that H maps an infinitesimal rectangle at (x, y) with area $dx\,dy$ into an infinitesimal parallelogram with area $|\det DH(x, y)dx\,dy|$. Thus, if $|\det DH(x, y)| < 1$, then H is area contracting. Notice that for the Hénon map

$$DH(x, y) = \begin{pmatrix} -2ax & 1 \\ b & 0 \end{pmatrix}$$

where $\det DH(x, y) = -b$. Hence if $|b| < 1$, the Hénon map is area contracting.

Second, H is invertible and

$$H^{-1}\begin{pmatrix} x \\ y \end{pmatrix} = \begin{pmatrix} \frac{1}{b}y \\ -1 + \frac{a}{b^2}y^2 + x \end{pmatrix}.$$

The simplest way to see this is to decompose H into three simple maps T_1, T_2, and T_3 as follows:

$$T_1\begin{pmatrix} x \\ y \end{pmatrix} = \begin{pmatrix} x \\ 1 - ax^2 + y \end{pmatrix}$$

$$T_2\begin{pmatrix} x \\ y \end{pmatrix} = \begin{pmatrix} bx \\ y \end{pmatrix}$$

$$T_3\begin{pmatrix} x \\ y \end{pmatrix} = \begin{pmatrix} y \\ x \end{pmatrix}$$

Observe that T_1 is an area-preserving bending map, T_2 contracts in the x direction (for $|b| < 1$), and T_3 rotates by $90°$ (see Fig. 4.16). The composite transformation $T_3 \circ T_2 \circ T_1$ yields the Hénon map H.

Now, we need to show that T_1, T_2, and T_3 are invertible, and hence $H^{-1} = T_1^{-1} \circ T_2^{-1} \circ T_3^{-1}$. The details are left to the reader as Problem 5.

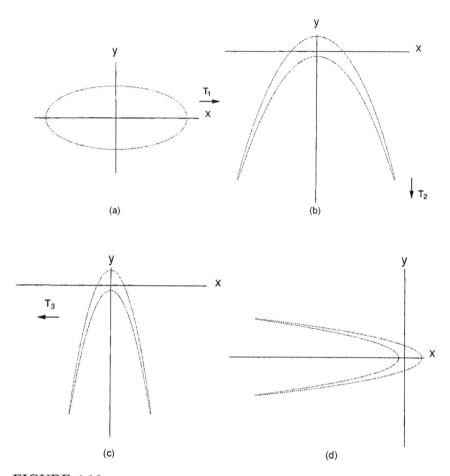

FIGURE 4.16
Decomposition of the action of the Henon map: $H = T_3 \circ T_2 \circ T_1$.

Third, H is one-to-one. This can easily be seen by showing that T_1, T_2, and T_3 are one-to-one.

Fourth, if $a \neq 0$, H has a fixed point $\begin{pmatrix} x^* \\ y^* \end{pmatrix}$ if $a \geq -\frac{1}{4}(1-b)^2$. If H has two fixed points $\begin{pmatrix} x_1^* \\ y_1^* \end{pmatrix}$ and $\begin{pmatrix} x_2^* \\ y_2^* \end{pmatrix}$, then we have

$$\begin{pmatrix} x_1^* \\ y_1^* \end{pmatrix} = \begin{pmatrix} \frac{1}{2a}(b-1+\sqrt{(1-b)^2+4a}) \\ \frac{b}{2a}(b-1+\sqrt{(1-b)^2+4a}) \end{pmatrix},$$

$$\begin{pmatrix} x_2^* \\ y_2^* \end{pmatrix} = \begin{pmatrix} \frac{1}{2a}(b-1-\sqrt{(1-b)^2+4a}) \\ \frac{b}{2a}(b-1-\sqrt{(1-b)^2+4a}) \end{pmatrix}.$$

Furthermore, if $a \neq 0$ and $a \in (-\frac{1}{4}(1-b)^2, \frac{3}{4}(1-b)^2)$, then $\begin{pmatrix} x_1^* \\ y_1^* \end{pmatrix}$ is asymptotically stable and $\begin{pmatrix} x_2^* \\ y_2^* \end{pmatrix}$ is a saddle. The details are left to the reader as Problem 5.

Finally, for a fixed value of the parameter b in $(0,1)$, $\begin{pmatrix} x_1^* \\ y_1^* \end{pmatrix}$ loses its stability and becomes a saddle at $a = \frac{3}{4}(1-b)^2$ and a new stable 2-cycle is born. The reason is that one of the eigenvalues of $DH\begin{pmatrix} x_1^* \\ y_1^* \end{pmatrix}$ will decrease and pass -1 (Problem 5). Hénon picked $b = 0.3$ and noticed that a period-doubling cascade starts at $a = 0.3675$ and ends at $a \approx 1.058$. Beyond $a \approx 1.06$, a strange attractor appears. In fact, there is a region R that gets mapped inside itself and contracts in area by the factor 0.3. This is called the trapping region, which is in the case of the Hénon map, a quadrilateral (see Fig. 4.17).

Forward iterates of the trapping region shrink down to a limit set called the Hénon attractor, which has zero area. Moreover, this attractor is indeed a strange attractor (see Fig. 4.18).

Figure 4.18 is generated for $a = 1.4$, $b = 0.3$ by computing 10,000 successive iterates of the Hénon map starting from the origin. Zooming into the strange attractor, we can see that there are six parallel curves. If we zoom further on the top three curves, we can see that they are really six curves grouped the same way as the first batch. This self-similarity continues to arbitrarily small scales. Benedick and Carleson [8] proved that this strange attractor is the closure of a branch of the unstable manifold. We will return to the Hénon map in the next chapter.

4.12.3   Discrete Epidemic Model for Gonorrhea [54]

Let us consider two distinct heterosexual populations P_1 and P_2. Let $x_i(n)$, $i = 1, 2$, be the fraction of the population P_i infected by gonorrhea at time period n. Then $1 - x_i(n)$ is the fraction of the population that is susceptible. It is assumed the populations P_1 and P_2 are constant and infected

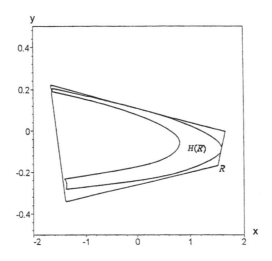

FIGURE 4.17
The trapping region R and its image $H(R)$.

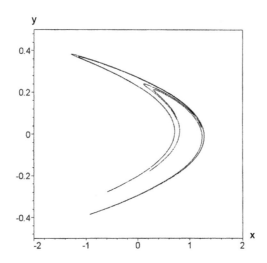

FIGURE 4.18
The Henon attractor.

members of the population can transmit gonorrhea to a susceptible person in the other population. To formulate our model, let us first assume there is no transmission of gonorrhea between the two populations. Then, the relation between the fraction of infected population P_i at time period $n + 1$ is given by

$$x_1(n+1) = (1 - b_1)x_1(n), \ 0 < b_1 < 1 \tag{4.79}$$
$$x_2(n+1) = (1 - b_2)x_2(n), \ 0 < b_2 < 1 \tag{4.80}$$

where $b_1 x_1(n)$ and $b_2 x_2(n)$ are the fractions of the populations P_1 and P_2, respectively, that have been cured. Clearly, $\lim_{n \to \infty} x_i(n) = 0$, which means that eventually there will be infected persons in either population. Unfortunately, this is not a realistic model and a modification is needed. Now, at time period n, part of the function of the susceptible $(1 - x_1(n))$ of population P_1 will be infected due to sexual engagements with population P_2. This is accounted for by adding the nonlinear term $a_1 x_2(n)(1 - x_1(n))$ to the right-hand side of Equation (4.79). Similarly, to account for the infected among susceptibles in population P_2, we add to the left-hand side of Equation (4.80) the nonlinear term $a_2 x_1(n)(1 - x_2(n))$. The more realistic model is now in the form

$$\begin{aligned} x_1(n+1) &= a_1 x_2(n)(1 - x_1(n)) + (1 - b_1)x_1(n) \\ x_2(n+1) &= a_2 x_1(n)(1 - x_2(n)) + (1 - b_2)x_2(n) \end{aligned} \tag{4.81}$$

where $0 < a_i < 1$, $0 < b_i < 1$, $i = 1, 2$. There are two fixed points, the origin O and the point $x^* = \begin{pmatrix} x_1^* \\ x_2^* \end{pmatrix}$, where

$$x_1^* = \frac{1 - \gamma_1 \gamma_2}{1 + \gamma_1}, \qquad x_2^* = \frac{1 - \gamma_1 \gamma_2}{1 + \gamma_2}$$

with $\gamma_i = \frac{b_i}{a_i}$.

Assume that $a_1 a_2 - b_1 b_2 \leq 0$. Then, taking $V\begin{pmatrix} x_1 \\ x_2 \end{pmatrix} = b_2 x_1 + a_1 x_2$ as our Liapunov function on the open set $G = \left\{ \begin{pmatrix} x_1 \\ x_2 \end{pmatrix} : 0 < x_i < 1, i = 1, 2 \right\}$, we get $\Delta V(x) = (a_1 a_2 - b_1 b_2)x_1 - a_1(a_2 + b_2)x_1 x_2 \leq 0$. Clearly, $E = \{x \in \overline{G} : \Delta V(x) = 0\}$ is the union of the coordinate axes. The largest invariant set M in E is the origin. Thus, by Theorem 4.6, the origin is globally asymptotically stable relative to G.

If $a_1 a_2 - b_1 b_2 > 0$, we obtain an unrealistic negative equilibrium point x^*. In this case, it may be shown (Problem 6) that the origin is unstable. Next, we consider the positive equilibrium point x^*. In this case, we choose the Liapunov function

$$V(x) = a_2 |x_1 - x_1^*| + a_1 \left(\frac{1 + \gamma_1}{1 + \gamma_2} \right) |x_2 - x_2^*|.$$

Then it may be shown that (Problem 6)

$$\Delta V(x) \le -a_1 a_2(x_1|x_2 - x_2^*| + \rho x_2|x_1 - x_1^*|) \le 0$$

with $\rho = \frac{1+\gamma_1}{1+\gamma_2}$. By the same reasoning as in the previous assumption, we conclude that x^* is globally asymptotically stable relative to $G - \{0\}$.

4.12.4   Perennial Grass

Let $x(n)$ denote the litter of a perennial grass per square meter on the ground in year (season) n. This occurs since underground roots sprout and produce new biomass, which then falls to the ground as litter (the dead plant material produced at the end of a growing season). A litter layer mulches the soil surface and intercepts light, thus inhibiting growth until it decays. Once litter decays, it stimulates growth by releasing nitrogen.

In [28], the following model was proposed.

$$x(n + 1) = ax(n) + (b + cx(n - 1))e^{-x(n)}, \quad n = 0, 1, \ldots \qquad (4.82)$$

with $a, c \in (0, 1)$ and $b > 0$ such that $x(-1)$ and $x(0)$ are positive initial data. Here $ax(n)$ is the fraction of the present litter that has not decayed yet, and $b + cx(n - 1))e^{-x(n)}$ is the litter accumulation, which depends on past and present litter. Writing Equation (4.82) as a system yields

$$\begin{pmatrix} y_1(n + 1) \\ y_2(n + 1) \end{pmatrix} = \begin{pmatrix} y_2(n) \\ (b + cy_1(n))e^{-y_2(n)} + ay_2(n) \end{pmatrix} \qquad (4.83)$$

The fixed points are obtained by solving the two equations $y_1 = y_2$ and $(b + cy_1)e^{-y_2} + ay_2 = y_2$. Hence, if $y^* = \begin{pmatrix} y_1^* \\ y_2^* \end{pmatrix}$ is an equilibrium point, then

$$h(y_1^*) = ay_1^* - y_1^* + (b + cy_1^*)e^{-y_1^*} = 0.$$

Note that $h(0) = b > 0$, and the origin is not an equilibrium point. Moreover, $h(\infty) = -\infty$, which implies by the intermediate value theorem the existence of a positive equilibrium point with $y_1^* = y_2^* > 0$. Furthermore, the positive fixed point is indeed unique since

$$\begin{aligned} h'(y_1^*) &= a - 1 + (c - b - cy_1^*)e^{-y_1^*} \\ &= \tfrac{1-a}{b+cy_1^*}(-b - by_1^* - cy_1^{*2}) \\ &< 0. \end{aligned}$$

The linearized system corresponding to System (4.83) is given by

$$u(n + 1) = Au(n) \qquad (4.84)$$

where

$$A = \begin{pmatrix} 0 & 1 \\ c\left(\frac{x^* - ax^*}{b + cx^*}\right) & a - (1 - a)x^* \end{pmatrix}$$

with $x^* = x_1^* = x_2^*$.

By Theorem 4.11 a sufficient condition for the local stability of x^* is given by

$$|a - (1 - a)x^*| < 1 - \frac{c(1-a)x^*}{b + cx^{*}} < 2. \qquad (4.85)$$

Now, Equation (4.85) is equivalent to the equation

$$(1-a)cx^{*2} + ((1-a)b - 2ax^*) - b(1+a) < 0,$$

that is

$$0 < x^* < \frac{a}{1-a} - \frac{b}{2c} + \frac{\sqrt{(2ac - (1-a)b^2 + 4(1-a^2)bc)}}{2(1-a)c}.$$

Hence, x^* is (locally) asymptotically stable if

$$x^* \leq \frac{a}{1-a}. \qquad (4.86)$$

Indeed, it was shown in [28], that under Condition (4.86), x^* attracts all positive solutions of Equation (4.83). For the case $x^* > \frac{a}{1-a}$, it may be shown [28] that Equation (4.83) is permanent, i.e., for any initial conditions $x(-1)$ and $x(0) \in (0, \infty)$, there exists $N \in \mathbb{Z}^+$ such that

$$D \leq x(n) \leq E, \quad \text{for } n \geq N$$

where D and E are positive numbers.

Further analysis of this model reveals a period-doubling bifurcation and the onset of chaos. Such behavior has been exhibited by computer analysis.

Exercises - (4.12)

1. In the epidemic Model (4.83), show by using linearization that the positive equilibrium point x^* is asymptotically stable if $\gamma_1 \gamma_2 < 1$.

2. In the epidemic Model (4.83), introduce a third population group that infects members of the group and those outside, and analyze the stability of the model.

3. Consider a predator–prey model in which predators search over a constant area and have unlimited capacity for consuming the prey:

$$N(t+1) = rN(t)\exp(-aP(t))$$
$$P(t+1) = N(t)\{1 - \exp(-aP(t))\}$$

where $t \in \mathbb{Z}^+$, $a > 0$, $r > 1$, $N(t)$ is the size of the prey population, and $P(t)$ is the size of the predator population at time period t.

(a) Find the positive equilibrium point (N^*, P^*).

(b) Determine the local stability of (N^*, P^*) using linearization.

4. A general form for models of insect predator (P)–prey (N), or insect parasitism, is

$$N(t+1) = rN(t)f(N(t), P(t)), \quad P(t+1) = N(t)[1 - f(N(t), P(t))]$$

where f is a nonlinear function that incorporates assumptions about predators searching, and $r > 0$ is the rate of increasing of the prey population. Furthermore, it is assumed that $0 < f < 1$, f is an increasing function.

(a) Find conditions on r so that a positive equilibrium point (N^*, P^*) exists.

(b) Determine under what conditions (N^*, P^*) is asymptotically stable.

5. (a) Show that the maps T_1, T_2, and T_3 that define the Hénon map are invertible.

(b) Show that for a fixed $b \in (0, 1)$ if $a > \frac{3}{4}(1 - b)^2$, then one of the eigenvalues of $DH(x_1^*)$ is less than -1.

(c) Show that if $a \in \left(\frac{1}{4}(1 - b)^2, \frac{3}{4}(1 - b)^2\right)$, $a \neq 0$, then x_1^* is asymptotically stable and x_2^* is a saddle.

6. (a) In the gonorrhea model, show that the origin is unstable if $a_1 a_2 - b_1 b_2 > 0$.

(b) In the gonorrhea model, show that the positive equilibrium point x^* is globally asymptotically stable relative to $G - \{0\}$, where

$$G = \left\{ \begin{pmatrix} x_1 \\ x_2 \end{pmatrix} : 0 < x_i < 1, \ i = 1, 2 \right\}.$$

7. [76] Consider the following mathematical model for plant competition:

$$p(n+1) = (1 - \alpha)\frac{cp(n)}{cp(n) + q(n)} + \alpha p(n)$$

$$q(n+1) = p(1 - \beta)\frac{dq(n)}{p(n) + dq(n)} + \beta q(n)$$

where $0 < \alpha$, $\beta < 1$.

(a) Find conditions under which the positive equilibrium point (p^*, q^*) exists.

(b) Use a Liapunov function to determine the stability of (p^*, q^*).

Appendix

THEOREM 4.14
[12] Let z be an attracting fixed point of a continuous map $f : I \to \mathcal{R}$, where I is an interval. Then z is stable.

PROOF   There is an open interval G containing z such that for all x in G, $f^n(x) \to z$ as $n \to \infty$. This implies that z is the only periodic point in G. In particular, $f^2(x) \neq x$ if $x \in G$ and $x \neq z$. Put

$$G_- = \{x \in G : x < z\}, \quad G_+ = \{x \in G : x > z\}.$$

By the Intermediate Value Theorem, $f^2(x) < x$ for all x in G_- or $f^2(x) > x$ for all x in G_-. Suppose $f^2(x) < x$ for all x in G_-. Then if $f^2(x)$ is also in G_-, $f^4(x) < f^2(x)$ and so $f^4(x) < x$. Now, for a given n, if $x \in G_-$ is close enough to z, $f^{2k}(x) \in G$ for $0 \leq k \leq n$ and then it follows by repetition of the above argument that for such x, $f^{2n}(x) < x$. However, since z is the only periodic point in G, it is also true that $f^{2n}(x) < x$ for all x in G_- or $f^{2n}(x) > x$ for all x in G_-. It follows that $f^{2n}(x) < x$ for all n and all $x \in G_-$. However this means that $f^{2n}(x)$ does not converge to z, a contradiction. Hence we must have $f^2(x) < x$ for all x in G_-. Similarly, we can show that $f^2(x) < x$ for all x in G_+.

It follows that $f(x) \neq x$ for $x \neq z$ in G. If $f(x) < x$ for all x in G_-, it would follow that $f^2(x) < f(x) < x$ if $x \in G_-$ is so close to z that $f(x) \in G$. So $f(x) > x$ for all x in G_-. Similarly, $f(x) < x$ for all x in G_+. Then if H is a sufficiently small open interval containing z, $\overline{H} \subset G$ and $f(\overline{H}) \subset G$.

Assume first that $f(x) \leq z$ for all $x \in H$ with $x < z$ and $f(x) \geq z$ for all $x \in H$ with $x > z$. Then for every closed interval $K \subset H$ with $z \in K$ we have $f(K) \subset K$. Thus z is stable.

Otherwise for some $x_0 \in H$, either $x_0 < z < f(x_0) = b$ or $b = f(x_0) < z < x_0$. Assume for definiteness that $x_0 < z < f(x_0) = b$. Define

$$a = \sup\{x_0 \leq a < z : f(a) = b\}.$$

Then $f(a) = b$, $a < z$ and $f(x) < b$ for $a < x \leq z$. Hence $f(x) > a$ for $z \leq x \leq b$, since $f(z) = z > a$ and $f(x) = a$ would imply $f^2(x) = b \geq x$. It follows that for $J = (a, b)$ we have $f(J) \subset J$, since $b > f(x) > x$ for $a < x < z$ and $a < f(x) < x$ for $z < x < b$. Then $f^n(\overline{J})$ is decreasing sequence of compact intervals all containing z. Let $K = [c, d]$ be the intersection. Note that $z \in K$ and $f(K) = K$. If $c = d = z$, it follows that z is stable.

Assume $c < d$. Then $z > c$, because if $z = c$, then $f(x) < x$ for $c < x \leq d$ so that $f(K) \neq K$. Also $z < d$, because if $z = d$, then $f(x) > x$ for $c \leq x < d$

so that $f(K) \neq K$. Since $f(x) > x$ for $c \leq x < z$ but $f(K) = K$, there must exist $x_2 \in (z, d]$ such that $f(x_2) = c$. Since $f(z) = z$ and there exists $x \in [c, z)$ such that $f(x) = d$ (since $f(K) = K$ but $f(x) < x$ if $z < x \leq d$), it follows from the Intermediate Value Theorem that there exists $x_1 \in [c, z)$ such that $f(x_1) = x_2$. However, then $f^2(x_1) = c \leq x_1$, which is a contradiction.   ∎

5

Bifurcation and Chaos in Two Dimensions

Lo! Thy dread empire, Chaos! Is restor'd; Light dies before thy uncreating word; Thy hand, great anarch! Lets the curtain fall, And universal darkness buries all.

Alexander Pope, The Duncaid

5.1 Center Manifolds

In Chapter 4, Theorem 4.11, we were able to give a complete determination of the stability of two-dimensional maps via linearization, when the fixed point is hyperbolic, i.e., the eigenvalues of the Jacobian matrix are off the unit circle. However, Theorem 4.11 failed to address the stability of non-hyperbolic fixed points. Here comes the center manifold theory to our rescue. Roughly speaking, a center manifold is a set M_c in a lower dimensional space, where the dynamics of the original system can be obtained by studying the dynamics on M_c. For instance, if a nonhyperbolic map is defined on $\mathbb{R}^2$, then its dynamics may be analyzed by studying the dynamics on an associated one-dimensional center manifold M_c. In light of our complete understanding of the stability of one-dimensional maps, the reduction procedure to center manifolds is one of the most powerful tools in dynamical systems. We begin our exploration by introducing the necessary notations, definitions, theory and examples.

Consider the s-parameter map $F(\mu, u)$, $F : \mathbb{R}^s \times \mathbb{R}^k \to \mathbb{R}^k$, with $u \in \mathbb{R}^k$, $\mu \in \mathbb{R}^s$, where F is C^r $(r \geq 3)$ on some sufficiently large open set in $\mathbb{R}^k \times \mathbb{R}^s$. Let (μ^*, u^*) be a fixed point of F, i.e.,

$$F(\mu^*, u^*) = u^*. \tag{5.1}$$

We have seen in Chapters 1 and 4 that the stability of hyperbolic fixed points of F is determined from the stability of the fixed points under the linear map

$$J = D_u F(\mu^*, u^*). \tag{5.2}$$

However, the situation is drastically different if one of the eigenvalues λ of J lies on the unit circle, that is, $|\lambda| = 1$. There are three separate cases in which the fixed point (u_0, μ_0) is nonhyperbolic.

1. J has one real eigenvalue equal to 1 and the other eigenvalues are off the unit circle.

2. J has one real eigenvalue equal to -1 and the other eigenvalues are off the unit circle.

3. J has two complex conjugate eigenvalues with modulus 1 and the other eigenvalues are off the unit circle.

When $k = 2$, cases 1 and 2 can be reduced to the one-dimensional cases that were discussed in Chapter 2. This task may be accomplished using the center manifold theory, which we will develop shortly. The third case is new and has no analogue in the one-dimensional theory. It will give rise to a new bifurcation, the **Neimark-Sacker** bifurcation.[1] We now present a version of center manifold theory that meets our needs.

By a change of variables, we may assume, without loss of generality, that $u_0 = 0$. Let us temporarily suppress the parameter μ. Then, the map F can be written in the form

$$
\begin{aligned}
x &\longmapsto Ax + f(x, y)\\
y &\longmapsto By + g(x, y)
\end{aligned}
\tag{5.3}
$$

where J in Equation (5.2) has the form $J = \begin{pmatrix} A & 0 \\ 0 & B \end{pmatrix}$. Moreover, all of the eigenvalues of A lie on the unit circle and all of the eigenvalues of B are off the unit circle. Furthermore,

$$
\begin{aligned}
f(0,0) &= 0, \quad g(0,0) = 0\\
Df(0,0) &= 0, \; Dg(0,0) = 0.
\end{aligned}
$$

Observe that System (5.3) corresponds to the system of difference equations

$$
\begin{aligned}
x(n+1) &= Ax(n) + f(x(n), y(n)),\\
y(n+1) &= By(n) + g(x(n), y(n)).
\end{aligned}
$$

From now on, we assume that A is a $t \times t$ matrix and B is an $s \times s$ matrix, with $t + s = k$. The following results are taken from [16].

---

[1] It is often called the Hopf bifurcation, a name borrowed mistakenly from the theory of differential equations.

THEOREM 5.1

There is a C^r-center manifold for System (5.3) that can be represented locally as

$$M_c = \{(r, y) \in \mathbb{R}^t \times \mathbb{R}^s : y - h(x), |x| < \delta, h(0) = 0,$$
$$Dh(0) = 0, \text{ for a sufficiently small } \delta\}. \tag{5.4}$$

Furthermore, the dynamics restricted to M_c are given locally by the map

$$x \longmapsto Ax + f(x, h(x)), x \in \mathbb{R}^t. \tag{5.5}$$

This theorem asserts the existence of a center manifold, i.e., a curve $y = h(x)$ on which the dynamics of System (5.3) is given by Equation (5.5). The next result states that the dynamics on the center manifold M_c determines completely the dynamics of System (5.3).

THEOREM 5.2

The following statements hold.

1. *If the fixed point $(0,0)$ of Equation (5.5) is stable, asymptotically stable, or unstable, then the fixed point (0,0) of System (5.3) is stable, asymptotically stable, or unstable, respectively.*

2. *For any solution $(x(n), y(n))$ of System (5.3) with an initial point (x_0, y_0) in a small neighborhood around the origin, there exists a solution $z(n)$ of Equation (5.5) and positive constants L, $\beta > 1$ such that*

$$|x(n) - z(n)| \le L\beta^n, \quad \text{and} \quad |y(n) - h(z(n))| \le L\beta^n \text{ for all } n \in \mathbb{Z}^+.$$

The question that still lingers is how to find the center manifold M_c or, equivalently, how to compute the curve $y = h(x)$. The first thing that comes to mind is to substitute for y in System (5.3) to obtain the system

$$x(n+1) = Ax(n) + f(x(n), h(x(n)))$$
$$y(n+1) = h(x(n+1))$$
$$= h[Ax(n) + f(x(n), h(x(n)))]$$
$$= Bh(x(n)) + g(x(n), h(x(n))).$$

Equating the two equations for $y(n+1)$ yields the functional equation

$$\mathcal{F}(h(x)) = h[Ax + f(x, h(x)] - Bh(x) - g(x, h(x)) = 0. \tag{5.6}$$

Solving Equation (5.6) is a formidable task, so at best one can hope to approximate its solution via power series. The next result provides the theoretical justification for our approximation.

THEOREM 5.3

Let $\psi : \mathbb{R}^t \to \mathbb{R}^s$ be a C^1-map with $\psi(0) = \psi'(0) = 0$. Suppose that[2] $\mathcal{F}(\psi(x)) = O(|x|^r)$ as $x \to 0$ for some $r > 1$. Then,

$$h(x) = \psi(x) + O(|x|^r) \quad as \quad x \to 0.$$

We now present examples to compute the center manifold M_c.

Example 5.1

Consider the map $F = \begin{pmatrix} f \\ g \end{pmatrix}$ given by

$$\begin{pmatrix} x \\ y \end{pmatrix} \longmapsto \begin{pmatrix} -1 & 0 \\ 0 & -\frac{1}{2} \end{pmatrix} \begin{pmatrix} x \\ y \end{pmatrix} + \begin{pmatrix} xy \\ x^2 \end{pmatrix}.$$

Then,

$$M_c = \{(x, y) \in \mathbb{R}^2 : y = h(x), h(0) = h'(0) = 0\}.$$

The function h must satisfy Equation (5.6)

$$h(Ax + f(x, h(x))) - Bh(x) - g(x, h(x)) = 0$$

or

$$h(-x + xh(x)) + \frac{1}{2}h(x) - x^2 = 0. \tag{5.7}$$

Let us assume that $h(x)$ takes the form

$$h(x) = c_1 x^2 + c_2 x^3 + O(x^4). \tag{5.8}$$

Then, substituting Equation (5.8) in Equation (5.7) yields

$$c_1 x^2 - c_2 x^3 + O(x^4) + \frac{1}{2}(c_1 x^2 + c_2 x^3 + O(x^4)) - x^2 = 0.$$

Hence,

$$c_1 + \frac{1}{2}c_1 - 1 = 0 \quad or \quad c_1 = \frac{2}{3}$$

$$-c_2 + \frac{1}{2}c_2 = 0 \quad or \quad c_2 = 0.$$

Consequently, $h(x) = \frac{2}{3}x^2 + O(x^4)$ and the map f on the center manifold is given by

$$x \longmapsto -x + \frac{2}{3}x^3 + O(x^5)$$

---

[2]$\mathcal{F}(\psi(x)) = O(|x|^r)$ as $x \to 0$, "read as big O," means that $\lim_{x \to 0} \frac{\mathcal{F}(\psi(x))}{O(|x|^r)} = M$, for some real number M.

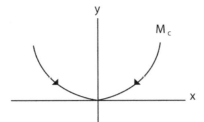

FIGURE 5.1
An asymptotically stable nonhyperbolic fixed point $(0,0)$. The curve $h(x) = -x + \frac{2}{3}x^3 + O(x^4)$ is the graph of the center manifold M_c. The orbits on the y-axis oscillate but converge to origin.

whose Schwarzian derivative at zero (see Chapter 1) is given by

$$S(f)(0) = \frac{f'''(0)}{f'(0)} - \frac{2}{3}\left(\frac{f''(0)}{f'(0)}\right)^2 = -4 < 0.$$

Thus, by Theorem 1.6, $x^* = 0$ is asymptotically stable under the map f. This implies by Theorem 5.2 that the origin is asymptotically stable under the map F (see Fig. 5.1).   ▯

Example 5.2

Consider the two dimensional map $F = \begin{pmatrix} f \\ g \end{pmatrix}$ given by

$$\begin{pmatrix} x \\ y \end{pmatrix} \longmapsto \begin{pmatrix} 0 & 1 \\ -\frac{1}{2} & \frac{3}{2} \end{pmatrix} \begin{pmatrix} x \\ y \end{pmatrix} + \begin{pmatrix} 0 \\ -y^3 \end{pmatrix}, \quad x, y \in \mathbb{R}^2. \tag{5.9}$$

The eigenvalues of the linear part are $\lambda_1 = 1$, $\lambda_2 = \frac{1}{2}$, with corresponding eigenvectors $V_1 = \begin{pmatrix} 1 \\ 1 \end{pmatrix}$ and $V_2 = \begin{pmatrix} 2 \\ 1 \end{pmatrix}$. In order to find the center manifold, we first need to diagonalize the matrix $\begin{pmatrix} 0 & 1 \\ -\frac{1}{2} & \frac{3}{2} \end{pmatrix}$ by using the matrix $T = \begin{pmatrix} 1 & 2 \\ 1 & 1 \end{pmatrix}$ whose columns are the eigenvectors V_1 and V_2. Letting $\begin{pmatrix} x \\ y \end{pmatrix} = T\begin{pmatrix} u \\ v \end{pmatrix}$, yields

$$T\begin{pmatrix} u \\ v \end{pmatrix} \longmapsto \begin{pmatrix} 0 & 1 \\ -\frac{1}{2} & \frac{3}{2} \end{pmatrix} T\begin{pmatrix} u \\ v \end{pmatrix} + \begin{pmatrix} 0 \\ -(u+v)^3 \end{pmatrix}.$$

Hence

$$\begin{pmatrix} u \\ v \end{pmatrix} \longmapsto T^{-1}\begin{pmatrix} 0 & 1 \\ -\frac{1}{2} & \frac{3}{2} \end{pmatrix} T\begin{pmatrix} u \\ v \end{pmatrix} + T^{-1}\begin{pmatrix} 0 \\ -(u+v)^3 \end{pmatrix}.$$

Substituting for $T^{-1} = \begin{pmatrix} -1 & 2 \\ 1 & -1 \end{pmatrix}$ we obtain

$$\mathcal{F}(h(u)) = h(AU + F(u, h(u)) - Bh(u) - g(u, h(u)) = 0 \qquad (5.10)$$

where $A = 1$, $B = \frac{1}{2}$, $f(u, v) = -2(u + v)^3$, $g(u, v) = (u + v)^3$. Let us assume that $h(x)$ takes the form

$$h(u) = au^2 + bu^3 + O(u^4). \qquad (5.11)$$

Substituting form Equation (5.11) into Equation (5.10) yields

$$a(u - 2(u + au^2 + bu^3 + O(u^4))^3)^2 + b(u - 2(u + au^2 + bu^3 + O(u^4))^3)^3 + \ldots$$
$$-\frac{1}{2}(au^2 + bu^3 + O(u^4)) - (u + au^2 + bu^3 + O(u^4))^3 = 0$$

or

$$au^2 + bu^3 - \frac{1}{2}au^2 - \frac{1}{2}bu^3 - u^3 + O(u^4) = 0.$$

Equating coefficients of like powers to zero yields

$$u^2 : a - \frac{1}{2}a = 0 \Rightarrow a = 0$$
$$u^3 : b - \frac{1}{2}b - 1 = 0 \Rightarrow b = 0.$$

Thus the center manifold is given by the graph of

$$h(u) = 2u^3 + O(u^4)$$

and the map f on the center manifold is given by

$$u \longmapsto u - 2u^3 + O(u^4).$$

Notice that $u^* = 0$ is a fixed point of f at which $f'(0) = 1$, $f''(0) = 0$, and $f'''(0) < 0$. This implies by Theorem 1.6 that the origin is asymptotically stable under the map $F = \begin{pmatrix} f \\ g \end{pmatrix}$ (see Figure 5.2).   ☐

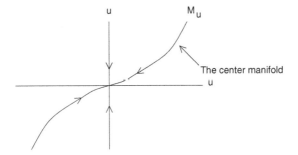

FIGURE 5.2
An asymptotically stable nonhyperbolic fixed point $(0,0)$. The curve $h(u) = 2u^3 + O(u^4)$ represents the center manifold M_c.

REMARK 5.1

1. The choice of the form of the map h is not unique. It can be shown (see [16]) that any two center manifolds of a given fixed point may differ only in transcendentally small terms. In other words, the Taylor series expansion of any two center manifolds must agree in all orders.

2. If the matrix J of the linear part is not in diagonal form, we must first diagonalize it before computing the center manifold as the next example will illustrate. ∎

Center Manifolds Depending on Parameters

Suppose now that System (5.3) depends on a vector of parameters, say, $\mu \in \mathbb{R}^m$. Then, System (5.3) takes the form

$$x(n+1) = Ax(n) + f(\mu, x(n), y(n))$$
$$y(n+1) = By(n) + g(\mu, x(n), y(n)) \qquad (5.12)$$

where

$$f(0,0,0) = 0, \quad g(0,0,0) = 0$$
$$Df(0,0,0) = 0, \quad Dg(0,0,0) = 0$$

where f and g are C^r functions $(r \geq 3)$ in some neighborhood of $(x, y, \mu) = (0,0,0)$.

The first step in handling Equation (5.12) is to increase the numbers of equations to $k + m$ by writing it in the form

$$x(n+1) = Ax(n) + f(\mu(n), x(n), y(n))$$
$$\mu(n+1) = \mu(n) \qquad (5.13)$$
$$y(n+1) = By(n) + g(\mu(n), x(n), y(n)).$$

The center manifold M_c now takes the form

$$M_c = \{(\mu, x, y) : y$$
$$= h(x, \mu), |x| < \delta_1, |\mu| < \delta_2, h(0,0) = 0, Dh(0,0) = 0\}. \quad (5.14)$$

Substituting for y into System (5.13) yields

$$x(n+1) = Ax(n) + f(\mu, x(n), h(x(n)))$$
$$y(n+1) = h[Ax(n) + f(\mu, x(n), h(x(n)))]$$
$$= Bh(\mu, x(n)) = g(\mu, x(n), h(\mu, x(n))).$$

The latter equations lead to the functional equation

$$\mathcal{F}(h(\mu, x)) = h[Ax + f(\mu, h(\mu, x), x)]$$
$$- Bh(\mu, x) - g(\mu, h(\mu, x), x) = 0. \quad (5.15)$$

Note that if μ is one-dimensional, it should be treated as a dependent variable, so terms like $x\mu$ or $y\mu$ will be viewed as nonlinear terms and will be absorbed by f and g. If μ is multidimensional, then we pick one component as a dependent variable and the remaining components are viewed as constants. For example, $h(\mu, x)$ takes the form

$$h(\mu, x) = c_1 x^2 + c_2 x\mu + c_3 \mu^2 + \ldots \quad (5.16)$$

Compare Equation (5.16) with Equation (5.8).

5.2 Bifurcation

5.2.1 Eigenvalues of 1 or -1

In this section we focus our attention on the bifurcation of two dimensional maps. The extension to higher dimensions should be apparent to the reader after comprehending the two dimensional case.

Let us consider the one-parameter family of maps

$$F(\mu, u) : \mathbb{R}^2 \times \mathbb{R} \to \mathbb{R} \quad (5.17)$$

with $u = (x, y) \in \mathbb{R}^2$, $\mu \in \mathbb{R}$ and $F \in C^r$, $r \geq 5$. If (μ^*, u^*) is a fixed point, then we make a change of variables, so that our fixed point is $(0,0)$. Let $J = D_u F(0,0)$. Then using the center manifold theorem, we find a one-dimensional map $f_\mu(u)$ defined on the center manifold M_c. By Theorem 5.1, we deduce the following statements:

1. Suppose that J has an eigenvalue equal to 1. Then we have

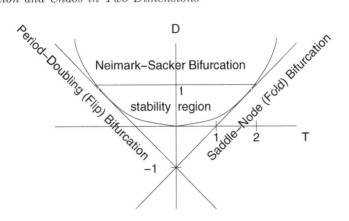

FIGURE 5.3
The occurrence of the three main types of bifurcation.

(a) a saddle-node bifurcation (fold bifurcation), if $\frac{\partial F}{\partial \mu}(0,0) \neq 0$ and $\frac{\partial^2 F}{\partial^2 u}(0,0) \neq 0$

(b) a pitchfork bifurcation, if $\frac{\partial f}{\partial \mu}(0,0) = 0$ and $\frac{\partial^2 f}{\partial^2 \mu}(0,0) = 0$

(c) a transcritical bifurcation, if $\frac{\partial f}{\partial \mu}(0,0) = 0$ and $\frac{\partial^2 f}{\partial^2 \mu}(0,0) \neq 0$.

2. If J has an eigenvalue equal to -1, then we have a period-doubling (flip) bifurcation.

3. If J has a pair of complex conjugate eigenvalues of modulus 1, a new type of bifurcation, called the Neimark-Sacker bifurcation appears. The Neimark-Sacker bifurcation will be discussed in more details in the sequel.

Let $T = tr\ J$, $D = det\ J$. Then the following trace-determinant diagram (Fig. 5.3) illustrate the main bifurcation phenomena (compare with Fig. 4.10).

5.2.2 A Pair of Eigenvalues of Modulus 1 - The Neimark-Sacker Bifurcation

We now turn our attention to the case when the Jacobian matrix $J = D_u F(u_0, \mu_0)$ has two complex conjugate eigenvalues. Let us start with an illustrative example.

Example 5.3

Consider the family of maps

$$F_\mu \begin{pmatrix} x_1 \\ x_2 \end{pmatrix} = (1 + \mu - x_1^2 - x_2^2) \begin{pmatrix} \cos\beta & -\sin\beta \\ \sin\beta & \cos\beta \end{pmatrix} \begin{pmatrix} x_1 \\ x_2 \end{pmatrix} \qquad (5.18)$$

where $\beta = \beta(\mu)$ is a smooth function of the parameter μ and $0 < \beta(0) < \pi$. Observe that the origin is a fixed point of the map F_μ for all μ with the Jacobian matrix

$$J = (1 + \mu) \begin{pmatrix} \cos\beta & -\sin\beta \\ \sin\beta & \cos\beta \end{pmatrix}.$$

The matrix J has eigenvalues $\lambda_{1,2} = (1+\mu)e^{\pm i\beta}$ with $|\lambda_{1,2}| = |1 + \mu|$. Hence, at $\mu = 0$, the eigenvalues cross the unit circle, a clear sign of the appearance of a Neimark-Sacker bifurcation. Clearly, the origin is asymptotically stable for $-2 < \mu < 0$. To analyze the bifurcation when $\mu = 0$, it is more convenient to write the map F_μ in polar coordinates (r, θ). To facilitate this change of coordinates, write Equation (5.18) as a two-dimensional system of difference equations:

$$\begin{pmatrix} x_1(n+1) \\ x_2(n+1) \end{pmatrix} = (1 + \mu - x_1^2(n) - x_2^2(n))$$

$$\begin{pmatrix} \cos\beta & -\sin\beta \\ \sin\beta & \cos\beta \end{pmatrix} \begin{pmatrix} x_1(n) \\ x_2(n) \end{pmatrix}. \qquad (5.19)$$

Now, putting $x_1(n) = r(n)\cos\theta(n)$, $x_2(n) = r(n)\sin\theta(n)$ in Equation (5.19), we obtain

$$r(n+1) = (1 + \mu)r(n) - r^3(n)$$
$$\theta(n+1) = \theta(n) + \beta. \qquad \square \qquad (5.20)$$

The form of Equation (5.20) enables us to detect easily the presence of an invariant circle by letting $(1 + \mu)r - r^3 = r$. Hence, the invariant circle is of radius $r^* = \sqrt{\mu}$. Thus, this invariant circle appears when μ crosses the value 0 as shown in Fig. 5.4. When $\mu = 0$, the map $r \longmapsto r - r^3$ is one-dimensional and its stability can be determined by using the techniques of Chapter 1.

Note also that the cobweb diagram indicates that the origin is (slowly) asymptotically stable (see Fig. 5.5).

Thus, for $\mu = 0$, the origin is asymptotically stable. When μ becomes positive, the origin loses its stability and gives rise to an attracting (asymptotically stable) circle with radius $r = \sqrt{\mu}$. The dynamics on this circle are determined by the map $\theta \longmapsto \theta + \beta$, which is a rotation by an angle β in the counterclockwise direction. This phenomenon is called a Neimark-Sacker bifurcation (or a Hopf bifurcation).

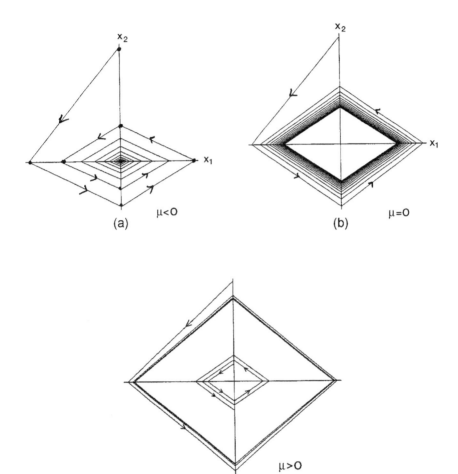

FIGURE 5.4

Supercritical Neimark-Sacker bifurcation of the map in (5.18), (a) $\mu < 0$, (b) $\mu = 0$, (c) $\mu > 0$.

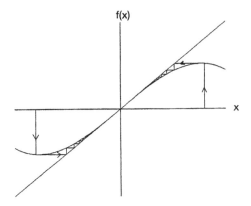

FIGURE 5.5

For $\mu = 0$, the Cobweb diagram for the one dimension map $r \to r - r^3$ shows that 0 is asymptotically stable.

REMARK 5.2   An analogous but different scenario may occur if we consider the map

$$\tilde{F}_\mu \begin{pmatrix} x_1 \\ x_2 \end{pmatrix} = (1 + \mu + x_1^2 + x_2^2) \begin{pmatrix} \cos\beta & -\sin\beta \\ \sin\beta & \cos\beta \end{pmatrix} \begin{pmatrix} x_1 \\ x_2 \end{pmatrix}. \qquad (5.21)$$

This map undergoes a Neimark-Sacker bifurcation at $\mu = 0$ but in a manner different from that of Equation (5.18) (see Fig. 5.6). The reader is asked to provide details in Problem 4.   ∎

To this end, we have described in detail the occurrences of the Neimark-Sacker bifurcation for the map in Example 5.3. In the sequel, we will show that the dynamics of this map are typical for a certain class of two-dimensional maps with one parameter. So, let us consider the family of C^r maps ($r \geq 5$) $F_\mu : \mathbb{R}^2 \times \mathbb{R} \to \mathbb{R}^2$ such that the following conditions hold:

1. $F_\mu(0) = 0$, i.e., the origin is a fixed point of F_μ.

2. $DF_\mu(0)$ has two complex conjugate eigenvalues $\lambda(\mu) = r(\mu)e^{i\theta(\mu)}$ and $\bar{\lambda}(\mu)$, where $r(0) = 1$, $r'(0) \neq 0$, $\theta(0) = \theta_0$. Thus, $|\lambda(0)| = 1$.

3. $e^{ik\theta_0} \neq 0$ for $k = 1, 2, 3, 4, 5$, i.e., $\lambda(0)$ is not a low root of unity.

Based on the above assumptions, we make the following claims.

1. By a change of basis in $\mathbb{R}^2$, we may assume, without loss of generality, that

$$J = DF_\mu(0,0) = (1 + \mu) \begin{pmatrix} \cos\beta(\mu) & -\sin\beta(\mu) \\ \sin\beta(\mu) & \cos\beta(\mu) \end{pmatrix}.$$

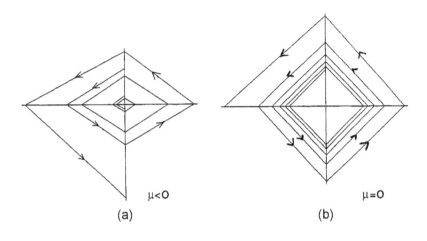

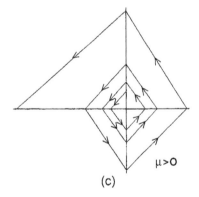

FIGURE 5.6

Supercritical Neimark-Sacker bifurcation of the map in (5.21), (a) $\mu < 0$, (b) $\mu = 0$, (c) $\mu > 0$.

2. From assumption 3, by a change of coordinates, we may assume that our map F_μ takes the form

$$F_\mu \begin{pmatrix} x_1 \\ x_2 \end{pmatrix} = N_\mu \begin{pmatrix} x_1 \\ x_2 \end{pmatrix} + O\left(\left\| \begin{pmatrix} x_1 \\ x_2 \end{pmatrix} \right\|^5 \right) \tag{5.22}$$

where

$$N_\mu \begin{pmatrix} r \\ \theta \end{pmatrix} = \begin{pmatrix} (1+\mu)r - f(\mu)r^3 \\ \theta + \beta(\mu) + g(\mu)r^2 \end{pmatrix}, \tag{5.23}$$

with $f(0) \neq 0$. Notice that the radius of the invariant circle is given by $r = \sqrt{\mu/f(\mu)}$.

THEOREM 5.4

(Neimark-Sacker). *Suppose that F_μ satisfies assumptions 1–3. Then, for sufficiently small μ, F_μ has an invariant closed curve enclosing the origin if $\mu/f(\mu) > 0$. Moreover, if $f(0) > 0$, this curve is attracting, and if $f(0) < 0$, it is repelling.*

PROOF   See [59, 72, 89].   ∎

Exercises - (5.1 and 5.2)

1. Consider the delayed logistic equation

$$x(n+1) = \mu x(n)(1 - x(n-1)).$$

   (a) Change the equation to a two-dimensional system.
   (b) Show that for $\mu > 1$, there exists a nontrivial positive fixed point.
   (c) Show that for $\mu = 2$, the nontrivial fixed point undergoes a Neimark-Sacker bifurcation.
   (d) Draw a phase space diagram for the system for
       i. $\mu < 2$.
       ii. $\mu = 2$.
       iii. $\mu > 2$.

2. Consider the discrete predator–prey system

$$x(n+1) = \alpha x(n)(1 - x(n)) - x(n)y(n)$$
$$y(n+1) = \frac{1}{\beta}x(n)y(n)$$

where $x(n)$ denotes the prey population at generation n and $y(n)$ denotes the predator population at generation n. Show that a nontrivial fixed point of the map undergoes a Neimark-Sacker bifurcation.

3. Discuss the existence of a Neimark-Sacker bifurcation for the Hénon map.

4. Analyze the bifurcation structure of the map in Equation (5.21).

5. Analyze the bifurcation structure of the map

$$r \longmapsto \mu r + \beta r^3$$
$$\theta \longmapsto \theta + \phi + \gamma r^2$$

with $\beta \neq 0$, $\gamma \neq 0$, and $\mu > 0$.

6. Consider the map

$$r \longmapsto r + (d\mu + ar^2)r$$
$$\theta \longmapsto \theta + \phi_0 + \phi_1\mu + br^2.$$

(a) Show that $r = \sqrt{-\mu d/a}$ is an invariant circle.

(b) Show that the invariant circle described in part (a) is asymptotically stable for $a < 0$ and unstable for $a > 0$.

(c) Draw the phase diagrams for the following cases:

   i. $d > 0$, $a > 0$.
   ii. $d > 0$, $a < 0$.
   iii. $d < 0$, $a > 0$.
   iv. $d < 0$, $a < 0$.

In Problems 7–13, compare center manifolds near the origin. Describe the bifurcations of the origin.

7.

$$x \longmapsto -\frac{1}{2}x - y - xy^2$$
$$y \longmapsto -\frac{1}{2}x + \mu y + x^2$$

8.

$$x \longmapsto x^2 + \mu y$$
$$y \longmapsto y + xy$$

9.

$$x \longmapsto -x + y - xy^2$$
$$y \longmapsto y + \mu y^2 + x^2 y$$

10.

$$x \longmapsto 2x + 3y$$
$$y \longmapsto x + x^2 + \mu y^2 + xy^2$$

11.

$$x \longmapsto \mu + \mu x + x^2 - y^2$$
$$y \longmapsto x^3 + y^2$$

12.

$$x \longmapsto \mu + x + x^2 - y^2$$
$$y \longmapsto \mu + x^2 + y^2$$

13.

$$x \longmapsto \mu + \frac{1}{2}x - y - x^2$$
$$y \longmapsto \frac{1}{2}x + y^2$$

14. Consider the map

$$F_\alpha \begin{pmatrix} x \\ y \end{pmatrix} = \begin{pmatrix} \cos\beta & -\sin\beta \\ \sin\beta & \cos\beta \end{pmatrix}$$
$$\left[(1+\alpha) \begin{pmatrix} x \\ y \end{pmatrix} + (x^2 + y^2) \begin{pmatrix} a & -b \\ b & -a \end{pmatrix} \begin{pmatrix} x \\ y \end{pmatrix} \right].$$

Let $z = x + iy$, $\overline{z} = x - iy$, $|z|^2 = z\overline{z} = x^2 + y^2$, $d = a + ib$. Show that the map F_α may be written in the complex form $z \longmapsto \mu z + cz|z|^2$.

15*. [25] Let F_μ be a map in the complex form $z \longmapsto \mu z + \alpha_1 z^2 + \alpha_2 z\overline{z} + \alpha_3 \overline{z}^2 + O(|z|^3)$, $\mu \neq 0$. Show that there exists a diffeomorphism h such that $h^{-1} \circ F_\mu \circ h = G_\mu = \mu z + O(|z|^3)$, provided that μ is not a kth root of unity for $k = 1$ or 3. (*Hint: Use* $h(z) = z + a_1 z^2 + a_2 z\overline{z} + a_3 \overline{z}^2$, *with* $a_1 = \frac{-\alpha_1}{\mu(1-\mu)}$, $a_2 = \frac{-\alpha_2}{\mu(1-\overline{\mu})}$, $a_3 = \frac{-\alpha_3}{\mu-\overline{\mu}^2}$.)

16*. [25] Let F_μ be a map in the complex form $z \longmapsto \mu z + \beta_1 z^3 + \beta_2 z^2 \overline{z} + \beta_3 z\overline{z}^2 + \beta_4 \overline{z}^3 + O(|z|^4)$, $\mu \neq 0$. Show that there exists a diffeomorphism h such that $h^{-1} \circ F_\mu \circ h = G_\mu = \mu z + \beta_2 z^2 \overline{z} + O(|z|^4)$, provided that μ is not a kth root of unity for $k = 2$ or 4. (*Hint: Use* $L(z) = z + b_1 z^3 + b_3 z\overline{z}^2 + b_4 \overline{z}^3$, *with* $b_1 = \frac{-\beta_1}{\mu(1-\mu^2)}$, $b_3 = \frac{-\beta_3}{\mu(1-\overline{\mu}^2)}$, $b_4 = \frac{-\beta_4}{\mu-\overline{\mu}^3}$.)

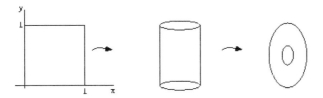

FIGURE 5.7
Construction of a torus.

17*. [25] Let F_μ be the map in the complex form $z \longmapsto \mu z + \beta_2 z^2 \overline{z} + O(|z|^4)$, $\mu \neq 0$. Show that there exists a diffeomorphism h such that $h^{-1} \circ F_\mu \circ h = G_\mu = \mu z + \beta_2 z^2 \overline{z} + O(|z|^5)$, provided that μ is not a kth root of unity for $k = 3$ or 5.

5.3 Hyperbolic Anosov Toral Automorphism

In Chapter 3, we used the double map on the circle S^1 to illustrate one-dimensional chaos. This was due mainly to the simplicity of its dynamics and the transparency of proving the existence of chaos. In this section, we initiate our exploration of two-dimensional chaos by introducing hyperbolic automorphisms on the torus. To construct a torus T, we start with a unit square in $\mathbb{R}^2$. Then, we identify the vertical sides by gluing them together. In the second step, we identify the horizontal sides, which are now in the form of circles, by gluing them together. In other words, a point of the form $(x, 0)$ is identified with the point $(x, 1)$ and a point of the form $(0, y)$ is identified with a point of the form $(1, y)$ (see Fig. 5.7).

Formally, the torus T is obtained from the plane by identifying all points whose coordinates differ by an integer. This defines an equivalence relation on $\mathbb{R}^2$, where $(x_1, y_1) \sim (x_2, y_2)$ if $x_2 - x_1 = n_1$ and $y_2 - y_1 = n_2$, for some $n_1, n_2 \in \mathbb{Z}$. This equivalence relation may be also described by the projection map $\pi : \mathbb{R}^2 \to T$ with $\pi(x, y) = [x, y]$. Clearly, for any pair of integers m and n, $\pi(x + n, y + m) = [x, y]$.

We are now ready to introduce the promised hyperbolic toral automorphism.

DEFINITION 5.1   *Let $A = (a_{ij})$ be a 2×2 matrix with the following properties:*

(a) A is hyperbolic, i.e., its eigenvalues are off the unit circle.

FIGURE 5.8

$$L_A \circ \pi \begin{pmatrix} x \\ y \end{pmatrix} = \pi \circ A \begin{pmatrix} x \\ y \end{pmatrix}.$$

(b) *The entries a_{ij}, $1 \le i$, $j \le 2$, are integers.*

(c) $\det A = \pm 1$.

It follows immediately from (a) and (c) that the eigenvalues of A are real with one eigenvalue inside the unit circle and the other outside the unit circle (Problem 3).

The matrix A induces a map $L_A : T \to T$, such that $L_A \circ \pi = \pi \circ A$. This map is called a <u>hyperbolic toral automorphism</u> (see Fig. 5.8). Since $\det A = \pm 1$, A^{-1} is also hyperbolic with integer entries (Problem 3). Thus A^{-1} induces the hyperbolic toral automorphism $(L_A)^{-1}$.

The map L_A on the torus T may be obtained by multiplying the matrix A times the vector $\begin{pmatrix} x \\ y \end{pmatrix}$, followed by taking the product vector modulus 1. Explicitly, $L_A \begin{pmatrix} x \\ y \end{pmatrix} = A \begin{pmatrix} x \\ y \end{pmatrix} \bmod 1$. Let us illustrate the dynamics of L_A by an example.

Example 5.4

(The Cat Map).   Consider the map C_A on the torus T induced by the matrix $A = \begin{pmatrix} 2 & 1 \\ 1 & 1 \end{pmatrix}$. This map is commonly called the **cat map**, for reasons that will be explained in the sequel. The eigenvalues of A are $\lambda_1 = \frac{3+\sqrt{5}}{2}$ and $\lambda_2 = \frac{3-\sqrt{5}}{2}$. Moreover, $\det A = 1$. Thus, by Definition 5.1, the map C_A on T is a hyperbolic toral automorphism.   ⬚

Let us now contemplate the image of a unit square under A. Note that

$$\begin{pmatrix} 2 & 1 \\ 1 & 1 \end{pmatrix} \begin{pmatrix} 1 \\ 0 \end{pmatrix} = \begin{pmatrix} 2 \\ 1 \end{pmatrix},$$

$$\begin{pmatrix} 2 & 1 \\ 1 & 1 \end{pmatrix} \begin{pmatrix} 0 \\ 1 \end{pmatrix} = \begin{pmatrix} 1 \\ 1 \end{pmatrix}, \qquad \begin{pmatrix} 2 & 1 \\ 1 & 1 \end{pmatrix} \begin{pmatrix} 1 \\ 1 \end{pmatrix} = \begin{pmatrix} 3 \\ 2 \end{pmatrix}.$$

Thus A takes the unit square into a parallelogram as shown in Fig. 5.9(a). In Fig. 5.9(b), we see that under A, the cat is smeared all over the parallelogram, and this is how the map acquired its name.

Note that the only fixed point of C_A is the origin. There are two cycles:

$$\left\{ \begin{pmatrix} 1/5 \\ 2/5 \end{pmatrix}, \begin{pmatrix} 4/5 \\ 3/5 \end{pmatrix} \right\} \text{ and } \left\{ \begin{pmatrix} 2/5 \\ 4/5 \end{pmatrix}, \begin{pmatrix} 3/5 \\ 1/5 \end{pmatrix} \right\}.$$

It is intriguing to observe that

$$A^n = \begin{pmatrix} F_{2n} & F_{2n-1} \\ F_{2n-1} & F_{2n-2} \end{pmatrix}$$

where F_n denotes the Fibonacci sequence with $F_0 = F_1 = 1$ (Problem 2). Moreover, the number of k-periodic points of C_A is given by

$$|(F_{2n} - 1)(F_{2n-2} - 1) - F_{2n-1}^2|. \tag{5.24}$$

After delving into the cat map as a warm-up, we now focus our attention on the general properties of hyperbolic toral automorphisms. The main results of this section follow.

THEOREM 5.5
Let $L_A : T \to T$ be a hyperbolic toral automorphism on the torus T. Then L_A is chaotic.

PROOF To prove that L_A is chaotic, it suffices, by virtue of Theorem 3.4, to show that the set of periodic points of L_A is dense in T, and L_A is topologically transitive. We complete the proof by establishing the following results.

LEMMA 5.1
The set of periodic points of L_A is dense in T.

PROOF For each positive integer n, let

$$U_n = \{ [i/n, j/n] \in T : 0 \le i, \ j < n, \ i, j \in \mathbb{Z}^+ \}.$$

Then U_n has n^2 points. Moreover, since the entries of A are integers, $L_A(U_n) \subset U_n$. Hence, for $x \in U_n$, there are positive integers $r < s$ such that $L_A^r(x) = L_A^s(x)$ and $|r - s| \le n^2$. This implies that $L_A^{s-r}(x) = x$; consequently, x is periodic with period less than or equal to n^2. Observe that

$$U = \bigcup_{n=1}^{\infty} U_n \tag{5.25}$$

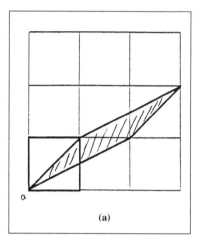

(a)

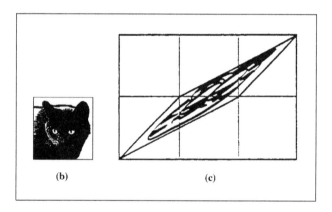

(b)   (c)

FIGURE 5.9

(a) Image of a unit square under the map A.   (b) A cat in a unit square.
(c) Image of a cat under the map A.

is dense in T (Problem 4).   ■

LEMMA 5.2
The map L_A is topologically transitive.

PROOF   See the Appendix at the end of the chapter.   ■

Exercises - (5.3)

1. Let $f_A : T \to T$ be the map induced by the matrix $A = \begin{pmatrix} 2 & 0 \\ 0 & 2 \end{pmatrix}$.

   (a) Show that the map f_A fails to be a diffeomorphism (*Hint: Show that f is noninvertible.*)

   (b) Prove that the set of periodic points of f_A is dense in T.

   (c) Show that the set of eventually fixed points of f_A is also dense in T.

   (d) Show that f is chaotic on T.

2. Let $A = \begin{pmatrix} 2 & 1 \\ 1 & 1 \end{pmatrix}$ be the matrix which induces the cat map C_A. Show that

$$A^n = \begin{pmatrix} F_{2n} & F_{2n-1} \\ F_{2n-1} & F_{2n-2} \end{pmatrix}$$

   where F_n is the Fibonacci sequence defined by $F_{n+1} = F_n + F_{n-1}$, with $F_0 = F_1 = 1$.

3. Let A be a 2×2 matrix that satisfies the assumptions in Definition 5.1.

   (a) Show that A has two real eigenvalues λ_1 and λ_2 such that $|\lambda_1| < 1$ and $|\lambda_2| > 1$.

   (b) Show that A^{-1} induces a toral automorphism $L_{A^{-1}} = (L_A)^{-1}$.

4. Prove that the set U defined by Formula (5.25) is dense in T.

5. Let L_A be a hyperbolic toral automorphism on the torus T. Then, a point $u \in T$ is called **nonwandering** if for every open neighborhood G of u there exists a positive integer n such that $L_A(G) \cap G \neq \phi$. Prove that every point in T is nonwandering.

6. A map f on a metric space X is called expansive if there is $\varepsilon > 0$ such that for every pair of points $x, y \in X$ there is an integer k such that $d(f^k(x), f^k(y)) > \varepsilon$. Show that a hyperbolic toral automorphism is expansive.

7. Given a map $f : T \to T$, there is a lift $F : \mathbb{R}^2 \to \mathbb{R}^2$ such that $\pi \circ F = f \circ \pi$ and $F(x+m) = F(x)+m$ for all $x \in \mathbb{R}^2$, $m \in \mathbb{Z} \times \mathbb{Z}$. If $f, g : T \to T$ are two C^1 maps with lifts $F, G : \mathbb{R}^2 \to \mathbb{R}^2$, then we define the $d_1(f, g) = \sup\{d(f \circ \pi(x), g \circ \pi(x)), \|F'(x) - G'(x)\| : x = (x_1, x_2) \in \mathbb{R}^2$, with $0 \le x_1, x_2 \le 1\}$.

   (a) Show that d_1 is a metric on the set of C^1 maps on the torus T.

   (b) A C^1 diffeomorphism f on T is said to be structurally stable if there is a ball $B_\delta(f)$ such that every $g \in B_\delta(f)$ is topologically conjugate to f. Prove that every hyperbolic toral automorphism is structurally stable.

8. A point x is recurrent for f if for any open neighborhood U of x there exists a positive integer n such that $f^n(x) \in U$. Prove that homoclinic points of a hyperbolic toral automorphism are not recurrent.

5.4   Symbolic Dynamics

In Chapter 3 we encountered the sequence space $\sum_2^+$ of all one-sided sequences of the form $x_1 x_2 \ldots$, where x_i is either 0 or 1. It was shown that the shift map σ on $\sum_2^+$ is chaotic. This, in turn, was used to show, via conjugacy, that several one-dimensional maps are chaotic. In an analogous fashion, we will introduce the space $\sum_2$ of all two-sided sequences $\ldots x_{-3} x_{-2} x_{-1} . x_0 x_1 x_2 x_3 \ldots$, where the decimal point separates both sides of the sequence. The space $\sum_2$ will be equipped with the metric defined by

$$d(x, y) = \sum_{i=-\infty}^{\infty} \frac{|x_i - y_i|}{2^{|i|}} \qquad (5.26)$$

for each $x, y \in \sum_2$.

It is easy to show that d is indeed metric on $\sum_2$. As in Sec. 3.7, one may easily show that if $d(x, y) < \frac{1}{2^n}$, then $x_i = y_i$ for $|i| \le n$. Conversely, if $x_i = y_i$ for $|i| \le n$, then $d(x, y) \le \frac{1}{2^{n-1}}$.

The shift map $\sigma : \sum_2 \to \sum_2$ moves the decimal point one place to the right. For instance, if $x = \ldots x_{-3} x_{-2} x_{-1} . x_0 x_1 x_2 x_3 \ldots$, then $\sigma(x) = \ldots x_{-3} x_{-2} x_{-1} x_0 . x_1 x_2 x_3 \ldots$

Note that there are two fixed points of σ, the doubly repeated sequence $\ldots \overline{0}.\overline{0} \ldots$ and $\ldots \overline{1}.\overline{1} \ldots$ The two cycles of σ are $\ldots \overline{10}.\overline{10} \ldots$, $\ldots \overline{01}.\overline{01} \ldots$ An n-cycle is of the form $\ldots \overline{x_{-n} \ldots x_{-2} x_{-1}.x_0 x_1 \ldots x_{n-1}} \ldots$, where $x_{-n} = x_0, \ldots, x_{-1} = x_{n-1}$.

LEMMA 5.3
The shift map $\sigma : \sum_2 \to \sum_2$ is a homeomorphism.

PROOF   The proof is similar to that of Lemma 3.6 and will hence be omitted. ∎

After the preceding preliminary work we are now ready to prove one of the main results in this section.

THEOREM 5.6
The shift map $\sigma : \sum_2 \to \sum_2$ is chaotic on $\sum_2$.

PROOF   To prove that σ is chaotic on $\sum_2$, then, according to Theorem 3.4, it suffices to show that it is transitive and the set of periodic points of σ is dense in $\sum_2$. In order to prove transitivity, we need to construct a dense orbit in $\sum_2$. For this purpose we select the two-sided sequence $\tilde{x}$, which has the form

$$\ldots 0011\ 0010\ 0001\ 0000\ 11\ 10\ 01\ 00.0\ 1\ 000\ 001\ 010\ 011\ 100\ 101\ 110\ 111 \ldots .$$

In the right side of the sequence $\tilde{x}$, for each positive integer n all possible n-tuples appear in a specific order, whereas in the left side of the sequence $\tilde{x}$, for each positive even integer n all possible n-tuples appear in the same backward order. We leave it to the reader to show that indeed the orbit of $\tilde{x}$ is dense in $\sum_2$. To show that the set of periodic points of σ is dense in $\sum_2$, let $x = \ldots x_{-3} x_{-2} x_{-1}.x_0 x_1 x_2 \ldots$ be an arbitrary sequence in $\sum_2$. Then, for each positive integer n, the sequence $\hat{x}$ given by

$$x_{-n} \ldots x_{-2} x_{-1} \underbrace{x_0 x_1 x_2 \ldots x_n}\ x_{-n} \ldots x_{-2} x_{-1}.\underbrace{x_0 x_1 x_2 \ldots x_n}\ x_{-n} \ldots$$

is of period $2n + 1$. Moreover, $d(x, \hat{x}) \leq \frac{1}{2^{n-1}}$, which implies that the set of periodic points of σ is dense in $\sum_2$. Hence, σ is chaotic on $\sum_2$. ∎

5.4.1   Subshifts of Finite Type

In Chapter 3, we have discussed $\sum_2^+$ and its extension $\sum_N^+$, the space of all one-sided sequences of positive integers between 1 and N. This sequence space

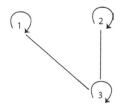

FIGURE 5.10
Graph of the partition in Example 5.5.

has a metric defined for $x = (x_0 x_1 x_2 \ldots)$ and $y = (y_0 y_1 y_2 \ldots)$ by

$$d_N(x, y) = \sum_{i=0}^{\infty} \frac{|x_i - y_i|}{N^i}.$$

Let $A = a_{ij}$ be an $N \times N$ matrix such that $a_{ij} = 0$ or 1 for all i and j, $\sum_j a_{ij} \geq 1$ for all i, and $\sum_i a_{ij} \geq 1$ for all j. Define

$$\sum_{A}^{+} = \{x \in \sum_{N}^{+} : a_{x_i x_{i+1}} = 1 \ \text{ for all } \ i \in \mathbb{Z}^+\}.$$

The matrix A is called the transition matrix. It generates a distinctive subspace $\sum_{A}^{+}$ of $\sum_{N}^{+}$. Moreover, the shift map $\sigma_A : \sum_{A}^{+} \to \sum_{A}^{+}$, which is the restriction of the shift map σ on $\sum_{N}^{+}$, is called a subshift of finite type for the matrix A.

Example 5.5
    Let

$$A = \begin{pmatrix} 1 & 0 & 0 \\ 0 & 1 & 1 \\ 1 & 0 & 1 \end{pmatrix}.$$

Then, $\sum_{A}^{+}$ consists of all sequences of 1's, 2's, and 3's such that 1 may follow 1, 2 may follow 2, 3 may follow 3, 3 may follow 2, but not vise versa, 1 may follow 3, but not vise versa. Fox example,

$$111\ldots, 222\ldots 333\ldots 111\ldots, 333\ldots 111\ldots, 222\ldots, 333\ldots$$

are all the types of sequences in $\sum_{A}^{+}$. A graphical representation is given in Fig. 5.10.   ▯

    Another interesting fact about σ_A is that we can compute the number of k-cycles of σ_A from knowing the trace of A^k, which will be denoted by $\mathrm{tr}(A^k)$, where $\mathrm{tr}(A^k)$ is the sum of diagonal entries of A^k. Moreover, it turns out that $\text{tr}(A^k)$ obeys the famous Fibonacci difference equation. These topics and more will be the focus of our discussion.

LEMMA 5.4
Let A be an $N \times N$ transition matrix. Then, the number of k-cycles of σ_A is equal to $\text{tr}(A^k)$.

PROOF   We first observe that $\sigma_A^k(x) = x$ for $x \in \Sigma_A^+$ if and only if x is in the form $(\overline{x_0 x_1 \ldots x_{k-1}} \ldots)$ such that $a_{x_0 x_1} = a_{x_1 x_2} = \cdots = a_{x_{k-1} x_0} = 1$. Hence, the product $a_{x_0 x_1} a_{x_1 x_2} \ldots a_{x_{k-1} x_0} = 1$ if and only if the string $x_0 x_1 \ldots x_{k-1}$ is a piece of a sequence in Σ_A^+, and equal to zero otherwise. This implies that

$$\sum_{x_0, x_1, \ldots, x_{k-1}} a_{x_0 x_1} a_{x_1 x_2} \ldots a_{x_{k-1} x_0} = \text{number of } k\text{-cycles of } \sigma_A$$

where the sum is taken over all positive values x_i between 1 and N. It may be shown that this sum is actually equal to $\text{tr}(A^k)$.   ∎

In general, σ_A does not have chaotic properties on Σ_A^+. However, for certain transition matrices, σ_A is transitive on Σ_A^+. Now, we introduce this type of transition matrices.

DEFINITION 5.2   *An $N \times N$ transition matrix A is **reducible** if there exists a pair i, j such that $(A^k)_{ij} = 0$ for all $k \geq 1$. Otherwise, A is said to be **irreducible**. In other words, A is irreducible if there is an allowable string $x_0 x_1 \ldots x_{N-1}$ from i to j for every pair i, j, that is, the transition from x_{j-1} to x_j is allowable for $j = 1, \ldots, N$, or $a_{x_{j-1} x_j} = 1$, $j = 1, \ldots, N$.*

For example, the matrix A in Example 5.5 is reducible because it is not possible to go from 1 to 2. The matrix

$$A = \begin{pmatrix} 1 & 0 & 1 \\ 0 & 1 & 1 \\ 1 & 1 & 0 \end{pmatrix}$$

is irreducible as you may see in Fig. 5.11.

Recall that a matrix $A = (a_{ij})$ is positive if $a_{ij} > 0$ for all i, j and eventually positive if the entries of A^k are positive for some $k > 1$. It follows that positive and eventually positive matrices are irreducible. The following result is key to our interest in irreducible matrices.

THEOREM 5.7
For a transition matrix A, the following are equivalent.

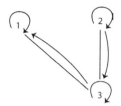

FIGURE 5.11

Graph of the partition for the matrix $A = \begin{pmatrix} 1 & 0 & 1 \\ 0 & 1 & 1 \\ 1 & 1 & 0 \end{pmatrix}$.

1. *A is irreducible.*

2. *σ_A is (topologically) transitive on $\sum_A^+$.*

PROOF   The proof will be left to the reader as Problem 14.   ∎

Exercises - (5.4)

1. Without appealing to Theorem 5.6, prove that $\sigma : \sum_2 \to \sum_2$ possesses sensitive dependence on initial conditions.

2. Show that the sequence $\tilde{x}$, given in the proof of Theorem 5.6, has a dense orbit.

3. Prove Lemma 5.3.

In Problems 4–7, use the following information:
For a sequence $x \in \sum_2$, define the stable set $W^s(x)$ and the unstable set $W^u(x)$ of x as follows:

$$W^s(x) = \{y : |\sigma^n(x) - \sigma^n(y)| \to 0 \text{ as } n \to \infty\}, \text{ and}$$

$$(5.27)$$

$$W^u(x) = \{y : |\sigma^n(x) - \sigma^n(y)| \to 0 \text{ as } n \to -\infty\}.$$

4. Let $z \in \sum_2$. Prove that $W^s(z)$ consists of all sequences whose terms agree with z to the right of some term in z.

5. A sequence $x \in \sum_2$ is said to be homoclinic to the sequence $0 = \ldots 000.000 \ldots$ if $x \in W^s(0) \cap W^u(0)$. Prove that the set of homoclinic sequences to 0 is dense in $\sum_2$.

6. A sequence $x \in \sum_2$ is said to be heteroclinic if $x \in W^s(0) \cap W^s(1)$, where $1 = \ldots 111.111\ldots$. Prove that the set of heteroclinic sequences is dense in $\sum_2$.

7. Define a map $h : \sum_2^+ \to \sum_2$ by $h(x_0x_1x_2 \ldots) = \ldots x_5x_3x_1.x_0x_2x_4 \ldots$. Prove that h is a homeomorphism.

8. Draw the graph of partition for the following matrices and then determine their reducibility.

   (a) $\begin{pmatrix} 1&1&0&0 \\ 1&1&1&0 \\ 1&0&1&1 \\ 0&0&1&0 \end{pmatrix}$

   (b) $\begin{pmatrix} 1&1&0&0 \\ 1&1&1&0 \\ 0&0&1&1 \\ 0&0&1&0 \end{pmatrix}$

9. Suppose the ij entry of A^k, where A is a transition matrix is b_{ij}. Show that there are b_{ij} allowable strings of length $k+1$ starting at i and ending at j, that is, $ix_1x_2 \ldots x_{k-1}j$.

10. A transition matrix is called a permutation if the sum of each row is 1. Show that a transition matrix $A = (a_{ij})$ is a permutation if and only if $\sum_j a_{ij} = 1$ for all i.

11. Let $A = \begin{pmatrix} 0&1 \\ 1&1 \end{pmatrix}$. Find all fixed points of σ_A, σ_A^2, σ_A^3, and σ_A^4. Give the least period of each cycle.

12. (a) Show that $\mathrm{tr}(A^{k+2}) = \mathrm{tr}(A^{k+1}) + \mathrm{tr}(A^k)$, where A is the matrix given in Problem 8.

    (b) Find a formula for the trace of B^k, where $B = \begin{pmatrix} 1&1&0 \\ 1&1&0 \\ 0&0&1 \end{pmatrix}$.

13. Let $F : \mathbb{R} \to \mathbb{R}$ be a C^1 function. Suppose that we have k closed and bounded intervals $I_1, I_2, \ldots, I_k$ and $M > 1$ such that $|F'(x)| \geq M$ for all $x \in \bigcup_{i=1}^k I_i = I$ and $F(I_i) \supset I_j$ whenever $F(I_i) \cap I_j \neq \phi$. Let A be the matrix of the subshifts of finite type defined by $a_{ij} = 1$ if $F(I_i) \supset I_j$ and $a_{ij} = 0$ if $F(I_i) \cap I_j = \phi$. Assuming that A is transitive and irreducible, show that F_{1_Λ} is conjugate to the subshift of finite type σ_A on $\sum_A$, where $\Lambda = \bigcup_{i=1}^k F^{-i}(I)$.

14.* Prove Theorem 5.7. (Hard)

15. (Project) Let A be an $N \times N$ transition matrix and let $\sum_A$ be the set of all two-sided sequences of 1's, 2's, and 3's. Develop parallel results such as Lemma 5.4 and Theorem 5.7.

5.5 The Horseshoe and Hénon Maps

In 1967, Steven Smale [98] introduced a map, which is often called Smale's horseshoe map. It will be the first example of a chaotic map that is topologically conjugate to the double-sequence space $\sum_2$. The map may be described geometrically as follows. We start with a square S with side length 1. Then, we attach to S two semicircles D_1 and D_2, to make up a region D (see Fig. 5.12). The horseshoe map $H : D \to D$ may be described as follows: contract S in the y direction by a factor of $\delta < \frac{1}{2}$ and expand it in the x direction by a factor of $\frac{1}{\delta}$.

The regions D_1 and D_2 are contracted to small semicircular regions with radius $\frac{1}{2}\delta$ and will be mapped inside the region D_1. The map H can be seen as a composition of two functions H_1 and H_2. The first map H_1 stretches out the region D to a region $\tilde{D}$ that is more than twice in length and is less than half in width. This is followed by the second map H_2, which folds or bends the region $\tilde{D}$ in the middle so that it crosses the square S twice. The main interest in the horseshoe map H is to describe its dynamics on the attractor:

$$\Lambda = \{X \in S : H^n(X) \in S \text{ for all } n \in \mathbb{Z}\}.$$

To facilitate our task, we first consider the set

$$\Lambda^+ = \{X : H^n(X) \in S \text{ for all } n \in \mathbb{Z}^+\}.$$

Note that in order for the positive orbit of X, $O^+(X)$, to be in S, X must belong to either V_0 or V_1. Now if $H^2(X) \in S$, then clearly $H(X) \in V_0 \cup V_1$ or $X \in H^{-1}(V_0) \cup H^{-1}(V_1)$. Since h preserves horizontal and vertical rectangles $H^{-1}(V_0)$ consists of two vertical rectangles V_{00} and V_{01}, each of width δ. Similarly, $H^{-1}(V_1)$ consists of two smaller vertical rectangles V_{11} and V_{10} each of width δ. Note that $X \in V_{ij}$ if $X \in V_i$ and $H(X) \in V_j$. We conclude that Λ^+ is the product of a Cantor set with a vertical interval.

Next, we consider the set

$$\Lambda^- = \{X : H^n(X) \in S \text{ for all } n \in \mathbb{Z}^-\}.$$

Observe first that in order for the negative orbit of X, $O^-(X)$, to be in S, u must belong to either $E_0 = H(V_0)$ or $E_1 = H(V_1)$. Now, if $H^{-1}(X) \in E_0 \cup E_1$, then $X \in H(E_0) \cup H(E_1)$.

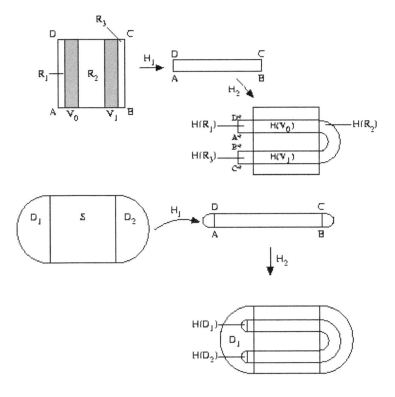

FIGURE 5.12
The construction of the horseshoe map $H = H_2 \circ H_1$.

Notice that $H(E_0) \cap S$ consists of two horizontal strips of width δ^2. Analogously, $H(E_1) \cap S$ consists of two horizontal strips of width δ^2 (see Fig. 5.13). One may conclude from this that Λ^- is the product of a Cantor set with a horizontal interval. Moreover, $\Lambda = \Lambda^+ \cap \Lambda^-$.

For each $X \in \Lambda^+$ we associate the forward sequence $x_0 x_1 x_2 \ldots$, where

$$x_n = \begin{cases} 0 \text{ if } H^n(X) \in V_0 \\ 1 \text{ if } H^n(X) \in V_1 \end{cases}.$$

Note that each forward sequence corresponds to a whole vertical line in T (see Fig. 5.14).

Similarly, we assign the backward sequence $\ldots x_{-3} x_{-2} x_{-1}$ to each $X \in \Lambda^-$, where

$$x_{-n} = \begin{cases} 0 \text{ if } H^{-n}(X) \in E_0 \\ 1 \text{ if } H^{-n}(X) \in E_1 \end{cases}.$$

Analogously, each backward sequence corresponds to a whole horizontal line in S (see Fig. 5.14). Now, combining the forward and backward sequences, we

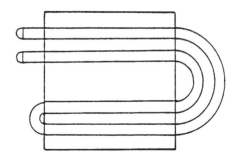

FIGURE 5.13
The graph of $H^2(S)$.

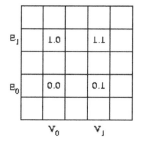
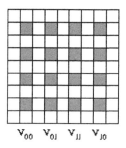

FIGURE 5.14
$H^{-1}(V_0)$ consists of two vertical rectangles V_{00} and V_{01}; $H^{-1}(V_1)$ consists of two vertical rectangles V_{11} and V_{10}.

obtain the two-sided sequence $\ldots x_{-3}x_{-2}x_{-1}.x_0x_1x_2\ldots$ which is a unique sequence in the sequence space $\sum_2$. By this we have already defined a conjugacy $g : \Lambda \to \sum_2$. We are now going to show that g is in fact a homeomorphism as we will demonstrate in the sequel.

THEOREM 5.8
$g : \Lambda \to \sum_2$ *is a homeomorphism.*

PROOF   We first show that g is one to one. So let $X_1, Y_2 \in \Lambda$ with $g(X_1) = g(Y_2)$. This implies that X_1 and Y_2 have the same forward and backward sequence representation, and consequently they must be on the same vertical and horizontal line. Hence, $X_1 = Y_2$ and g is one to one. Next, we show that g is onto. Let $x = \ldots x_{-3}x_{-2}x_{-1}.x_0x_1x_2\ldots$ be a sequence in $\sum_2$. We need to produce a point $U \in \Lambda$ such that $g(U) = x$. To accomplish this task, we define for each $n \in \mathbb{Z}$ the set

$$A_n = \{v \in V_0 \cup V_1 : g(v) = \ldots y_{-3}y_{-2}y_{-1}.y_0y_1y_2\ldots$$

with

$$y_0y_1y_2\ldots y_n = x_0x_1x_2\ldots x_n\}.$$

Note that each set A_n is closed. Moreover, $\bigcap_{n\geq 0} A_n$ is a vertical line in S and $\bigcap_{n<0} A_n$ is a horizontal line in S. Hence, $\bigcap_{n\in\mathbb{Z}} A_n$ is a single point U for which $g(U) = x$. Thus, g is onto. We leave it to the reader to show that both g and g^{-1} are continuous (Problem 1).   ∎

COROLLARY 5.1
The Horseshoe map H *is chaotic on* Λ.

PROOF   We observe that the map g establishes a topological conjugacy between the horseshoe map H and the shift map σ (Fig. 5.15). For if $g(U) = \ldots x_{-3}x_{-2}x_{-1}.x_0x_1x_2\ldots$, then $g(H(U)) = \ldots x_{-2}x_{-1}x_0.x_1x_2x_3\ldots = \sigma(g(U))$. Since σ is chaotic on $\sum_2$ (Theorem 3.8), it follows by Theorem 3.9 that H is chaotic on Λ.   ∎

REMARK 5.3   Observe that since H is a contraction on D_1, it follows by the contraction mapping principle (Theorem 6.1) that H has a unique fixed point $W \in D_1$. Moreover, $\lim_{n\to\infty} H^n(U) = W$ for all $U \in D_1$. Now, since $H(D_2) \subset D_1$, $\lim_{n\to\infty} H^n(U) = W$ for all $U \in D_2$. Moreover, if $H^n(U) \notin S$ for some $U \in S$, then $\lim_{n\to\infty} H^n(U) = W$.   ∎

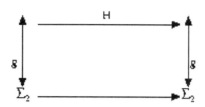

FIGURE 5.15
Topological conjugacy $\sigma \circ g = g \circ H$.

5.5.1   The Hénon Map

Example 5.6
(The Hénon Map). Consider the Hénon map

$$H_{ab}\begin{pmatrix} x \\ y \end{pmatrix} = \begin{pmatrix} 1 - ax^2 + y \\ bx \end{pmatrix}.$$

As we have seen in Chapter 4, the Hénon map has two fixed points if $a > -\frac{1}{4}(1-b)^2$. These fixed points are

$$u_1^* = \left(\frac{1}{2a}(b-1+\sqrt{(1-b)^2+4a}), \ \frac{b}{2a}(b-1+\sqrt{(1-b)^2+4a}) \right)^T,$$

$$u_2^* = \left(\frac{1}{2a}(b-1-\sqrt{(1-b)^2+4a}), \ \frac{b}{2a}(b-1-\sqrt{(1-b)^2+4a}) \right)^T.$$

Recall that for $|b| < 1$ and $a \in (-\frac{1}{4}(1-b)^2, \frac{3}{4}(1-b)^2) = I$, the fixed point u_1^* is asymptotically stable while u_2^* is a saddle. Now at the left end point $a_1 = -\frac{1}{4}(1-b)^2$ of the parameter interval I, we have $u_2^* = u_1^* = \left(\frac{-(1-b)}{2a}, \frac{-b(1-b)}{2a} \right)$. Moreover, the Jacobian matrix $J = D_u H(u_1^*, a_1) = \begin{pmatrix} 1-b & 1 \\ b & 0 \end{pmatrix}$ has an eigenvalue equal to $+1$. Hence, by the center manifold theorem, there is a saddle node bifurcation at u_1^*.

On the other hand, at the right end point $a_2 = \frac{3}{4}(1-b)^2$ of I, we have $u_1^* = \left(\frac{(1-b)}{2a}, \frac{b(1-b)}{2a} \right)^T$. Furthermore, the Jacobian matrix $J = D_u H(u_1^*, a_2) = \begin{pmatrix} -(1-b) & 1 \\ b & 0 \end{pmatrix}$ has an eigenvalue equal to -1. Again, by the center manifold theorem, we have a period-doubling bifurcation. For a fixed b, we may plot a bifurcation diagram showing the x components of an orbit and the parameter a on the x axis. Figure 5.16 shows the bifurcation diagram of the Hénon map for $a \in [0, 1.4]$, and $b = 0.3$. We see a 4-cycle going to a chaotic region and then when $a = 2$ we have two pieces of a chaotic attractor [2]. ☐

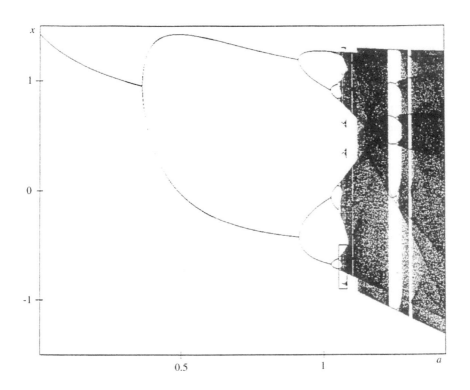

FIGURE 5.16
The bifurcation diagram of the Henon map for a fixed $b = 0.3$.

In this section, we turn our attention to the existence of a horseshoe for the Hénon map. The main result, due to Devaney and Nitecki [27], asserts the existence of a horseshoe for large values of a.

THEOREM 5.9

Assume that $a \geq (5 + 2\sqrt{5})(1 + |b|^2)/4$, $b \neq 0$ and let $\ell = \frac{1}{2}[1 + |b| + ((1 + |b|)^2 + 4a)^{\frac{1}{2}}]$ be the side length of a square $S = \{(x, y) : |x| \leq \ell, |y| \leq \ell\}$.

Define $\Lambda := \bigcap_{i=-\infty}^{\infty} H^i(S)$. *Then, the following hold true:*

1. Λ *is a cantor set.*

2. H *has a hyperbolic structure on* Λ.

3. $H_{|\Lambda}$ *is topologically conjugate to the two-sided shift* $\sum_2$. *Thus, H has an embedded horseshoe.*

PROOF   The interested reader may find the proof of this theorem in [27].

REMARK 5.4   In 1995, Brown [14] discussed the existence of a horseshoe in the Hénon map for the case when $b = -1$, i.e., when

$$H \begin{pmatrix} x \\ y \end{pmatrix} = \begin{pmatrix} 1 - ax^2 + y \\ -x \end{pmatrix}.$$

In this case, H is area preserving and orientation preserving as well. However, before introducing his results, we need to introduce the notion of transverse homoclinic points. Let x^* be a fixed point saddle of a map f. Then a point p is said to be a **transverse** homoclinic point for x^* if $p \in W^s(x^*) \cap W^u(x^*)$ and $W^s(X^*)$ and $W^u(x^*)$ cross-transversally, i.e., the two manifolds intersect with a positive angle between them, where $W^s(x^*)$ and $W^s(x^*)$ are as defined in (5.27). In other words, the angle between the lines tangent to the two manifolds at the point of crossing is nonzero (see Fig. 5.17).   ∎

We are now ready to state a remarkable result by Steven Smale [98].

THEOREM 5.10
Let f be a diffeomorphism (C^1 homeomorphism) on the plane and x^* be a fixed point saddle. If f has a transverse homoclinic point p for x^*, then there is a horseshoe for f.

Using the above result of Smale, Brown was able to establish the following interesting result.

THEOREM 5.11
If $a > 0$ and $b = -1$, then H has an embedded horseshoe.

The main idea in the proof is to show the existence of a transverse homoclinic point for the Hénon map H. Then the result follows immediately by Smale's theorem.

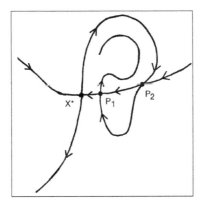

FIGURE 5.17

P_1 is a transverse homoclinic point, while P_2 is not a transverse homoclinic point since the angle between $W^s(P_2)$ and $W^u(P_2)$ is zero.

Exercises - (5.5)

1. Complete the proof of Theorem 5.8 by showing that both g and g^{-1} are continuous maps.

2. Show that homoclinic points map to homoclinic points under f and its inverse f^{-1}.

3. Construct the period table for the Hénon map H for periods up to six as follows:

Period k	Number of fixed points of H^k	Number of fixed points of H^k due to lower period orbits	Number of cycles of period k

4. [40] Let the region D in the plane be made up of three disks, A, B, and C, and two strings, S_1 and S_2, connecting them as in Fig. 5.18. The map F takes D inside itself. The disks A, B, and C are permuted with $F(A) \subset B$, $F(B) \subset C$, $F(C) \subset A$. The map F on these sets is a

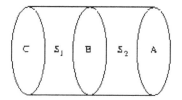

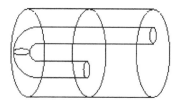

FIGURE 5.18
Exercise 4.

contraction and there is a unique attracting cycle of period 3, $\{\bar{x}_1, \bar{x}_2, \bar{x}_3\}$ with $\bar{x}_1 \in A$, $\bar{x}_2 \in B$, and $\bar{x}_3 \in C$. The strip S_1 is stretched across S_1, B, and S_2 with a contraction in the vertical direction. The strip S_2 is stretched across S_1 (see Fig. 5.18). Let $\Lambda = \bigcap_{i=-\infty}^{\infty} F^i(S_1 \cup S_2)$. Prove that $F|_\Lambda$ is conjugate to the two sided subshift of finite type $\sum_A$ for the transition matrix

$$A = \begin{pmatrix} 1 & 1 \\ 1 & 0 \end{pmatrix}.$$

5. Consider the map defined on a region D as in Fig. 5.19, where F contracts vertical lengths and expands horizontal lengths in the square S as in the case of the Smale horseshoe. Let

$$\Lambda = \{x \in D : F^n(x) \in S \text{ for all } n \in \mathbb{Z}\}.$$

Show that F is (topologically) conjugate to a two-sided subshift of finite type generated by a 3×3 matrix A. What is A?

6. Rework Problem 5 with the map defined geometrically in Fig. 5.20.

In Problems 7–9, use the following information:

Let f be a C^1 map on $\mathbb{R}^2$ and $J_n = Df^n(x_0)$. Let $B = B_1(x_0)$ be the unit disk around x_0. Then $J_n B$ is an ellipse with orthogonal axes of length r_1^n and

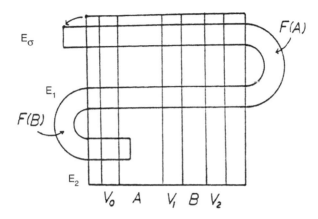

FIGURE 5.19
Exercise 5.

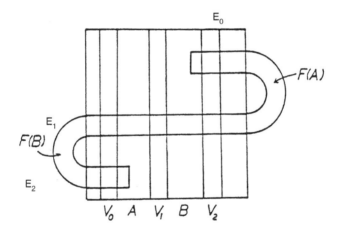

FIGURE 5.20
Exercise 6.

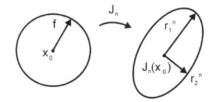

FIGURE 5.21
Exercises 7–10. A disc is mapped under J_n into an ellipse.

r_2^n (see Fig. 5.21). Let r_k^n be the length of the kth longest orthogonal axis of the ellipse $J_n B$.

DEFINITION 5.3   *The kth Liapunov number of x_0 is defined by*

$$L_k = \lim_{n \to \infty} (r_k^n)^{\frac{1}{n}}$$

if the limit exists. The kth Liapunov exponent of x_0 is defined by

$$h_k = \ln L_k.$$

DEFINITION 5.4   *Let $\{x_0, x_1, x_2, \ldots\}$ be the orbit of $x_0 \in \mathbb{R}$, which is assumed to be bounded. Then the orbit is said to be chaotic [2] if*

a) No Liapunov number is exactly 1

b) $h_1(x_0) > 0$

c) The orbit does not converge to a periodic orbit, i.e., it is not asymptotically periodic.

   7. Show that the horseshoe map has chaotic orbits.

   8. Show that the set of chaotic orbits of the horseshoe map is uncountable.

   9. Show that the horseshoe map has a dense chaotic orbit.

   10. Consider the map $f(r, \theta) = (r^2, 2\theta)$ in polar coordinates $[f(z) = z^2$ in the complex domain]. Show that f has chaotic orbits.

5.6   A Case Study: The Extinction and Sustainability in Ancient Civilizations

The following discrete population model was developed in [7] to investigate the population dynamics of Easter Island

$$P(n + 1) = P(n) + aP(n)\left(1 - \frac{P(n)}{R(n)}\right) \tag{5.28}$$

$$R(n + 1) = R(n) + cR(n)\left(1 - \frac{R(n)}{K}\right) - hP(n)$$

where $P(n)$ is the size of the population in year n, and $R(n)$ is the amount of resource, say trees, animals, or crops, present in year n. The parameters a, c,

K, and h represent, respectively, the intrinsic growth rate of the population, the intrinsic growth rate of the resource, the resource carrying capacity, and the harvesting rate. It follows that a, c, h, $K > 0$. The following assumptions have been made in this model.

(a) The resource is governed by a logistic equation in the absence of people.

(b) The per capita consumption of the resource by people does not depend on the amount of the resource. Hence instead of having $-hP(n)R(n)$ in the second equation, we have $-hP(n)$ for the harvesting effect. This is more appropriate if the population has easy access to resources.

(c) The population is governed by a logistic equation with the resources $R(n)$ comprising the carrying capacity of the population.

Using estimations from archeology, the initial population in 400 A.D., $P_0 = 50$ and the number of trees then, $R_0 = 70,000$. Moreover, the parameters are $a = 0.044$, $c = 0.001$, $h = 0.018$, and $K = 70,000$. With these values of the parameters and initial data, model (5.28) is successful as illustrated in Fig. 5.22 which depicts the population predicted by model (5.28) together with data points estimated from archeology.

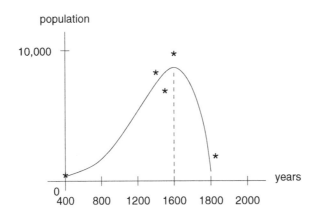

FIGURE 5.22
The graph of population versus time (time-series) using model (5.28). Each "*" is a data point approximated through archeological evidence.

To simplify our notation, we let $K = 1$, which is the same as taking the units of P as percentage of total sustainable population. Hence our model can be written as a two-dimensional map

$$F\begin{pmatrix} P \\ R \end{pmatrix} = \begin{pmatrix} P + aP\left(1 - \frac{P}{R}\right) \\ R + cR(1 - R) - hP \end{pmatrix}. \tag{5.29}$$

Local stability and trace-determinant analysis

The map F has three fixed points $(P_1^*, R_1^*) = (0,0)$, $(P_2^*, R_2^*) = (0,1)$ and $(P_3^*, R_3^*) = \left(1 - \frac{h}{c}, 1 - \frac{h}{c}\right)$. Now the Jacobian matrix of F at a fixed point (P^*, R^*)

$$J = DF\begin{pmatrix} P^* \\ R^* \end{pmatrix} = \begin{pmatrix} 1 + a - \frac{2aP^*}{R^*} & \vdots & \frac{2aP^*}{R^*} \\ \cdots & \cdots & \cdots \\ -h & \vdots & 1 + c - 2cR^* \end{pmatrix}.$$

(i) The fixed point $(P_1^*, R_1^*) = (0,0)$. Notice that the fixed point $(0,0)$ is a singularity of the matrix J since we have the term $0/0$ appearing in its entries. Thus we need a different type of analysis that does not rely on the properties of J. Notice that when $P = 0$ (in the absence of population), Equation (5.28) reduces to

$$F\begin{pmatrix} 0 \\ R \end{pmatrix} = \begin{pmatrix} 0 \\ R + cR(1 - R) \end{pmatrix},$$

which is the one-dimensional map $f(R) = R + cR(1 - R)$ defined on the y-axis. Since $f'(0) = 1 + c > 0$, it follows by Theorem 1.3 that the fixed point 0 is unstable. Hence, the fixed point $(0,0)$ is unstable. This makes sense, since in the absence of a human population, the resource does not go extinct.

(ii) The fixed point

$$(P_2^*, R_2^*) = (0,1) : J|_{(0,1)} = \begin{pmatrix} 1 + a & 0 \\ -h & 1 - c \end{pmatrix}.$$

The eigenvalues of J are $\lambda_1 = 1 + a$ and $\lambda_2 = 1 - c$. Since $\lambda_1 > 1$ and $0 < \lambda_2 < 1$, the fixed point (P_2^*, R_2^*) is an unstable "saddle." The interpretation of this is that when settles, first arrive on the island, their small colony has a chance to grow. Their arrival on the island may be viewed as a perturbation from the fixed point (P_2^*, R_2^*).

(iii) The fixed point

$$(P_3^*, R_3^*) = \left(1 - \frac{h}{c}, 1 - \frac{h}{c}\right) : J|_{(P_3^*, R_3^*)} = \begin{pmatrix} 1 - a & a \\ -h & -c + 2h + 1 \end{pmatrix}.$$

Now $tr\, J = 2 - a - c + 2h$, and $det\, J = h(2 - a) - c(1 - a) + 1 - a = h(2 - a) + (1 - a)(1 - c)$.

Using Fig. 5.3, we make the following conclusions.

(a) The upper side of the stability triangle which corresponds to a Neimark-Sacker bifurcation is given by $det\, A < 1$. Hence

$$h(2 - a) + (1 - a)(1 - c) < 1. \tag{5.30}$$

(b) The right side of the stability triangle which corresponds to a saddle-node (fold) bifurcation is given by $det\ A > tr\ A - 1$. Hence

$$2 - a - c + 2h - 1 < h(2 - a) + (1 - a)(1 - c). \qquad (5.31)$$

(c) The left side of the stability triangle which corresponds to a period-doubling (flip) bifurcation is given by $det\ A > -tr\ A - 1$. Hence

$$-(2 - a - c + 2h) < h(2 - a) + (1 - a)(1 - c). \qquad (5.32)$$

Inequalities (5.30), (5.31), and (5.32) simplify to

$$h(2 - a) + (1 - a)(1 - c) < 1 \qquad (5.30')$$
$$h < c \qquad (5.31')$$
$$(2 - a)(c - h - 2) < 2h. \qquad (5.32')$$

Notice condition (5.31)$'$ has to be assumed in order to have the fixed point, $(P_3^*, R_3^*) = \left(1 - \frac{h}{c}, 1 - \frac{h}{c}\right)$ positive. Now combining (5.30)$'$ and (5.32)$'$ yields

$$\frac{(a - 2)(c - a)}{a - 4} < h < \frac{1 - (1 - a)(1 - c)}{2 - a}. \qquad (5.33)$$

Using (5.33) and (5.31)$'$ we drew Fig 5.23, which shows the stability zone in the $c - h$ plane for (i) $a = 0.0045$, (ii) $a = 0.3$, and (iii) $a = 1$. In (i) for $a = 0.0045$, we have the stability zone defined by $\frac{c}{2} - 1 < h < \frac{c}{2}$. In (ii) for $a = 0.3$, we have the stability zone defined by $\frac{17}{37}c - \frac{51}{370} < h < \frac{7}{17}c + \frac{3}{17}$. And in (iii) for $a = 1$, we have the stability zone defined by $\frac{c}{3} - \frac{1}{3} < h < 1$.

The Neimark-Sacker bifurcation occurs when $det\ J = 1$, $det\ J > tr\ J - 1$ and $det\ J < -tr\ J - 1$. These conditions reduce to

$$det\ J = 1, \quad \text{and} \quad -2 < tr\ j < 2. \qquad (5.34)$$

Thus we have $h(2 - a) + (1 - a)(1 - c) = 1$, and consequently,

$$h = \frac{a + c - ac}{2 - a}. \qquad (5.35)$$

From the second part of (5.34) we have

$$-2 < 2 - a - c + 2h < 2. \qquad (5.36)$$

Substituting (5.37) into (5.36) yields

$$\frac{a + c}{2} - 2 < \frac{a + c - ac}{2 - a} < \frac{a + c}{2}.$$

This may be simplified further to

$$a < c < a - 4 + \frac{8}{a}. \qquad (5.37)$$

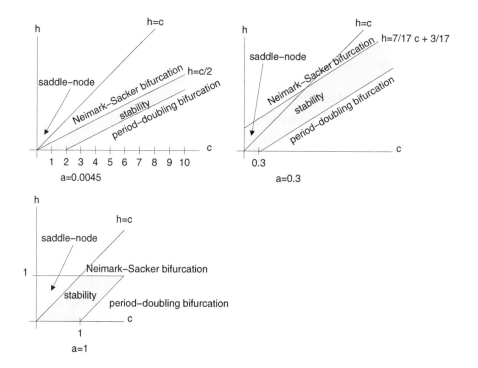

FIGURE 5.23
Stability regions.

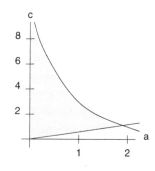

FIGURE 5.24
The shaded area is the region at which Neimark-Sacker bifurcation occurs.

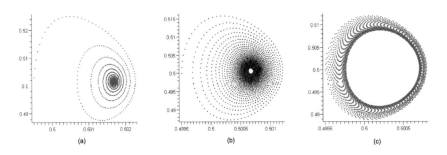

FIGURE 5.25
The Niemark-Sacker invariant closed curves for $a = 0.0045$, $c = 3$, and $h = 1.495$, $h = 1.498$, $h = 1.499$.

Figure 5.23 illustrates the occurrence of the Neimark-Sacker bifurcation in the $a - c$ plane.

The archeological estimation for a in the Easter Island is $a \approx 0.0045$. This leaves a fairly wide window for c at which the Neimark-Sacker bifurcation can be present.

Notice that the fixed point $(R_3^*, P_3^*) = \left(1 - \frac{h}{c}, 1 - \frac{h}{c}\right)$ loses its stability as long as $h = \frac{a+c-ac}{2-a}$ and inequality (5.37) holds. By Theorem 5.4 we have a closed invariant curve around (R_3^*, P_3^*). Figure 5.25 illustrates the phenomenon for $a = 0.0045$, $c = 3$, and with the parameter values of $h = 1.495$, $h = 1.498$, and $h = 1.499$.

CONCLUDING REMARKS

Notice that as h increases (and thus the increase in harvesting), the positive fixed point looses it stability. This makes sense as too much harvesting of the resources in the island would definitely destabilize the population. The only surprise, if any, is that destabilization occurs if the growth rate c increases. Such increase in c may occur through new improved farming techniques, for instance, leading to period-doubling (flip) bifurcation. For resource growth rate c greater than the population growth rate a, increasing the harvesting rate h pushes the positive fixed point to a Neimark-Sacker bifurcation.

Research Projects for Graduate and Undergraduate Students

Project A

Consider the Neimark-Sacker bifurcation depicted in Fig. 5.25.

1. Increase the value of h until the closed invariant curve develops cusps.

2. Keep increasing the value of h and observe that a periodic orbit inside the invariant closed curve undergoes a secondary Neimark-Sacker

bifurcation.

3. Show that beyond a certain parameter value h, a chaotic region appears.

4. Simulate the attractors for the following parameter values:

    (i) $a = 0.3$, $c = 3$, $h = 0.17$

    (ii) $a = 0.3$, $c = 3$, $h = 1.53$

    (iii) $a = 0.3$, $c = 3$, $h = 1.5335$

    (iv) $a = 0.3$, $c = 3.5$, $h = 1.27, 1.278$

    (v) $a = 0.3$, $c = 3.5$, $h = 1.5$

5. Provide a mathematical analysis for the phenomena that occurred in your simulations.

Project B

Study the stability and bifurcation of the predator-prey model

$$x(n + 1) = ax(n)(1 - x(n)) - x(n)y(n) \tag{5.38}$$

$$y(n + 1) = \frac{1}{\beta}x(n)y(n)$$

where $x(n)$ is the population size of the prey and $y(n)$ is the population size of the prey in generation n. This model is obtained from Maynard Smith [99]

$$u(n + 1) = Ru(n) - (R - 1)u^2(n)/u^* - cu(n)v(n) \tag{5.39}$$

$$v(n + 1) = ru(n)v(n)/u^*$$

where $u(n)$ and $v(n)$ are the sizes of the prey and predator populations in year n whose maximum reproductive rates are R and r, respectively, and u^* is the fixed point of $u(n)$ in the absence of $v(n)$. System (5.38) is obtained from (5.39) by letting $x(n) = (R - 1)u(n)/Ru^*$ and $y(n) = cv(n)$.

Project C

Triangular Maps. A map $F : I^2 \to I^2$ on $I^2 = [0, 1] \times [0, 1]$ is called a triangular map if it is of the form $F(x, y) = (f(x), g(x, y))$. Such maps have applications to neural networks. The main question here is which of the properties of f holds also for F.

(a) Prove that if F is continuous and (x, y) is a periodic point of F, then x is a periodic point of f. Moreover, if x is a periodic point of f, then there exists $y \in I$ such that (x, y) is a periodic point of F.

(b) Show that if an orbit of (x, y) under F is dense in I^2, then the orbit of x under f is dense in I.

(c) Consider the map $F(x, y) = (T(x), g(x, y))$ on I^2, where T is our old friend, the tent map,

$$T(x) = \begin{cases} 2x & \text{for } x \le \frac{1}{2} \\ 2(1 - x) & \text{for } x > \frac{1}{2} \end{cases}$$

and

$$g(x, y) = \begin{cases} \left(\frac{1}{2} + x\right) y & \text{for } x \le \frac{1}{2} \\ \left(\frac{3}{2} - x\right) y + \left(x - \frac{1}{2}\right) & \text{for } x > \frac{1}{2}. \end{cases}$$

Show that F is chaotic on I^2 by showing that F is transitive and the set of periodic points is dense in I^2.

Appendix: Topologial Transitivity of the Toral Automorphism

Consider a 2×2 matrix A that satisfies the hypotheses in Definition 5.1. Let λ_s be the eigenvalue of A with $|\lambda_s| < 1$, and λ_u be the eigenvalue of A with $|\lambda_u| > 1$. If V_s and V_u are the corresponding eigenvectors of A, respectively, then the stable $W^s(0)$ and unstable $W_u(0)$ subspaces of A are straight lines through the origin in $\mathbb{R}^2$. Explicitly,

$$W^s(0) = \{tV_s : t \in \mathbb{R}\}, \ W^u(0) = \{tV_u : t \in \mathbb{R}\}.$$

Now, for $[x, y] \in T$, let ℓ_s and ℓ_u be lines in $\mathbb{R}^2$ that intersect at (x, y) and are parallel to $W^s(0)$ and $W^u(0)$, respectively. We define two subspaces $W^s[x, y]$ and $W^u[x, y]$ of the torus T as the projection of these lines in T, i.e.:

$$W^s[x, y] = \pi(\ell_s), \ W^u[x, y] = \pi(\ell_u).$$

In the next lemma, we will show that these two subspaces are indeed the stable and unstable sets of L_A, respectively. In other words, we will show that

$$W^s[x, y] = \{[\tilde{x}, \tilde{y}] : d(L_A^n[\tilde{x}, \tilde{y}], L_A^n[x, y]) \to 0 \text{ as } n \to \infty\}$$
$$W^u[x, y] = \{[\tilde{x}, \tilde{y}] : d(L_A^{-n}[\tilde{x}, \tilde{y}], L_A^{-n}[x, y]) \to 0 \text{ as } n \to \infty\}$$

where d is the distance in T induced by the Euclidean distance in $\mathbb{R}^2$.

Now let $[x, y] \in T$ be a periodic point under L_A. We say that a point $p \ne [x, y]$ is <u>homoclinic</u> to $[x, y]$ if $p \in W^s[x, y] \cap W^u[x, y]$. If, in addition, $W^s[x, y]$ and $W^u[x, y]$ meet at a nonzero angle, then p is said to be a <u>transverse</u> homoclinic point.

Since this is the case for L_A, all homoclinic points here are transverse. It follows from Lemma 5.1 that the set of transverse homoclinic points is dense in T.

LEMMA 5.5
The following statements hold.

1. *For each* $[x,y] \in T$, $W^s[x,y]$ *is the stable set of* L_A *associated to* $[x,y]$ *and* $W^u[x,y]$ *is the unstable set of* L_A, *associated to* $[x,y]$.

2. *For each* $[x,y] \in T$, *the sets* $W^s[x,y]$ *and* $W^u[x,y]$ *are dense in* T.

3. *Transverse homoclinic points are dense in* T.

PROOF

1. Let (x,y) and $(\tilde{x},\tilde{y})$ be two points on a line parallel to W^s in $\mathbb{R}^2$. Let the distance between these two points be Euclidean. Then $|A^n \begin{pmatrix} x \\ y \end{pmatrix} - A^n \begin{pmatrix} \tilde{x} \\ \tilde{y} \end{pmatrix}| = |\lambda_s^n| \to 0$ as $n \to \infty$. This implies that $|L_A^n[x,y] - L_A^n[\tilde{x},\tilde{y}]| \to 0$ as $n \to \infty$. The proof that $W^u[x,y]$ is the unstable set of L_A is similar and will be omitted.

2. We claim that $W^s(0)$ is a line with an irrational slope. For if not, then $W^s(0)$ must pass through a point (k, ℓ), whose coordinates are integers. But, then $A^n \begin{pmatrix} k \\ \ell \end{pmatrix}$ would have integer coordinates which contradicts the fact that $A^n \begin{pmatrix} k \\ \ell \end{pmatrix} \to 0$ as $n \to \infty$.

   Let x_j be the x coordinate of the intersection of the line $y = j$ and $W^s(0)$, $j = 1,\, 2,\, 3,\, \dots$.
   Since the slope of $W^s(0) = \frac{1}{x_1}$ is irrational, it follows that x_1 is irrational. Moreover, $x_n = nx_1$ is also irrational. Note that $\pi(x_j, j) = [\alpha_j, 0]$, $0 \le \alpha_j < 1$. Now the line $y = 0$ defines a circle in T and the α_j's are the successive images of $[0]$ under an irrational translation of this circle. Hence these points are dense in the circle from which the proof easily follows. ∎

PROOF OF LEMMA 5.2   Let U and V be two open sets in T. Let $[r] \in U$ and $[s] \in V$ be two points in T that are homoclinic to $[0]$. Let $\varepsilon > 0$. Choose an open interval I_u of length $\delta > 0$ in $W^u[0]$ and containing $[r]$ and I_s of length $\delta > 0$ in $W^s[0]$ containing $[s]$. Choose n sufficiently large such that

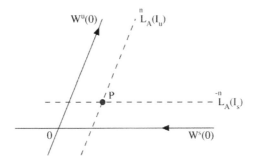

FIGURE 5.26

$W^u(0)$ is parallel to $L_A^n(I_u)$, $W^s(0)$ is parallel to $L_A^{-1}(I_s)$.

1. $d(L_A^n[r],\ [0]) < \varepsilon/2$

2. $d(L_A^{-n}[s],\ [0]) < \varepsilon/2$

3. $|\lambda_u|^n \delta > \varepsilon$

Since $L_A^n(I_u)$ and $L_A^{-n}(I_s)$ are parallel to $W^u[0]$ and $W^s[0]$, respectively, $L_A^n(I_u) \cap L_A^{-n}(I_s) \neq \phi$ (see Fig. 5.26). Let $[q] \in L_A^n(I_u) \cap L_A^{-n}(I_s)$. Then $[p] = L_A^{-n}[q] \in U$ and $L_A^n[q] \in V$. Hence $L_A^{2n}[p] \in V$, which completes the proof. ∎

6

Fractals

Fractal geometry will make you see everything differently. There is danger in reading further. You risk the loss of your childhood vision of clouds, forests, galaxies, leaves, feathers, flowers, rocks, mountains, torrents of water, carpets, bricks, and much else besides. Never again will your interpretation of these things be quite the same.

Michael Barnsley

6.1 Examples of Fractals

The word *fractal* was coined from the Latin adjective *fractus* by Benoit Mandelbrot in 1977 [67]. The corresponding Latin verb *frangere* means "to break," to create irregular fragments.

Before delving into the mathematical foundation of fractals, we will introduce, as a warm-up, some of the more popular examples of fractals. Intuitively, a fractal is a geometrical figure that consists of an identical motif repeating itself on ever-reducing scales, i.e., self-similar. A more specific definition requires more mathematical vocabulary and will be given in later sections.

Example 6.1
(The Sierpinski Triangle or Gasket). Our first example of a fractal is due to the Polish mathematician Waclaw Sierpinski (1882–1969) who introduced it in 1916 [95]. The construction goes as follows: we start with an equilateral triangle with sides of unit length (for simplicity), thought of as a solid object. By connecting the midpoints of the three sides of the triangle, we obtain four smaller equilateral triangles, then remove the middle one. This leaves three smaller equilateral triangles whose sides are of length $\frac{1}{2}$. In the next step, we repeat the same procedure with the three remaining triangles to obtain 9 equilateral triangles with sides of length $(\frac{1}{2})^2$. If we continue this construction, then at the nth stage we will have 3^n equilateral triangles with sides of length $(\frac{1}{2})^n$. Fig. 6.1 illustrates our construction for $n = 1, 2, 3$. $\quad\Box$

Side length = 1
Step 0: Initiator

3^1 triangles: Side length = 1/2
Step1: Generator

3^2 triangles: Side length = (1/2)^2
Step 2

3^5 triangles: Side length = (1/2)^5
Step 5

FIGURE 6.1
Construction of the Sierpinski triangle.

Basic Construction of the Sierpinski Triangle

If we carry out this process indefinitely, then we obtain, as the limit, the Sierpinski triangle or gasket. Note that the Sierpinski gasket is self-similar since every part of it is similar to the whole.

Moreover, the sum of the areas of the parts removed in the construction of the Sierpinski triangle is equal to the area of the original triangle. Let A be the area of the original triangle, the initiator. In the first step, we remove $\frac{1}{4}A$, and in the next step we remove three triangles each of area $\left(\frac{1}{4}\right)^2 A$, etc. Hence, the sum of the areas removed is

$$\frac{1}{4}A + 3\left(\frac{1}{4}\right)^2 A + 3^2\left(\frac{1}{4}\right)^3 A + \cdots = A.$$

Thus, the remaining set, the Sierpinski gasket, has a zero area.

It is worthwhile to mention that the Sierpinski gasket can be produced in many different ways. One such construction is shown in Fig. 6.2. Following [37], the figure in step 0 will be called the **initiator** and the figure in step 1 will be called the **generator**.

Example 6.2
(The Koch Curve). The second popular fractal was introduced by the Swedish mathematician Helge Von Koch in 1904 [105] and is named after him. The **initiator** of the Koch curve is a straight line. The **generator** is obtained by partitioning the initiator into three equal segments. Then, we remove the middle third and replace it with an equilateral triangle as shown in Fig. 6.3.

If the initiator has length 1, then the generator will consist of four line segments, each of length $\frac{1}{3}$. Hence, the total length of the generator is $\frac{4}{3}$. In the second step, each one of the four line segments will act as an initiator and is replaced by the corresponding scaled-down generator. The resulting curve will have 16 line segments, each of length $\left(\frac{1}{3}\right)^2$. Furthermore, the length of the whole curve is $\left(\frac{4}{3}\right)^2$. If we repeat this process indefinitely, the limiting curve is called the Koch curve. It is worth mentioning that the Koch curve was the first example of a curve that is not differentiable anywhere, i.e., a curve without a tangent anywhere. In addition, the length of the Koch curve is ∞ since the length of the curve at the nth stage is $\left(\frac{4}{3}\right)^n$, which clearly goes to ∞ when $n \to \infty$. Moreover, it is undoubtedly self-similar since every part, however small, is, itself, a miniature of the whole. ▯

An interesting variation of the Koch curve is the **Koch snowflake** or island. The initiator this time is an equilateral triangle, where each side represents the initiator in the Koch curve construction (Fig. 6.4). One may show, as in the Koch curve, that the encompassing curve is of infinite length. The Koch snowflake is an example of a finite area encompassed by a curve of infinite

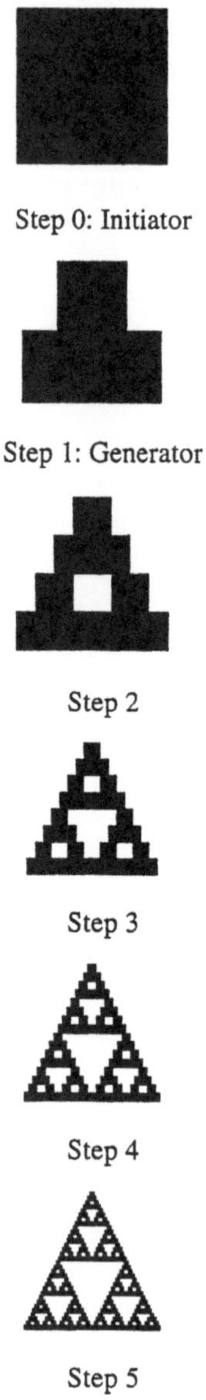

Step 0: Initiator

Step 1: Generator

Step 2

Step 3

Step 4

Step 5

FIGURE 6.2
Another construction of the Sierpinski triangle.

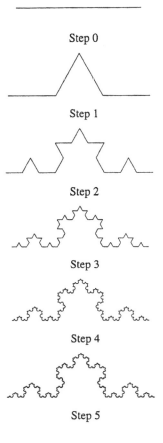

Step 0

Step 1

Step 2

Step 3

Step 4

Step 5

FIGURE 6.3
Construction of the Koch curve.

length. This means that we can paint the inside of the Koch island, but we can never wrap a length of string around its boundary. In Problem 17 you will be asked to find the exact area enclosed by the Koch island.

Example 6.3
(The Pythagorean Tree). A beautiful fractal, called the Pythagorean tree, was discovered in 1957 by the German mathematician A. E. Bosman (1891–1961). We start with a square as our **initiator**. To the top side, we attach an isosceles right triangle. The **generator** is obtained by attaching two squares along the free sides of the triangle, and so forth (Fig. 6.5). ⬚

Let us number the squares as in Fig. 6.5 where a square of index n supports two smaller squares; the one on the left has index $2n$ and the one on the

Step 0

Step 1: Generator

Step 2

Step 3

Step 4

FIGURE 6.4
Construction of the Koch snowflake.

Step 0: Initiator

Step 1: Generator

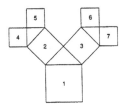

Step 2

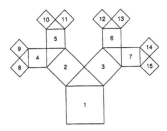

Step 3

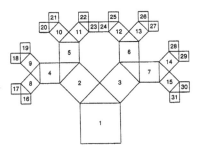

Step 4

FIGURE 6.5
Construction of the Pythagorean tree.

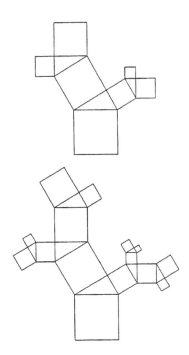

FIGURE 6.6
Construction of Pythagorean pine tree.

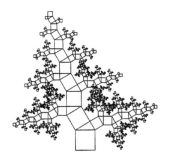

FIGURE 6.7
A pine tree generated as a fractal in 50 steps.

right has index $2n+1$. Using this numbering system, the position of a square depends on the index divisibility by 2. If we write $n = a_r a_{r-1} \ldots a_1$ in the binary system, where $a_i = 0$ or 1, taking 1 to represent right square and 0 to represent left square, we may locate easily the square with index n. For example, $26 = 1 \times 2^4 + 1 \times 2^3 + 0 \times 2^2 + 1 \times 2 + 0 = (1\ 1\ 0\ 1\ 0)_2$ in the binary system. Reading from left to right (the first number 1 represents the first square, the initiator)

$$1(\text{right}), 0(\text{left}), 1(\text{right}), 0(\text{left}).$$

Many variations of the Pythagorean tree may be easily constructed. The Pythagorean pine tree (Fig. 6.6) is constructed by following the squares alternately by a scalene triangle and its mirror image (i.e., we flip the orientation of the triangle after each step).

A Mathematical Curiosity: The Pascal Triangle

Pascal's triangle may be constructed as follows:
  The top row is 1 and will be labeled as row 0.
  The next row is 1 1 and will be labeled as row 1.
  Now each entry in the next row is the sum of the two entries above it (i.e., the one above it and to the right, and the one above and to the left). If there is no number to the left or to the right above the entry, you place a zero there.
  Figure 6.8 shows all rows of the Pascal triangle from row 0 to row 12.
  Pascal's triangle has always seemed to pop up in the strangest places: number theory, probability theory, and polynomial expansion. We will treat the latter, the expansion of $(1+x)^n$.

$$
\begin{array}{llllllll}
(1+x)^0 = & & & & 1 & & & \\
(1+x)^1 = & & & 1 & + & x & & \\
(1+x)^2 = & & 1 & + & 2x & + & x^2 & \\
(1+x)^3 = & & 1 & + & 3x & + & 3x^2 & + & x^3 \\
(1+x)^4 = & 1 & + & 4x & + & 6x^2 & + & 4x^3 & + & x^4 \\
(1+x)^5 = & 1 + 5x & + & 10x^2 & + & 10x^3 & + & 5x^4 & + & x^5
\end{array}
$$

A quick glance at the coefficients of the polynomials, one recognizes at once the Pascal's triangle. And because of this connection, the entries in Pascal's traingle are called the binomial coefficients. Notice that in row 5 $[(x+1)^5]$, the coefficients are

$$\binom{5}{0} = 1, \quad \binom{5}{1} = 5, \quad \binom{5}{2} = 10, \quad \binom{5}{3} = 10, \quad \binom{5}{4} = 5, \quad \binom{5}{5} = 1$$

where $\binom{n}{k} = \frac{n!}{k!(n-k)!}$.

  Let us go back to the Pascal's traingle in Fig. 6.8. Draw a grid square around each integer in the Pascal triangle. Color the grid black if it contains an odd number and leave it uncolored if it contains an even number. The result is what you see in Fig. 6.9, the beginning of Sierpinski's triangle.

```
                              1
                           1     1
                        1     2     1
                     1     3     3     1
                  1     4     6     4     1
               1     5    10    10     5     1
            1     6    15    20    15     6     1
         1     7    21    35    35    21     7     1
      1     8    28    56    70    56    28     8     1
   1     9    36    84   126   126    84    36     9     1
1    10    45   120   210   252   210   120    45    10     1
1    11    55   165   330   462   462   330   165    55    11     1
1    12    66   220   495   792   924   792   495   220    66    12     1
```

FIGURE 6.8

The first 13 rows of Pascal's triangle.

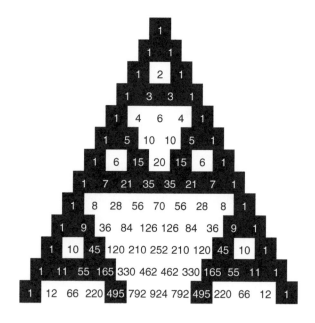

FIGURE 6.9

The beginning of Sierpinski's triangle.

6.2 L-system

In 1968, Aristid Lindenmayer [83] introduced the concept of L systems. He was trying to describe the growth process of living organisms such as branching pattern of plants. The graphical implementation of L-systems is based on "turtle" graphics. The following code words are used in the movement of the turtle.

> F : move one step forward (of a certain fixed length L),
>   drawing the path of motion
> f : the same as F but do not draw the path of motion
> $+$: turn left (counterclockwise) by a fixed angle of θ
> $-$: turn right (clockwise) by the angle of θ

The size of the step length L and the angle θ must be specified before implementing an L-system. In addition, as in IFS, we need to specify an initiator, called *axiom* here, which is the initial figure that we start with. Moreover, the generator in IFS will be called here as *the Production Rule*. Our first example is the classical Koch curve.

Example 6.4
(The Construction of the Koch Curve by an L-system) .
   The following specifications of the basic elements of our construction are as follows.

- $\theta = \frac{\pi}{3}$

- $L = \frac{1}{3}$

- Axiom: F a line segment of a unit length

- Production Rule: $F + F - -F + F$

What the Production Rule is saying is to replace every line (or every F) with the following sequence $F + F - -F + F$. Here is a graphical description of the Production Rule.

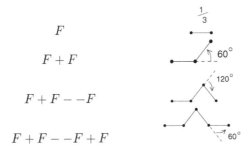

F

$F + F$

$F + F - -F$

$F + F - -F + F$

TABLE 6.1
The construction process of the Koch curve.

Iteration	Fractal Image	Describing String and Rules
Axiom		F ; $\theta = 60°$, $L = \frac{1}{3}$
First Iteration		$F + F - - F + F$
Second Iteration		$F + F - - F + F +$
		$F + F - - F + F - -$
		$F + F - - F + F +$
		$F + F - - F + F$

Table 6.1 shows the iteration process of the L-system of the Koch curve.

□

Example 6.5
(The Peano Curve). Table 6.2 summerizes the construction of the Peano curve.

□

Our last example requires a little more sophistication in our coding. The new codes are:

1. Left square bracket "[" which indicates a branch point.

2. Right square bracket "]" which indicates a point at which the branch is complete.

Example 6.6
(A Simple Bush).
Table 6.3 describes the construction of a bush.

□

6.3   The Dimension of a Fractal

In this section we address the question of defining a fractal. In Sec. 6.1, we have seen that self-similarity is an important feature of fractals. But such a property is shared by nonfractals such as lines, squares, cubes, etc.

TABLE 6.2

The construction of the Peano curve.

Iteration	Fractal Image	Describing String and Rules
Axiom		F ; $\theta = 90°$, $L = \frac{1}{2}$
First Iteration	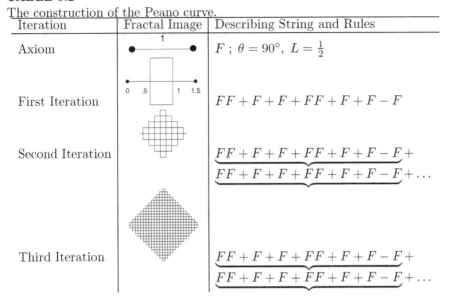	$FF + F + F + FF + F + F - F$
Second Iteration		$FF + F + F + FF + F + F - F +$ $FF + F + F + FF + F + F - F + \cdots$
Third Iteration		$FF + F + F + FF + F + F - F +$ $FF + F + F + FF + F + F - F + \cdots$

Moreover, there are fractals that do not possess the property of self-similarity (see Example 6.8). Hence, we need to find another characteristic of fractals that is not shared by nonfractals. This leads us to Mandelbrot's definition of a fractal. According to Mandelbrot, a fractal is a set whose fractal dimension is strictly greater than its topological dimension. So, we need to know what those dimensions are. In general, a topological dimension of a set agrees with our intuition about dimension of known sets: a smooth curve is one-dimensional, a disc in the plane is two-dimensional, a solid cube is three-dimensional, etc. The formal definition is done inductively as follows.

DEFINITION 6.1   *A set A has a **topological dimension 0** if every point in A has an arbitrarily small neighborhood whose boundary does not intersect A. A set A has a **topological dimension** $k > 0$ denoted by $D_t(A)$, if every point in A has an arbitrarily small neighborhood whose boundary intersects A in a set of topological dimension $k - 1$, and k is the least positive integer for which this holds.*

Sets of topological dimension 0 are in abundance. The most famous set of topological dimension 0 is the Cantor set. Clearly, the set of integers, the set of rational numbers, and the set of irrational numbers are all of topological dimension 0. In the plane, any set that consists of scattered isolated points is of topological dimension 0 (see Fig. 6.10).

Note that a line is of topological dimension 1; so is the circle. The boundary

TABLE 6.3

The construction of a bush.

Iteration	Fractal Image	Describing String and Rules
Axiom		F ; $\theta = 25°$, $L = \frac{1}{2}$
First Iteration		$FF + [+F - F - F] - [-F + F + F]$
Tenth Iteration		. . .

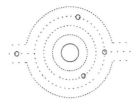

FIGURE 6.10

A set of topological dimension 0.

of a neighborhood of points in either the line or the circle intersects either set in a set of two points that is of topological dimension zero (Fig. 6.11). Now, what about the topological dimension of the Koch snowflake? Is it 2 or 1 (Fig. 6.12)?

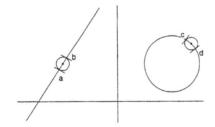

FIGURE 6.11

Two sets of topological dimension 1.

Examples of sets of topological dimension 2 include regions in the plane such as discs, solid squares, etc. (see Fig. 6.13).

Next, we introduce the notion of **fractal dimension**. This concept of a fractal dimension was introduced in 1977 by Mandelbrot [67]. It corresponds to the notion of "capacity" used in 1958 by Kolmogrov [57].

Suppose that a line segment of length 1 is divided into N equal subsegments with a scaling ratio h. Then, we have $Nh = 1$. On the other hand, if a square region is divided into N equal subsquares with a scaling ratio h, then the relation is $Nh^2 = 1$. Similarly, if a cube is divided into N equal subcubes by having its sides scaled by a factor h, then $Nh^3 = 1$. Hence, it is reasonable to use the exponent d in the formula $Nh^d = 1$ as the similarity dimension D_s (see Fig. 6.14).

Note that since $Nh^d = 1$, $d = \frac{\ln N}{\ln(1/h)}$. Thus, for self-similar sets we adopt

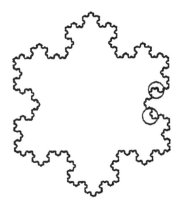

FIGURE 6.12
The snowflake is of topological dimension 1.

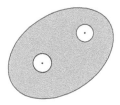

FIGURE 6.13
A set of topological dimension 2.

the following definition for the similarity dimension $D_s(A)$ of a set A:

$$D_s(A) = \frac{\ln N}{\ln (1/h)}. \qquad (6.1)$$

where N is the number of pieces and $1/h$ is the magnification factor.

Example 6.7
Find the similarity dimension of (a) the Koch snowflake H and (b) the Cantor set K. □

SOLUTION

1. The generator of the Koch snowflake H is made up of four equal line segments scaled down by a factor of $\frac{1}{3}$ (see Fig. 6.15). Hence, $N = 4$, $h = \frac{1}{3}$.

N = 3, h = 1/3, d = 1

N = 9, h = 1/3, d = 2

N = 27, h = 1/3, d = 3

FIGURE 6.14

The division of a cube into N equal subcubes.

It follows from Eq. (6.1) that

$$D_s(H) = \frac{\ln 4}{\ln 3} \approx 1.26.$$

2. The generator of the Cantor set C is made up of two equal line segments scaled down by a factor of $\frac{1}{3}$ (see Fig. 6.16). Hence, $N = 2$, $h = \frac{1}{3}$.

It follows from Eq. (6.1) that

$$D_s(C) = \frac{\ln 2}{\ln 3} \approx 0.63. \quad \blacksquare$$

REMARK 6.1 It may be shown, as in Example 6.1, that the length of the Cantor set is zero. In other words, the sum of the lengths of the removed intervals is equal to 1 (Problem 16). $\blacksquare$

Box Dimension

When a set is not self-similar, then our definition of a fractal dimension is not adequate anymore; hence the need for a more general definition of dimension. This new definition is realized by the use of special sets commonly known as k-dimensional boxes. A k-dimensional box is a subset of $\mathbb{R}^k$ defined as

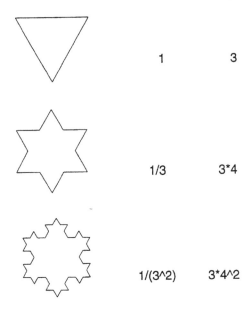

	1	3
	1/3	3*4
	1/(3^2)	3*4^2

FIGURE 6.15
The similarity dimension of the Koch snowflake.

$\{x = (x_1, x_2, \ldots, x_k) : 0 \leq x_i \leq \ell$ for all $1 \leq i \leq k$, for some $\ell > 0\}$. A box of dimension 1 is a closed interval, of dimension 2 a solid square, of dimension 3 a solid cube. Let $N(h)$ be the smallest number of k-dimensional boxes of side length h required to cover G. Then, the box dimension of G is defined to be

$$D_b(G) = \lim_{h \to 0} \frac{\ln N(h)}{\ln(\frac{1}{h})}. \tag{6.2}$$

Example 6.8
(A Non-Self-Similar Fractal). (see Fig. 6.17) This fractal is constructed as follows: divide a square region into nine equal squares and then delete one of them at random. Repeat this process on each of the remaining eight squares. The limiting set A is a fractal that is not self-similar. Now

$$D_b(A) = \lim_{h \to 0} \frac{\ln N(h)}{\ln(\frac{1}{h})} = \lim_{n \to \infty} \frac{\ln 8^n}{\ln 3^n} = \frac{\ln 8}{\ln 3} \approx 1.89. \quad \square$$

We now define the fractal dimension $D_f(A)$ of a set A whether it is self-similar or not as either its similarity dimension $D_s(A)$ or its box dimension $D_b(A)$. This occurs despite the fact that the box dimension is more general

	h	N
	1	1
	1/3	2
	1/(3^2)	2^2
	1/(3^3)	2^3

FIGURE 6.16
The construction of a Cantor set.

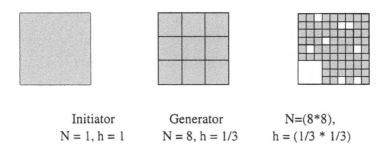

Initiator	Generator	N=(8*8),
N = 1, h = 1	N = 8, h = 1/3	h = (1/3 * 1/3)

FIGURE 6.17
A non-self similar fractal.

than the similarity dimension; the fractal dimension of most of the fractals in the book can be computed using the similarity dimension. It is generally true that for any set A in $\mathbb{R}^k$, $D_t(A) \leq D_f(A)$. The equality $D_t(A) = D_f(A)$ is attained only for Euclidean figures such as lines, squares, discs, cubes, etc. As we mentioned earlier, if $D_t(A) < D_f(A)$, then A is a fractal.

REMARK 6.2 There are some pitfalls of the box dimension. For example, the box dimension of the set

$$A = \{0, \frac{1}{2}, \frac{1}{3}, \ldots, \frac{1}{n}, \ldots\}$$

is $\frac{1}{2}$. To show this, observe that the difference between two consecutive numbers in A is $\frac{1}{k-1} - \frac{1}{k} = \frac{1}{k(k-1)}$. Choose k such that $\frac{1}{k(k-1)} \leq h < \frac{1}{(k-1)(k-2)}$. To cover the subset $\{1, \frac{1}{2}, \ldots, \frac{1}{k-1}\}$ we need $(k-1)$ intervals of length $2h$ and centered at these points. The remaining points $\{\frac{1}{k}, \frac{1}{k+1}, \ldots, 0\}$ can be covered by $\frac{1}{2kh}$ intervals of length $2h$. Hence, the total number of intervals of length $2h$ needed to cover the set A is given by $N(2h) = (k-1) + \frac{1}{2kh}$. A first-order approximation of h is $\frac{1}{k^2}$. Hence we may approximate $(k-1)$ by $\frac{1}{\sqrt{h}}$ and $\frac{1}{2kh}$

by $\frac{1}{2\sqrt{h}}$. Then $N(2h) = \frac{3}{2\sqrt{h}}$. It follows from Formula (6.2) that

$$D_b(A) = \lim_{h \to 0} \frac{\ln \frac{3}{2\sqrt{h}}}{\ln \left(\frac{1}{2h} \right)} = \frac{1}{2}. \quad \blacksquare$$

Another striking example is the set of rational numbers G between 0 and 1. It can be shown that $D_b(G) = 1$ (Problem 22).

Exercises - (6.1, 6.2 and 6.3)

In Problems 1–9 the initiator and the generator of a fractal are provided.

a) Use five iterations to generate the shown picture.

b) Find the topological dimension of the fractal.

c) Find the similarity dimension of the fractal.

1.

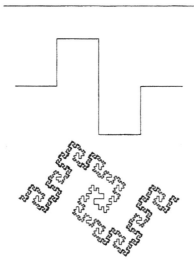

2.

3.

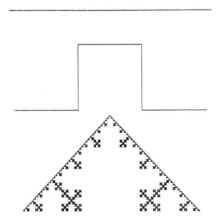

4.

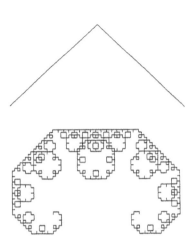

5.

6.

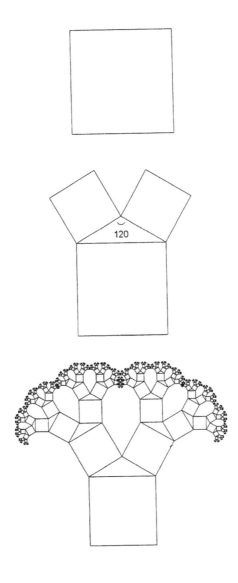

7. The Menger sponge is obtained from a unit cube by boring out the middle ninth of the cube in each direction and then continuing the process indefinitely on the remaining subcubes.

8.

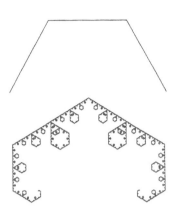

9.

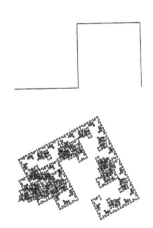

In Problems 10–13

(a) Find the second iteration.

(b) Draw its graphical representation of first and second iterations.

(c) Use the computer to generate the tenth iteration.

10. Axiom: $F + F + F + F$, F is a line segment of length 1
    Production Rule: $F + F - F - FF + F + F + F - F$, $\theta = 90°$, $L = \frac{1}{3}$

11. Axiom: $F + +F + +F$, F is a line segment of length 1
    Production Rule: $F - F + +F - F$, $\theta = \frac{\pi}{2}$, $L = \frac{1}{2}$

12. Axiom: F, a line segment of length 1
    Production Rule: $F[+F]F[-F]F$, $\theta = \frac{\pi}{7}$, $L = \frac{1}{2}$

13. Axiom: F, a line segment of length 1
    Production Rule: $-F + F[+F - F-] - [-F + F + F]$, $\theta = \frac{\pi}{8}$, $L = \frac{1}{2}$

14. Axiom: F, is a line segment of length 1
    Production Rule: $F + F - F - FFF + F + F - F$, $\theta = \frac{\pi}{2}$, $L = \frac{1}{2}$

15. Axiom: $F + F + F + F$, F is a line segment of length 1
    Production Rule: $F \rightarrow F + f - F - FFF + F + f - F$, $f \rightarrow fff$, $\theta = \frac{\pi}{2}$,
    $L = \frac{1}{2}$

16. Show that the length of the Cantor middle-third set is zero.

17.* Find the area enclosed by the Koch island if the initiator is an equilateral
    of side length 1.

18. Assume that $D_f(A)$ and $D_f(B)$ exist and that $A \subset B$. Show that
    $D_f(A) \leq D_f(B)$.

19. Let C_5 be the Cantor set in $[0, 1]$ obtained by deleting the middle fifth
    segment of each remaining line segment and continuing the process in-
    definitely. Find $D_t(C_5)$ and $D_f(C_5)$.

20. Generalize the previous problem to find $D_t(C_{2k+1})$ and $D_f(C_{2k+1})$, for
    $k > 2$.

21. Divide the interval $[0,1)$ into five equal pieces, delete the second and
    fourth subintervals, and then repeat the process indefinitely to obtain
    the "even-fifth" Cantor set $\tilde{C}_s$. Find $D_t(\tilde{C}_s)$ and $D_f(\tilde{C}_s)$.

22. Show that the box dimension of the set of rational numbers between 0
    and 1 is 1.

6.4   Iterated Function System

In this section, we lay the mathematical foundation of the construction of
fractals. There are two types of algorithms to generate fractals, deterministic
and random. Our focus will be mainly on deterministic iterated function
systems (IFS). They will be explored in detail in this section.

6.4.1 Deterministic IFS

Let us begin by recalling the definition of a linear transformation on the plane. A map $F : \mathbb{R}^2 \to \mathbb{R}^2$ is a linear transformation if

1. $F(p_1 + p_2) = F(p_1) + F(p_2)$, for any two points $p_1, p_2 \in \mathbb{R}^2$.

2. $F(\alpha p) = \alpha F(p)$, for $\alpha \in \mathbb{R}$, $p \in \mathbb{R}^2$.

A linear transformation f may be represented by a matrix $A = \begin{pmatrix} a & b \\ c & d \end{pmatrix}$, i.e.,

$$F \begin{pmatrix} x \\ y \end{pmatrix} = \begin{pmatrix} a & b \\ c & d \end{pmatrix} \begin{pmatrix} x \\ y \end{pmatrix} = \begin{pmatrix} ax + by \\ cx + dy \end{pmatrix}. \tag{6.3}$$

(Note that F and A are indistinguishable for a given coordinate system.) Given a point $\begin{pmatrix} x \\ y \end{pmatrix} \in \mathbb{R}^2$, we would like to determine the location of the point $F \begin{pmatrix} x \\ y \end{pmatrix}$. To facilitate this task, we may write the matrix A in the following convenient form.

$$A = \begin{pmatrix} r\cos\theta & -s\sin\phi \\ r\sin\theta & s\cos\phi \end{pmatrix} \tag{6.4}$$

where

$$r = \sqrt{a^2 + c^2}, \quad \cos\theta = a/\sqrt{a^2 + c^2}$$
$$s = \sqrt{b^2 + d^2}, \quad \cos(\pi - \phi) = b/\sqrt{b^2 + d^2}.$$

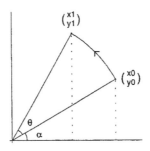

FIGURE 6.18
A rotation by an angle θ.

We now list some important special cases for A.

1. $s = r = 1, \theta = \phi$. Here we have

$$A = \begin{pmatrix} \cos \theta & -\sin \theta \\ \sin \theta & \cos \theta \end{pmatrix}$$

$$\begin{pmatrix} x(1) \\ y(1) \end{pmatrix} = F \begin{pmatrix} x_0 \\ y_0 \end{pmatrix} = \begin{pmatrix} x_0 \cos \theta - y_0 \sin \theta \\ x_0 \sin \theta + y_0 \cos \theta \end{pmatrix}$$

From Fig. 6.18 we have

$$x(1) = \cos(\theta + \alpha)$$
$$= \cos \alpha \cos \theta - \sin \theta \sin \alpha$$
$$= x_0 \cos \theta - y_0 \sin \theta$$
$$y(1) = \sin(\theta + \alpha)$$
$$= x_0 \sin \theta + y_0 \cos \theta$$

The point $\begin{pmatrix} x(1) \\ y(1) \end{pmatrix} = F \begin{pmatrix} x_0 \\ y_0 \end{pmatrix}$ is thus obtained by rotating clockwise by an angle θ. As an example, let $A = \begin{pmatrix} \frac{1}{\sqrt{2}} & \frac{-1}{\sqrt{2}} \\ \frac{-1}{\sqrt{2}} & \frac{1}{\sqrt{2}} \end{pmatrix}$. Then, $r = s = 1, \theta = \phi = \frac{\pi}{4}$, and the transformation rotates every point in the plane by an angle $\frac{\pi}{4}$.

2. $r = s > 0, \theta = \phi$ defines a transformation given by rotating by an angle θ and scaling by a factor r; $0 < r < 1$ gives a contraction, and $r > 1$ gives a dilation [see Fig. 6.19(a) and (b)].

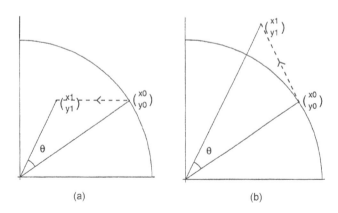

(a)   (b)

FIGURE 6.19
(a) A rotation and a contraction, (b) a rotation and an expansion.

3. $r = s > 0, \theta = \phi = 0$. This occurs if $b = c = 0$ and $a = d$, i.e.,

$$A = \begin{pmatrix} a & 0 \\ 0 & a \end{pmatrix}, \qquad F\begin{pmatrix} x_0 \\ y_0 \end{pmatrix} = \begin{pmatrix} ax_0 \\ ay_0 \end{pmatrix}.$$

This transformation produces a contraction if $0 < a < 1$ or a dilation if $a > 1$.

4. $r = a, s = b, 0 \le a \le 1, 0 \le b \le 1, \phi = \theta = 0$. The transformation is thus given by

$$A = \begin{pmatrix} a & 0 \\ 0 & b \end{pmatrix}, F\begin{pmatrix} x_0 \\ y_0 \end{pmatrix} = \begin{pmatrix} ax_0 \\ by_0 \end{pmatrix}.$$

This transformation reduces by a factor of a in the x direction and by a factor of b in the y direction (see Fig. 6.20).

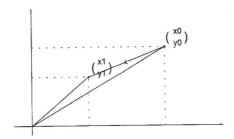

FIGURE 6.20
A reduction by a factor of a in the x-direction and by a factor of b in the y-direction.

5. $s = r, 0 \le r \le 1, \theta = \pi, \phi = 0$ gives

$$A = \begin{pmatrix} -r & 0 \\ 0 & r \end{pmatrix}, F\begin{pmatrix} x_0 \\ y_0 \end{pmatrix} = r\begin{pmatrix} -x_0 \\ y_0 \end{pmatrix}.$$

This transformation reduces x_0, y_0 by a factor r, which simultaneously reflects around the y axis (Fig. 6.21).

Next we define an **affine** linear transformation as a map $F : \mathbb{R}^2 \to \mathbb{R}^2$, which can be represented by

$$F\begin{pmatrix} x_0 \\ y_0 \end{pmatrix} = A\begin{pmatrix} x_0 \\ y_0 \end{pmatrix} + \begin{pmatrix} e \\ f \end{pmatrix}, \qquad (6.5)$$

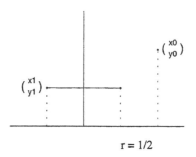

r = 1/2

FIGURE 6.21
A reduction by a factor of r followed by a reflection around the y-axis.

or in vector notation $F(z_0) = Az_0 + t$, where $z_0 = (x_0, y_0)^T, t = (e, f)^T, A = \begin{pmatrix} a & b \\ c & d \end{pmatrix}, e, f \in \mathbb{R}$. Explicitly,

$$\begin{pmatrix} x(1) \\ y(1) \end{pmatrix} = \begin{pmatrix} x_0 \\ y_0 \end{pmatrix} = \begin{pmatrix} ax_0 + by_0 + e \\ cx_0 + dy_0 + f \end{pmatrix}. \tag{6.6}$$

Note that an affine linear transformation involves a translation by the vector $\begin{pmatrix} e \\ f \end{pmatrix}$.

Recall that a map $F : S \to S$ is said to be a **contraction** if for some $0 < \alpha < 1$:

$$||F(z_1) - F(z_2)|| \leq \alpha ||z_1 - z_2|| \quad \text{for all } z_1, z_2 \in S. \tag{6.7}$$

The constant α is called the **contraction factor** of f.

For the affine transformation defined by (6.3),

$$||F(z_1) - F(z_2)|| = ||Az_1 - Az_2|| \leq ||A|| ||z_1 - z_2||. \tag{6.8}$$

It follows from (6.7) that f is a contraction of $\mathbb{R}^2$ if $||A|| < 1$ or if the eigenvalues λ_1, λ_2 of A satisfy $|\lambda_1| < 1, |\lambda_2| < 1$ (see Chapter 4). This is the case if the matrix A is given in the form

$$A = \begin{pmatrix} r\cos\theta & -r\sin\theta \\ r\sin\theta & r\cos\theta \end{pmatrix}, \ |r| < 1.$$

Here the eigenvalues of A are given by $\lambda_{1,2} = r\cos\theta \pm ir\sin\theta$ and thus $|\lambda_{1,2}| = |r| < 1$.

So, if $F\begin{pmatrix} x \\ y \end{pmatrix} = A\begin{pmatrix} x \\ y \end{pmatrix} = \begin{pmatrix} r\cos\theta & -r\sin\theta \\ r\sin\theta & r\cos\theta \end{pmatrix}\begin{pmatrix} x \\ y \end{pmatrix}$, then the effect of F will be a rotation of the given figure by an angle θ counterclockwise and the reduction of distance by a factor of r (Fig. 6.22).

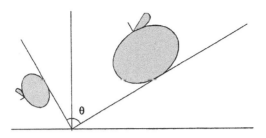

FIGURE 6.22
A rotation by an angle θ followed by a reduction by a factor r.

$$\theta = \frac{\pi}{2}, \qquad r = \frac{1}{2}$$

We now turn our attention to the notion of **iterated function system**. Let H be the collection of all closed and bounded subsets of $\mathbb{R}^2$. Let $F_1, F_2, \ldots, F_N$ be a family of contraction of $\mathbb{R}^2$. Let the function F be defined by

$$F(S) = F_1(S) \cup F_2(S) \cup \ldots \cup F_N(S) \tag{6.9}$$

for $S \in H$.

We call F the **union** of the functions $F_1, F_2, \ldots, F_N$ and it is commonly known as the Hutchinson operator [51]. If $F_1(S), F_2(S), \ldots, F_N(S)$ are disjoint (except possibly for boundaries), and $S = F(S)$, then S is said to be **self-similar**. In the sequel, we will show that all fractals enjoy the property of self-similarity.

Finally, we define the notion of an iterated function system (IFS), which will provide us with codes necessary to build up fractals new and old.

DEFINITION 6.2 *Let $F_1, F_2, \ldots, F_N$ be a family of contractions on $\mathbb{R}^k$ and S a closed bounded subset of $\mathbb{R}^k$. Then the system $\{S : F = \cup_{i=1}^{N} F_i\}$ is called an **iterated function system** (IFS).*

It will be shown later that the union map F is also a contraction (Theorem 6.2). Furthermore, the sequence of iterations $\{F^n(S) : n \in \mathbb{Z}^+\}$ converges to a closed and bounded set A_F (see Theorem 6.3) which is called the **attractor** for F. This attractor set A_F is what we call a fractal. Since $F^{n+1}(S) \to A_F$, too, $F(A_F) = A_F$, i.e., A_F is an invariant set.

We now give some examples to illustrate the concept of IFS. The first example will be our familiar Sierpinski gasket.

Example 6.9

Let $S = \triangle$ be a solid equilateral triangle along with the family of contractions.

$$F_1\begin{pmatrix} x \\ y \end{pmatrix} = \begin{pmatrix} \frac{1}{2} & 0 \\ 0 & \frac{1}{2} \end{pmatrix}\begin{pmatrix} x \\ y \end{pmatrix},$$

$$F_2\begin{pmatrix} x \\ y \end{pmatrix} = \begin{pmatrix} \frac{1}{2} & 0 \\ 0 & \frac{1}{2} \end{pmatrix}\begin{pmatrix} x \\ y \end{pmatrix} + \begin{pmatrix} \frac{1}{2} \\ 0 \end{pmatrix},$$

$$F_3\begin{pmatrix} x \\ y \end{pmatrix} = \begin{pmatrix} \frac{1}{2} & 0 \\ 0 & \frac{1}{2} \end{pmatrix}\begin{pmatrix} x \\ y \end{pmatrix} + \begin{pmatrix} \frac{1}{4} \\ \frac{\sqrt{3}}{4} \end{pmatrix},$$

then $\{S : F = \cup_{i=1}^{3} F_i\}$ is an IFS. Note that $||F_i(u) - F_i(v)|| = \frac{1}{2}||u - v||$ for $i = 1, 2, 3$. Now $F_1(S) = S_1, F_2(S) = S_2, F_3(S) = S_3$, and thus $F(S)$ is the union of the three sets S_1, S_2, S_3 as shown in Fig. 6.23(b). Moreover, repeating this process again we get $F_1(S_1) = S_{11}, F_2(S_1) = S_{21}, F_3(S_1) = S_{31}; F_1(S_2) = S_{12}, F_2(S_2) = S_{22}, F_3(S_2) = S_{32}; F_1(S_3) = S_{13}, F_2(S_3) = S_{23}, F_3(S_3) = S_{33}$ [Fig. 6.23(c)]. ▯

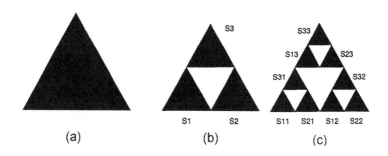

(a)   (b)   (c)

FIGURE 6.23

Construction of the Sierpinski triangle using the maps

	a	b	c	d	e	f
F_1	$\frac{1}{2}$	0	0	$\frac{1}{2}$	0	0
F_2	$\frac{1}{2}$	0	0	$\frac{1}{2}$	$\frac{1}{2}$	0
F_3	$\frac{1}{2}$	0	0	$\frac{1}{2}$	$\frac{1}{4}$	$\frac{\sqrt{3}}{4}$.

The sequence $\{F^n(S)\}_{n=1}^{\infty}$ converges to the Sierpinski gasket $G = \lim_{n\to\infty} F^n(S)$. If we write the code of the transformation given in Eq. (6.6) as

$$a \quad b \quad c \quad d \quad e \quad f,$$

then the code of the Sierpinski gasket is

$$
\begin{array}{ccccccc}
 & a & b & c & d & e & f \\
F_1 & \tfrac{1}{2} & 0 & 0 & \tfrac{1}{2} & 0 & 0 \\[2mm]
F_2 & \tfrac{1}{2} & 0 & 0 & \tfrac{1}{2} & \tfrac{1}{2} & 0 \\[2mm]
F_3 & \tfrac{1}{2} & 0 & 0 & \tfrac{1}{2} & \tfrac{1}{4} & \tfrac{\sqrt{3}}{4}.
\end{array}
$$

Example 6.10
(The Koch Curve). Find the IFS that generates the Koch curve K as its attractor. ▯

SOLUTION The initiator is the interval $I = [0,1]$, and the generator [see Fig. 6.24(b)] consists of four line segments K_1, K_2, K_3, K_4 each of length $\tfrac{1}{3}$.

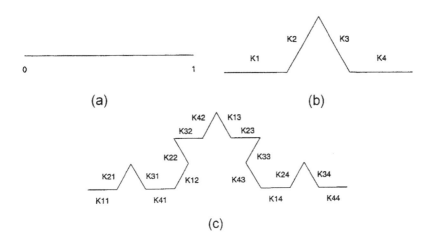

FIGURE 6.24
Construction of the Koch curve using the maps F_1, F_2, F_3, F_4.

Hence, we need four contractions F_1, F_2, F_3, F_4 to produce these line segments. We now give descriptions of these contractions [see Fig. 6.24(b)].

1. To obtain K_1, we shrink the line segment $[0, 1]$ by a factor $\frac{1}{3}$. Thus,

$$F_1 \begin{pmatrix} x \\ y \end{pmatrix} = \begin{pmatrix} \frac{1}{3} & 0 \\ 0 & \frac{1}{3} \end{pmatrix} \begin{pmatrix} x \\ y \end{pmatrix}.$$

2. To obtain K_2, we contract the initiator $[0, 1]$ by a factor of $\frac{1}{3}$, rotate counterclockwise by an angle $\frac{\pi}{3}$, and then translate by the vector $\begin{pmatrix} \frac{1}{3} \\ 0 \end{pmatrix}$. Thus,

$$F_2 \begin{pmatrix} x \\ y \end{pmatrix} = \begin{pmatrix} 1/3 \cos \pi/3 & -1/3 \sin \pi/3 \\ 1/3 \sin \pi/3 & 1/3 \cos \pi/3 \end{pmatrix} \begin{pmatrix} x \\ y \end{pmatrix} + \begin{pmatrix} 1/3 \\ 0 \end{pmatrix}$$

$$= \begin{pmatrix} \frac{1}{6} & -\frac{\sqrt{3}}{6} \\ \frac{\sqrt{3}}{6} & \frac{1}{6} \end{pmatrix} \begin{pmatrix} x \\ y \end{pmatrix} + \begin{pmatrix} 1/3 \\ 0 \end{pmatrix}.$$

3. To obtain K_3, contract by a factor of $\frac{1}{3}$, rotate clockwise by $\frac{\pi}{3}$ and translate by the vector $\begin{pmatrix} 1/2 \\ \frac{\sqrt{3}}{6} \end{pmatrix}$. Thus,

$$F_3 \begin{pmatrix} x \\ y \end{pmatrix} = \begin{pmatrix} \frac{1}{3} \cos -\frac{\pi}{3} & -\frac{1}{3} \sin -\frac{\pi}{3} \\ \frac{1}{3} \sin -\frac{\pi}{3} & \frac{1}{3} \cos -\frac{\pi}{3} \end{pmatrix} \begin{pmatrix} x \\ y \end{pmatrix} + \begin{pmatrix} 1/2 \\ \frac{\sqrt{3}}{6} \end{pmatrix}$$

$$= \begin{pmatrix} \frac{1}{6} & \frac{\sqrt{3}}{6} \\ -\frac{\sqrt{3}}{6} & \frac{1}{6} \end{pmatrix} \begin{pmatrix} x \\ y \end{pmatrix} + \begin{pmatrix} 1/2 \\ \frac{\sqrt{3}}{6} \end{pmatrix}.$$

4. To generate K_4, contract by a factor of $\frac{1}{3}$ and then translate by the vector $\begin{pmatrix} 2/3 \\ 0 \end{pmatrix}$. Thus,

$$F_4 \begin{pmatrix} x \\ y \end{pmatrix} = \begin{pmatrix} 1/3 & 0 \\ 0 & 1/3 \end{pmatrix} \begin{pmatrix} x \\ y \end{pmatrix} + \begin{pmatrix} 2/3 \\ 0 \end{pmatrix}.$$

We conclude this step by giving the code of this IFS:

	a	b	c	d	e	f
F_1	1/3	0	0	1/3	0	0
F_2	1/6	$\frac{-\sqrt{3}}{6}$	$\frac{\sqrt{3}}{6}$	1/6	1/3	0
F_3	1/6	$\frac{\sqrt{3}}{6}$	$\frac{-\sqrt{3}}{6}$	1/6	1/2	$\frac{\sqrt{3}}{6}$
F_4	1/3	0	0	1/3	2/3	0

Fig. 6.24(c) is obtained by iterating, i.e., by applying the transformations F_i on the new sets K_j. For example, $F_i(K_j) = K_{ij}, 1 \leq i, j \leq 4$. We observe that the Koch curve K is obtained as the attractor of the set I, i.e., the limit of the sequence $\{F^n(I)\}$, where F is the union of contractions $F_1, F_2, F_3,$ and F_4.

It is often more convenient to write an affine transformation

$$F\begin{pmatrix} x \\ y \end{pmatrix} = \begin{pmatrix} a & b \\ c & d \end{pmatrix}\begin{pmatrix} x \\ y \end{pmatrix} + \begin{pmatrix} e \\ f \end{pmatrix}$$

or

$$\quad\quad a \quad b \quad c \quad d \quad e \quad f$$

in the polar coordinate form given in (6.4) as

$$F\begin{pmatrix} x \\ y \end{pmatrix} = \begin{pmatrix} r\cos\theta & -s\sin\phi \\ r\sin\theta & s\cos\phi \end{pmatrix}\begin{pmatrix} x \\ y \end{pmatrix} + \begin{pmatrix} e \\ f \end{pmatrix}.$$

Figure 6.25 illustrates the effect of applying the affine transformation defined below on a square.

r	s	θ	ϕ	e	f
1/4	1/2	$\pi/6$	$\pi/3$	0.5	0.5

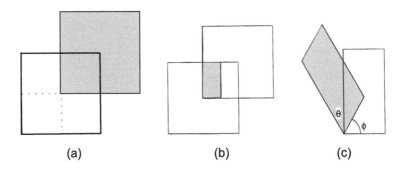

(a)   (b)   (c)

FIGURE 6.25
The effect of applying the affine transformation $\frac{1}{4}$, $\frac{1}{2}$, $\pi/6$, $\pi/3$, 0.5, 0.5.

We now use the above representation to give our last example in this section, namely, the Barnsley's fern.

Example 6.11
(The Barnsley's Fern). Barnsley [5] was able to generate the fern, one of the most celebrated fractals, using only four transformations given below:

	r	s	θ	ϕ	e	f
F_1	.85	.85	−2.5°	−2.5°	0	1.6
F_2	.3	.34	49°	49°	0	1.6
F_3	.3	.37	120°	−50°	0	.44
F_4	0	.16	0	0	0	0

The initiator S is chosen as a rectangle [see Fig. 6.26(a)].

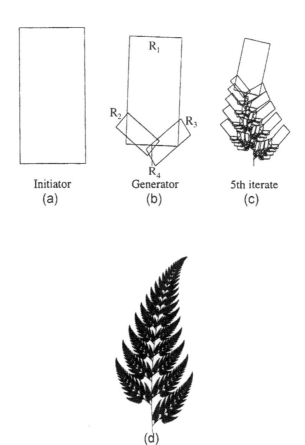

FIGURE 6.26
The construction of Barnsley's fern.

Note that $F_1(S)$ = region 1 $(R_1), F_2(S)$ = region 2 $(R_2), F_3(S)$ =
region 3 $(R_3), F_4(S)$ = region 4 (R_4), (the vertical line) [see Fig. 6.26(b)].
Furthermore, Barnsley's fern [see Fig. 6.26(c)] can be grouped into four parts
[see Fig. 6.26(b)]: R_1, R_2, R_3, R_4. This may help us to the conclusion that we
need four transformations to generate Barnsley's fern [Fig. 6.26(d)].

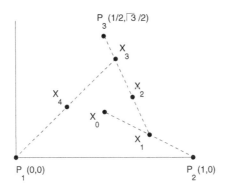

FIGURE 6.27
Illustration of the chaos game.

6.4.2 The Random Iterated Function System and the Chaos Game.

Thus far, we have only discussed the "deterministic" iterated function system (DIFS). The random iterated function system (RIFS) is often used for computers having the capability of displaying graphic images one pixel at a time on screen. Let us first describe the chaos game which generates the Sierpinski triangle using RIFS. Let $P_1 = (0,0)$, $P_2 = (0,1)$, $P_3 = \left(\frac{1}{2}, \frac{\sqrt{3}}{2}\right)$. Pick a starting point X_0 (preferably inside the triangle $P_1 P_2 P_3$). Now we roll a die, if 1 or 2 shows we move halfway toward P_1, if 3 or 4 shows we move halfway toward P_2, and if 5 or 6 shows we move halfway toward P_3. Call the new point X_1. By repeating this process indefinitely, we generate a sequence X_0, X_1, X_2,.... If we discard the first 100 transient points in the sequence, the remaining points consistutes the Sierpinski triangle (Fig. 6.27).

Exercises - (6.4)

1. Find an affine transformation F on the triangle S such that $F(S)$ is the given triangle.

    (a)

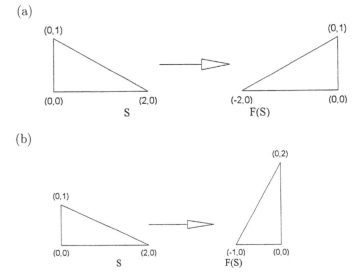

    (b)

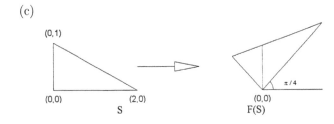

    (c)

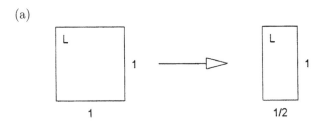

2. Find an affine transformation F on the square S such that $F(S)$ is the given figure.

    (a)

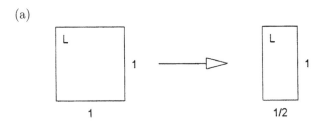

(b)

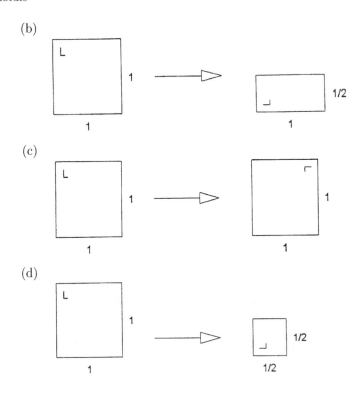

(c)

(d)

3. Find an IFS for the following initiator and generator.

Initiator   Generator

In Problems 4–7 find an IFS for the given initiators and generators.

4.

5.

 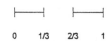

6.

7.

In Problems 8–11 use the computer program IFS, available at CRC's website www.crcpress.com/, to generate the attractor of the given system.

8.

	a	b	c	d	e	f
F_1	0.5	0	0	0.5	0	0
F_2	0.5	0	0	0.5	0.5	0
F_3	0.5	0	0	0.5	0.25	0.5

Sierpinski triangle

9.

	a	b	c	d	e	f
F_1	0	−0.5	0.5	0	0.5	0
F_2	0	0.5	−0.5	0	0.5	0.5
F_3	0.5	0	0	0.5	0.25	0.5

Twin Christmas tree

10.

	a	b	c	d	e	f
F_1	0	0.577	−0.577	0	0.0951	0.5893
F_2	0	0.577	−0.577	0	0.4413	0.7893
F_3	0.5	0.577	−0.577	0	0.0952	0.9893

Dragon

11.

	a	b	c	d	e	f
$\mathbf{F_1}$	0.195	−0.488	0.344	0.443	0.4431	0.2452
$\mathbf{F_2}$	0.462	0.414	−0.252	0.361	0.2511	0.5692
$\mathbf{F_3}$	−0.058	−0.070	0.453	−0.111	0.5976	0.0969
$\mathbf{F_4}$	−0.035	−0.070	0.469	−0.022	0.4884	0.5069
$\mathbf{F_5}$	−0.637	0	0	0.501	0.8562	0.2513
Tree						

In Problems 12–15 determine an IFS whose attractor is the given figure. In all the problems, the initiator is a unit square with an inscribed letter L as shown below.

12.

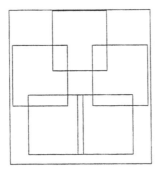 

13.

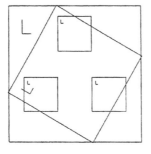 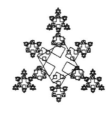

14.

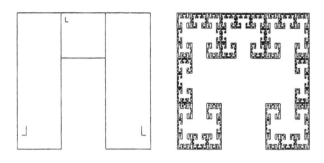

15.

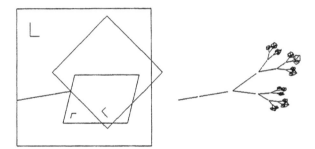

6.5   Mathematical Foundation of Fractals

In this section, we restrict our presentation to $\mathbb{R}^2$. The results obtained are easily extended to $\mathbb{R}^k$ or metric spaces with minor modifications. The object is H, the set of all closed and bounded subsets of $\mathbb{R}^2$. We would like to define a norm $\| \ \|$ or a metric D on H that measures the distance between two sets in H. Such a norm would enable us to precisely make the statement that a sequence of iterates $\{F^n(S)\}$ in H converges to a set E in H. Let $A, B \in H$. Then, the distance between a point $a \in A$ and B is given by

$$d(a, B) = \inf\{\|a - b\| : b \in B\}.$$

The distance between the set A and the set B is given by

$$d(A, B) = \sup\{d(a, B) : a \in A\}.$$

The main problem with this definition of the distance between the sets A and B is that $d(A, B) \neq d(B, A)$ as may be seen from the following example.

Example 6.12

Find $d(A, B), d(B, A)$ for the sets $A = \{(x, y) : (x + 1)^2 + y^2 \leq 1\}, B = \{(x, y) : (x - 3)^2 + y^2 \leq 4\}$ (see Fig. 6.28).

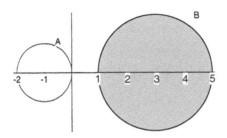

FIGURE 6.28

$d(A, B) \neq d(B, A)$.

SOLUTION   From Fig. 6.28, it is clear that $d(A, B) = \max\{d(a, B) : a \in A\} = 3$, $d(B, A) = \max\{d(b, A) : b \in B\} = 5$. Hence, $d(A, B) \neq d(B, A)$.

From the above example, we conclude that the above notion of distance between sets is not suitable since it is not symmetrical and thus does not fit our intuition about distances. Hence, we are led to the introduction of the **Hausdorff distance.** ∎

DEFINITION 6.3   *Let $A, B \in H$. Then, the **Hausdorff distance** between A and B is defined as*

$$D(A, B) = \max\{d(A, B), d(B, A)\}.$$

Note that in Example 6.12, $D(A, B) = \max\{3, 5\} = 5$. For the convenience of the reader, we present a more geometrical perspective for defining the Hausdorff distance. First, we define the ε-neighborhood $N_\varepsilon(A)$ of a set A (see Fig. 6.29) as

$$N_\varepsilon(A) = \{x \in \mathbb{R}^2 : d(x, a) \leq \varepsilon \text{ for some } a \in A\}. \tag{6.10}$$

Note that for two sets $A, B \in H$, we have (see Fig. 6.30)

$$d(A, B) = \inf\{\varepsilon > 0 : A \subset N_\varepsilon(B)\}.$$
$$d(B, A) = \inf\{\varepsilon > 0 : B \subset N_\varepsilon(A)\}.$$

It can be shown that the pair (H, D) is a metric space (see Section 3.5) (Problem 4).

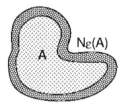

FIGURE 6.29
A set A and its neighborhood $N_\epsilon(A)$.

The most important property of (H, D) is that it is a complete metric space, i.e., every Cauchy sequence[1] in H converges to an element in H (for a proof, see Dugundji [30]). This completeness property is what is needed to apply the contractions mapping principle stated below.

THEOREM 6.1
The contraction mapping principle.
   Let $F : X \to X$ be a contraction mapping, with a contraction factor α, on a complete metric space X. Then, F has a unique fixed point $x^* \in X$, which is a global attractor, i.e., for every $x \in X$, $F^n(x) \to x^*$ as $n \to \infty$.

PROOF   Let x be an arbitrary point in X. Then for $m < n$,

$$\begin{aligned}
d(F^m(x), F^n(x)) &= d(F^m(x), F^m(F^{n-m}(x))) \\
&\leq \alpha^m d(x, F^{n-m}(x)) \\
&\leq \alpha^m [d(x, F(x)) + d(F(x), F^2(x)) + \cdots \\
&\quad + d(F^{n-m-1}(x), F^{n-m}(x)) \\
&\leq \alpha^m d(x, F(x))[1 + \alpha + \ldots + \alpha^{n-m-1}] \\
&< \frac{\alpha^m d(x, F(x))}{1 - \alpha}
\end{aligned}$$

(6.11)

Since $\alpha < 1$, it is evident from (6.11) that $\{F^n(x)\}$ is a Cauchy sequence, and by the completeness of X, $\lim_{n \to \infty} F^n(x) = x^*$, for some $x^* \in X$. We now use the continuity of F to infer that x^* is a fixed point.

$$F(x^*) = F(\lim F^n(x)) = \lim_{n \to \infty} F^{n+1}(x) = x^*.$$

We leave it to the reader to show that x^* is the only fixed point of F (Problem 9). ∎

---

[1] A sequence $\{x_n\}$ in a metric space X is said to be Cauchy if for every $\varepsilon > 0$ there exists a positive integer N such that $d(x_n, x_m) < \varepsilon$ for every $n, m \geq N$.

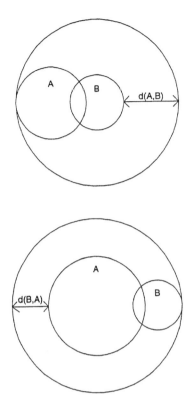

FIGURE 6.30
$d(A, B)$ and $d(B, A)$.

To complete our program, we need to prove two results. First, we need to show that a contraction on $\mathbb{R}^2$ is also a contraction on H. Second, the union of contraction is also a contraction.

LEMMA 6.1
If F is a contraction on $\mathbb{R}^2$, then it is also a contraction on H.

PROOF Let $A, B \in H$. Claim that for $a \in A$ there exists $b \in B$ such that $||a - b|| = d(a, B)$. To prove this, we note that the set $G = \{d(a, x) : x \in B\}$ is bounded below by 0. Hence, by the well-ordering principle[2] it has a greatest

---

[2] Well-ordering principle: If a subset of the real numbers is bounded below, then it has a greatest lower bound.

lower bound $\delta \geq 0$. Thus, $d(a, B) = \delta$. Furthermore, there is a sequence $\{b_n\}$ in B such that $d(a, b_n) \to \delta$ as $n \to \infty$.

Since B is closed and bounded, there is a subsequence $\{b_{n_i}\}$ of $\{b_n\}$, which converges to $b \in B$.

Now,

$$d(a, b) \leq d(a, b_{n_i}) + d(b_{n_i}, b).$$

Since $d(a, b_{n_i}) \to \delta, d(b_{n_i}, b) \to 0$, we have $d(a, b) \leq \delta$. But by the definition of $\delta, d(a, b) \geq \delta$.

Hence, $d(a, b) = \delta = d(a, B)$. Let F be a contraction on $\mathbb{R}^2$ with a contraction factor α. Then

$$||F(a) - F(b)|| \leq \alpha ||a - b|| = \alpha d(a, B) \leq \alpha D(A, B).$$

Hence, $d(F(a), F(b)) \leq \alpha D(A, B)$ and consequently, $D(F(A), F(B)) \leq \alpha D(A, B)$.
∎

LEMMA 6.2

If $F_1, F_2, \ldots, F_N$ are contractions, with contraction factors $\alpha_i, 1 \leq i \leq N$, on H, then so is their union $F = \cup_{i=1}^N F_i$, with a contraction factor $\alpha = \max\{\alpha_i : 1 \leq i \leq N\}$.

PROOF   We note first that if $A, B \in H$, then $D\{(F_1(A) \cup F_2(A)), (F_1(B) \cup F_2(B))\} \leq \max\{D(F_1(A), F_1(B)), D(F_2(A), F_2(B))\}$ (see Problems 2 and 3). Hence,

$$D(F_1(A) \cup F_2(A), F_1(B) \cup F_2(B)) \leq \alpha D(A, B)$$

where $\alpha = \max\{\alpha_1, \alpha_2\}$.

We now use mathematical induction on N to complete the proof of the lemma. Suppose that for $N = k, \cup_{i=1}^k F_i$ is a contraction, i.e.,

$$D\left(\cup_{i=1}^k F_i(A), \cup_{i=1}^k F_i(B)\right) \leq \alpha(k) D(A, B),$$

where $\alpha(k) = \max\{\alpha_1, \alpha_2, \ldots, \alpha_k\}$, then,

$$D\left(\cup_{i=1}^{k+1} F_i(A), \cup_{i=1}^{k+1} F_i(B)\right) = D\left(\cup_{i=1}^k F_i(A) \cup F_{k+1}(A),\right.$$
$$\left. \cup_{i=1}^k F_i(B) \cup F_{k+1}(B)\right)$$
$$\leq \alpha(k+1) D(A, B) \qquad (6.12)$$

where $\alpha(k+1) = \max\{\alpha(k), \alpha_{k+1}\}$.

Hence, $F = \cup_{i=1}^N F_i$ is a contraction for all N. Finally we are able to apply the contraction mapping principle to establish the main result in this section.
∎

THEOREM 6.2

If $F_1, F_2, \ldots, F_N$ are contractions on $\mathbb{R}^2$, then there exists a unique global attractor $A \in H$ for the union map $F = \cup_{i=1}^{N} F_i$. Explicitly, for every $B \in H, F^n(B)$ converges to A in the Hausdorff metric.

PROOF The proof is left to the reader as Problem 5. ∎

Exercises - (6.5)

1. Find the Hausdorff distance between the sets A and B shown in the figures.

   a.

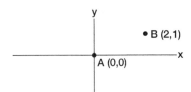

   b.

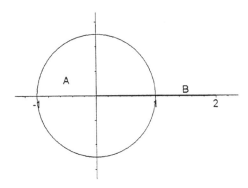

c.

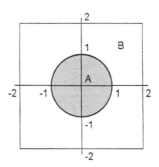

d.

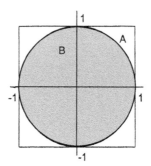

e.

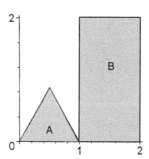

2. Let $A, B, C \in H$. Show that $D(A \cup B, C) \leq \max\{D(A, C), D(B, C)\}$.

3. Let $A, B, C, E \in H$. Show that $D(A \cup B, C \cup E) \leq \max\{D(A, C), D(B, E)\}$.

4. Prove that the Hausdorff distance is a metric on H, that is, for $A, B, C \in H$ we have:

   (a) $D(A, B) \geq 0$, $D(A, B) = 0$ if and only if $A = B$.

   (b) $D(A, B) = D(B, A)$.

(c) $D(A, B) \le D(A, C) + D(C, B)$.

5. Prove Theorem 6.2.

6. Prove that the attractor set A in Theorem 6.2 is invariant, i.e., $F(A) = A$.

7. Let $F : X \to X$ be a contraction with a contraction factor $\alpha \in (0, 1)$, on a complete metric space X and x^* be the unique fixed point of F. Show that

$$d(F^n(x), x^*) \le \frac{\alpha^n}{1 - \alpha} d(x, x^*).$$

8. Let $F : X \to X$ be a contraction with a contraction factor $\alpha \in (0, 1)$, on a complete metric space X. Prove that for any $x \in X, n \in \mathbb{Z}^+$,

$$d(x, F^n(x)) \le \frac{1 - \alpha^n}{1 - \alpha} d(x, F(x)).$$

9. Complete the proof of Theorem 6.1 by showing that the fixed point x^* of F is unique.

10. Let A be a closed and bounded subset of $\mathbb{R}^2$ and $\varepsilon > 0$. Prove that the set $N_\varepsilon(A)$ defined in (6.10) is also closed and bounded.

    (*Hint: Use the Bolzano-Weierstrass theorem: every bounded sequence in $\mathbb{R}^2$ has a convergent subsequence.*)

11. Prove that in a metric space X every convergent sequence is Cauchy.

In Problems 12–15, let $\{A_n\}_{n=1}^\infty$ be a Cauchy sequence in H. Define $A = \{x \in \mathbb{R}^2 : \text{there is a Cauchy sequence } \{a_n\} \text{ in } \mathbb{R}^2, \text{ with } a_n \in A_n \text{ and } a_n \to X \text{ as } n \to \infty\}$.

12. Suppose that $\{a_{n_k}\}$ is a Cauchy sequence in $\mathbb{R}^2$ with $a_{n_k} \in A_{n_k}$ for every k. Prove that $\{a_{n_k}\}$ is a subsequence of a Cauchy sequence $\{a_n\}$ in $\mathbb{R}^2$ with $a_n \in A_n$ for every n.

13. Prove that $A \ne \phi$.

14. Prove that A is closed and bounded.

15. Prove that $\lim_{n \to \infty} D(A_n, A) = 0$ and conclude that H is a complete metric space.

6.6   The Collage Theorem and Image Compression

In Sec. 6.4, we have seen how an iterated function system IFS: $\{S : F = \cup_{i=1}^{N} F_i\}$ generated a fractal. In this section, we are interested in the inverse problem. Here, we are given a figure G, such as a tree or a fern, and the question is to find an IFS that will generate a figure as close as we wish from G. The importance of this process is evident in the field of image compression. Suppose we are interested in sending the Barnsley's fern in Example 6.11 on a 512×512 screen via a high-definition television (HDTV) signals. This would then require sending 262,144 pieces of data where each pixel is represented by either a 0 or a 1. However, using the IFS: $\{S : F = F_1 \cup F_2 \cup F_3 \cup F_4\}$ used in the above-mentioned example, we only need the 24 coefficients in F; 6 coefficients for each affine transformation F_i, $1 \le i \le 4$. Then, using our IFS, one can construct the Barnsley's fern at the other end of the transmission. The compression ratio is then 262,144 to 24 or approximately 10,922 to 1. In practice, the situation is much more involved since the IFS of a given figure is unknown for most of the figures we are interested in.

However, Barnsley and Hurd [6] have developed a patented algorithm called the IFS compression, which is an interactive image modeling method based on the following theorem commonly known as the collage theorem.

THEOREM 6.3 (The Collage Theorem)
Let $\{S : \cup_{i=1}^{N} F_i\}$ be an IFS, with contraction factors

$$\alpha_1, \alpha_2, \ldots, \alpha_N, \alpha = max\,\{\alpha_i : 1 \le i \le n\},$$

for which A is the attractor. If for any $\varepsilon > 0$,

$$H(S, \cup_{i=1}^{N} F_i(S)) < \varepsilon,$$

then

$$H(S, A) < \tfrac{\varepsilon}{1-\alpha}.$$

PROOF   Let $F = \cup_{i=1}^{N} F_i$. Then $\lim_{k \to \infty} F^k(S) = A$. Hence,

$$
\begin{aligned}
D(S, A) &= D(S, \lim_{k \to \infty} F^k(S)) \\
&= \lim_{k \to \infty} D(S, F^k(S)) \\
&\le \lim_{k \to \infty} [D(S, F(S)) + D(F(S), F^2(S)) + \cdots D(F^{k-1}(S), F^k(S)] \\
&\le \lim_{k \to \infty} D(S, F(S))[1 + \alpha + \alpha^2 + \cdots + \alpha^{N-1}] \\
&\le \frac{\varepsilon}{1 - \alpha} \quad \blacksquare
\end{aligned}
$$

Real Fern   Approximate Collage   Fractal

FIGURE 6.31

Steps of Barnsley's IFS compression algorithm.

We now give a brief description of Barnsley's IFS compression algorithm [6]. Initially, we begin with a target image S, which may be either a digitized image or a polygonalized approximation of a given configuration. Second, the set S is rendered on a computer graphics monitor. Then it is acted on by *a* contraction F_1 defined as

$$F_1 \begin{pmatrix} x \\ y \end{pmatrix} = \begin{pmatrix} a_1 & b_1 \\ c_1 & d_1 \end{pmatrix} \begin{pmatrix} x \\ y \end{pmatrix} + \begin{pmatrix} e_1 \\ f_1 \end{pmatrix}$$

with $a_1 = b_1 = c_1 = d_1 = 0.25$. The image $F_1(S)$ is then displayed on the monitor, but with a different color from S. Now, we interactively adjust the coefficients by the use of a mouse so that $F_1(S)$ is translated, rotated, or sheared on the screen in such a way that $F_1(S)$ lies over a part of S. Once this is done, we record the new coefficient of F_1. In a similar fashion, we introduce another contraction F_2 such that $F_2(S)$ covers another part of S with little or no overlap with $F_1(S)$. We continue this process until we find a set of contracting affine transformations $F_1, F_2, \ldots, F_n$ such that the set

$$S^N = \cup_{i=1}^{N} F_i(S)$$

is visually close to S, that is, $H(S, S^N)$ is quite small. The collage theorem asserts that the attractor A, which is determined by $IFS : \{S : \cup_{i=1}^{N} F_i(S)\}$, will be visually close to S.

An application of Barnsley's IFS compression algorithm is illustrated in Fig. 6.31.

For more details, interested readers may consult Barnsley and Hurd [6].

7

The Julia and Mandelbrot Sets

A manifesto: There is a fractal face to the geometry of nature.

Benoit Mandelbrot

7.1 Introduction

You may have already seen beautiful and facinating pictures of the Julia set and the Mandelbrot set. These pictures become even more intriguing if we zoom in at finer scales. Images of Julia sets and Mandelbrot sets can be found on posters, book covers, T-shirts, carpets, screen savers, and web pages. Moreover, these images have captured the imagination of mathematicians and the public at large alike. The mathematics behind the beauty of these pictures is part of a branch of mathematics called complex dynamics (see for example [15] for a readable account). Remarkable progress on iterating complex functions, not just real functions, was made by Gaston Julia [52] and Pierre Fatou [38]. This is indeed a remarkable feat considering that all this was done before computer graphics were available to them.

The subject stayed moribund, however, until Benoit Mandelbrot made it popular after the appearance of his seminal book "The Fractal Geometry of Nature" in 1982. Taking advantage of the availability of computer graphics, Mandelbrot has created some of the most facinating pictures ever produced mathematically. More importantly, he was able to lay down the foundation of his new objects "Mandelbrot sets" and to give a great impetus to an otherwise dormant subject.

In this chapter we will take you on a short tour to explore this remarkable field of mathematics and unveil the mystery behind the beauty of "complex"

fractals.

7.2   Mapping by Functions on the Complex Domain

Let $z = x + iy$ be a complex number and let $\mathbb{C}$ denote the set of complex numbers. Then x is called the real part of z, $\Re(z)$, and y is called the imaginary part of z, $\Im(z)$. Note that both x and y are real numbers. If we let the x axis to be the real axis and the y axis to be the imaginary axis, then the complex number $z = x + iy$ is represented by the point (x, y) in this complex plane (see Fig. 7.1). The modulus $|z|$ of z is defined as $|z| = \sqrt{x^2 + y^2}$; it is the distance between z and the origin. A complex number $z = x + iy$ may be represented in polar coordinates. Let $r = |z|$, and $\theta = \tan^{-1}(\frac{y}{x})$. Then θ is called the argument of z, denoted by $\arg(z)$. Moreover, $z = re^{i\theta}$. It is noteworthy to observe that $|z| = r|e^{i\theta}| = r$, since $|e^{i\theta}| = |\cos\theta + i\sin\theta| = 1$.

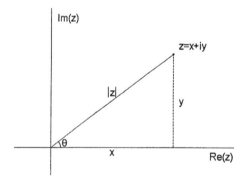

FIGURE 7.1
The modulus of a complex number $|z|$ and its argument θ.

The triangle inequality that we encountered in the real number system still holds for complex numbers.

Triangle Inequality for Complex Numbers

Let $z_1, z_2 \in \mathbb{C}$. Then

(a) $|z_1 + z_2| \leq |z_1| + |z_2|$

(b) $|z_1 + z_2| \geq |z_1| - |z_2|$

Consider a linear map $f : \mathbb{C} \to \mathbb{C}$, where $\mathbb{C}$ is the set of complex numbers, of the form $f(z) = \alpha z$, where $\alpha = a + ib$, and $z = x + iy$.

Now, α and z may be written in the following exponential forms:

$$\alpha = se^{i\beta}, \quad \text{with} \quad s = \sqrt{a^2 + b^2}, \quad \text{and} \quad \beta = \tan^{-1}\left(\frac{b}{a}\right)$$

$$z = re^{i\theta}, \quad \text{with} \quad r = \sqrt{x^2 + y^2}, \quad \text{and} \quad \theta = \tan^{-1}\left(\frac{y}{x}\right)$$

We may write $f(z)$ as $f(z) = sre^{i(\theta+\beta)}$.
Note also that $f^2(z) = s^2 re^{i(\theta+2\beta)}$, and generally

$$f^n(z) = s^n re^{i(\theta+n\beta)}. \tag{7.1}$$

Clearly, we have three cases to consider:

1. $s < 1$: In this case, it follows from Equation (7.1) that the orbit of z will spiral toward the origin. We may say, then, that the origin is asymptotically stable.

2. $s > 1$: From Equation (7.1) we conclude that the orbit of z spirals further away from the origin and thus the origin is unstable.

3. $s = 1$: In this case the orbit of z stays on the circle of radius r_0 and the map is a rotation on the circle. Recall that we have discussed this map in Chapter 3. It was shown that if β is rational, then every point on a circle of radius r_0 is periodic, and if β is irrational, then the map on each circle of radius r_0 is transitive, with the set of periodic points dense but not chaotic.

Next, we consider more complicated nonlinear maps.

Example 7.1
Consider the squaring map $Q_0(z) = z^2$. Then for $z = re^{i\theta}, Q_0(z) = r^2 e^{i2\theta}$. Note that this function maps the upper half plane $r \geq 0, 0 \leq \theta \leq \pi$ onto the entire complex plane (Fig. 7.3). □

Now, if we let $z = x + iy$ and $w = Q_0(z) = u + iv$, then $u + iv = x^2 - y^2 + i2xy$. Thus,

$$u = x^2 - y^2, \quad v = 2xy. \tag{7.2}$$

Hence, each branch of the hyperbola $x^2 - y^2 = a, (a > 0)$ is mapped in a one-to-one manner onto the vertical line $u = a$. To see this, we note from the first part of Equation (7.2) that $u = a$ if (x, y) is a point on one of the two branches of the hyperbola. When in particular it lies on the right-hand

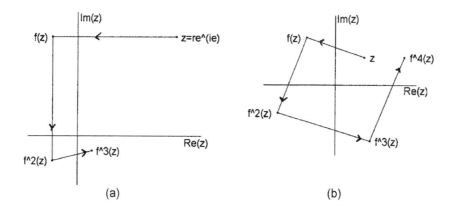

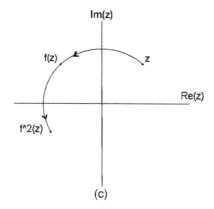

FIGURE 7.2

Iteration of a point $z = re^{i\theta}$ under the map $f(z) = \alpha z$, $\alpha = a + ib$, $s = \sqrt{a^2 + b^2}$. (a) $s < 1$: origin is asymptotically stable, (b) $s > 1$: origin is unstable, (c) $c = 1$: origin is stable.

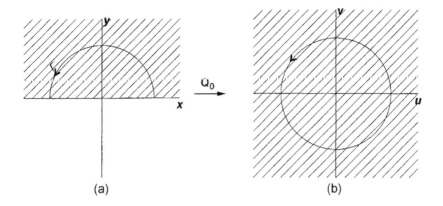

(a)   (b)

FIGURE 7.3
The map $Q_0(z) = z^2$ maps the upper half plane onto the entire complex plane.

branch, the second part of Equation (7.2) tells us that $v = 2y\sqrt{y^2 + a}$. Thus, the image of the right-hand branch can be expressed parametrically as

$$u = a, \qquad v = 2y\sqrt{y^2 + a}, \qquad -\infty < y < \infty$$

and is evident that the image of a point (x, y) on that branch moves upward along the entire line as (x, y) traces out the branch in the upward direction (Fig. 7.4).

Similarly,

$$u = a, \qquad v = -2y\sqrt{y^2 + a}$$

furnishes a parametric representation for the image of the left-hand branch of the hyperbola. Thus, the left-hand branch of the hyperbola is mapped to the line $u = a$.

Let us now turn our attention to the analysis of the dynamics of the map Q_0. Clearly, $Q_0^n(z) = r^{2^n}\, e^{i\, 2^n \theta}$. Furthermore, $|Q_0^n(z)| = r^{2^n}$. Consequently, we conclude that (see Fig. 7.5)

1. $|Q_0^n(z)| \to 0$ as $n \to \infty$ if $r < 1$ or $(|z| < 1)$.

2. $|Q_0^n(z)| \to \infty$ as $n \to \infty$ if $r > 1$ or $(|z| > 1)$.

3. $|Q_0^n(z)| = 1$ if $r = 1$ or $(|z| = 1)$.

Next we consider the square root map.

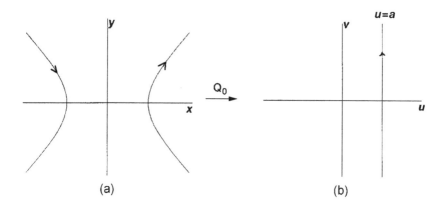

FIGURE 7.4

Q_0 maps a hyperbola to the line $u = a$.

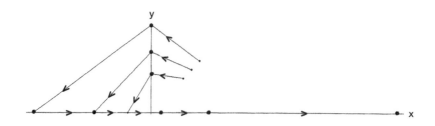

FIGURE 7.5

Orbits of the points $0.8e^{\frac{i\pi}{4}}$, $e^{\frac{i\pi}{4}}$, and $1.2e^{\frac{i\pi}{4}}$ under iteration of $Q_0(z) = z^2$.

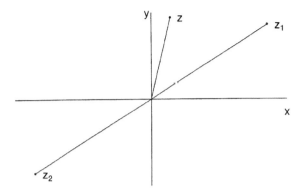

FIGURE 7.6
$z = re^{i\theta}$ and its two square roots z_1 and z_2.

Example 7.2
Consider the function $f(z) = z^{1/2}$.
If $z = r\,e^{i\theta}$ $(r > 0, \quad -\pi < \theta \le \pi)$, then

$$z^{1/2} = \sqrt{r}\; e^{i\frac{(\theta + 2k\pi)}{2}}, \qquad k = 0, 1. \quad \Box \tag{7.3}$$

The principal branch of the double-valued function $z^{1/2}$ is given by $f_0(z) = \sqrt{r}\; e^{i\,\theta/2}$, $-\pi < \theta \le \pi$, $r > 0$. Note that the origin and the ray $\theta = \pi$ form the branch cut for f_0, and the origin is the branch point.

From Equation (7.3) the two square roots of z are

$$z_1 = \sqrt{r}\,(\cos(\theta/2) + i\sin(\theta/2))$$
$$z_2 = \sqrt{r}\,\left(\cos\left(\frac{\theta}{2} + \pi\right) + i\sin\left(\frac{\theta}{2} + \pi\right)\right)$$
$$= -\sqrt{r}\,(\cos(\theta/2) + i\sin(\theta/2))$$

(see Fig. 7.6).

If S^1 is a circle of radius r and center at the origin, then $f(S^1)$ is another circle of radius $\sqrt{r}$ centered at the origin [Fig. 7.7a(a)].

The situation is entirely different when the circle S^1 does not contain the origin. Here the circle lies in a wedge $\theta_1 \le \theta \le \theta_2$. Hence, the argument of each point in $f(S^1)$ lies in the wedge $\theta_1/2 \le \theta \le \theta_2/2$ and its reflection with respect to the origin. Hence, $f(S)$ is the union of two closed curves as shown in Fig. 7.7a(b). Observe that when the circle S^1 touches the origin, $f(S^1)$ looks like figure eight [Fig. 7.7c(c)]. Finally, when S^1 encircles the origin, $f(S^1)$ looks like a peanut shell [Fig. 7.7(d)].

A set D in the complex plane $\mathbb{C}$ is called a **domain** if it is open and connected.

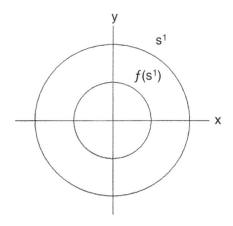

FIGURE 7.7a
(a) The image of S^1 when it is centered at the origin.

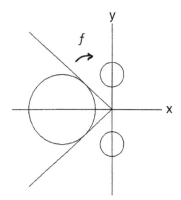

FIGURE 7.7b
(b) The image of S^1 when it is not centered at the origin.

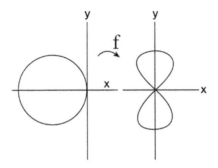

FIGURE 7.7c
(c) The image of S^1 when it passes through the origin.

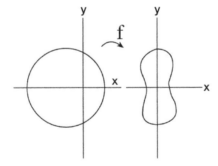

FIGURE 7.7d
(d) The image of S^1 when it encircles the origin.

DEFINITION 7.1   *A function f is said to be analytic in a domain D if it has a derivative at every point in D.*

We now state the main stability theorem for complex functions.

THEOREM 7.1
Let z^* be a fixed point of an analytic complex function f. Then, the following statements hold:

1. *If $|f'(z^*)| < 1$, then z^* is asymptotically stable.*

2. *If $|f'(z^*)| > 1$, then z^* is unstable.*

PROOF   The proof is similar to that of Theorem 1.3 in Chapter 1 and will be left to the reader as Problem 10.   ∎

As an immediate consequence of Theorem 7.1, we have the following result:

COROLLARY 7.1
Let z be a k-periodic point of an analytic function f. Then the following statements hold:

1. *If $|f'(z)f'(f(z))\ldots f'(f^{k-1}(z))| < 1$, then z is asymptotically stable.*

2. *If $|f'(z)f'(f(z))\ldots f'(f^{k-1}(z))| > 1$, then z is unstable.*

Example 7.3
Consider the map $f(z) = z^3$, $z \in C$.

(a) Find the fixed points of f and determine their stability.

(b) Find the 2-cycles of f and determine their stability.   ☐

SOLUTION

(a) Fixed points: $z^3 - z = z(z^2 - 1) = 0$. Hence, the fixed points are: $z_1^* = 0$, $z_2^* = 1$, $z_3^* = -1$. Since $|f'(z_1^*)| = 0$, $|f'(z_2^*)| = 3$, and $|f'(z_3)| = 3$, it follows from Theorem 7.1, that 0 is asymptotically stable, while 1 and -1 are unstable.

(b) To find the 2-cycles, we solve the equation $f^2(z) = z$. Hence, $z^9 - z = z(z^8 - 1) = 0$. Since 0 is a fixed point, we have $z^8 = 1$. Thus, the 2-cycles of $f(z)$ are the eighth roots of 1, excluding 1 and -1 since they are fixed points of f. The eighth roots of 1 are w, w^2, w^3, w^5, w^6, w^7, where

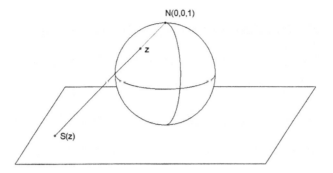

N(0,0,1)

S(z)

FIGURE 7.8
The Riemann Sphere: Stereographic projection from S^2 into $\mathbb{C}$.

$w = \cos\frac{\pi}{4} + i\sin\frac{\pi}{4} = \frac{\sqrt{2}}{2} + i\frac{\sqrt{2}}{2}$. Note that $w^4 = -1$ is excluded from this list. Hence, the 2-cycles are $\{w, w^3\}$, $\{w^2, w^6\}$, and $\{w^5, w^7\}$. It follows from Corollary 7.1 that all 2-cycles are unstable. ∎

7.3   The Riemann Sphere

To simplify the study of the dynamics of analytic maps, it is beneficial to consider the extended complex plane $\mathbb{C} \cup \{\infty\}$. To describe the topology of this space, we introduce a special representation of the complex plane. Consider the sphere S^2 with radius $\frac{1}{2}$ and center $(0, 0, \frac{1}{2})$ that is tangent to the complex plane $\mathbb{C}$ at the origin $(0, 0, 0)$. The point $N(0, 0, 1)$ will be referred to as the north pole of S^2 (see Fig. 7.8). We now introduce the stereographic projection S.

Let $P(a, b, c) \varepsilon S^2 \backslash \{N\}$. The line joining N to P will pierce $\mathbb{C}$ at the point $Q(a/(1-c), b/(1-c), 0)$, which corresponds to the complex number $z = (a + ib)/(1-c)$ (Problem 6). Conversely, any point $Q(x, y, 0)$ in $\mathbb{C}$ corresponding to the complex number $z = x + iy$ lies on a line passing through the point N and intersecting the sphere S^2 at a point $P(\alpha, \beta, \gamma)$ with

$$\alpha = \frac{x}{x^2 + y^2 + 1}, \quad \beta = \frac{y}{x^2 + y^2 + 1}, \quad \gamma = \frac{x^2 + y^2}{x^2 + y^2 + 1}$$

(see Problem 11).

Note that this gives a correspondence W from $\mathbb{C}$ onto $S^2 \backslash \{N\}$. We then let $W(\infty) = N$. Hence, the extended complex plane $\overline{\mathbb{C}} = \mathbb{C} \cup \{\infty\}$ is identified with the sphere S^2, and either one will be called a Riemann sphere.

For any $z_0 \neq \infty$, we define an open ball $B_\varepsilon(z_0)$ as $B_\varepsilon(z_0) = \{z \varepsilon \mathbb{C} : |z - z_0| < \varepsilon\}$. We define open balls around ∞ as follows: For any $\varepsilon > 0$, we let $B_\varepsilon(\infty) = \{z \varepsilon \mathbb{C} : |z| > \frac{1}{\varepsilon}\}$. Note that the W takes $B_\varepsilon(\infty)$ to an open neighborhood of the north pole $N(0, 0, 1)$. The above description of open balls determines a metric on the extended complex place $\overline{\mathbb{C}}$.

Linear Fractional Transformation (Möbius Transformation)

The transformation $T(z) = \frac{az+b}{cz+d}$, $ad - bc \neq 0$, where a, b, c, d are complex constants, is called a linear fractional transformation (or Möbius transformation). The map T may be extended to $\overline{\mathbb{C}}$ by letting $T(\infty) = a/c$, and $T(z_0) = \infty$ when $cz_0 + d = 0$. An important property of the map T is that it maps circles in $\overline{\mathbb{C}}$ to circles. Note that a line in the complex plane $\mathbb{C}$ becomes a circle through ∞ in the extended complex place $\overline{\mathbb{C}}$. Hence, T maps lines and circles in $\mathbb{C}$ to lines and circles in $\overline{\mathbb{C}}$.

Exercises - (7.2 and 7.3)

1. If $z = x + iy$, then we may write $z = r(\cos\theta + i\sin\theta)$, where $r = \sqrt{x^2 + y^2}$, $\theta = \arctan(y/x)$. The two square roots of z are given by

$$\pm\sqrt{r}(\cos(\theta/2) + i\sin(\theta/2)).$$

Find the square roots and draw them on the complex plane.

   (a) i
   (b) $-1 + i$
   (c) $-1 + \sqrt{3}i$
   (d) $1 + i$
   (e) $-2 + 2i$
   (f) -6

2. Plot the orbit of $z = 0$ under the following maps:

   (a) $g(z) = z - 1$
   (b) $f(z) = z + 2$

3. Let $f(z) = az + b$, $a, b \in C$.

   (a) Under what conditions does f have fixed points? Find the fixed points of f if they exist.

(b) Show that if $a \neq 1$, then f is topologically conjugate to a map of the form $z \rightarrow cz$.

4. Let $g(z) = az$, with $a = \frac{4}{5}e^{\pi i/3}$.

   (a) Show that the orbit of 1 under g looks like a spiral. Then find the equation of this spiral.

   (b) Show that if z_1 and z_2 are two points on the spiral that lie on the same ray extending from the origin, then there exists $k \in \mathbb{Z}^+$ such that $g^k(z_1) = z_2$ or $g^k(z_2) = z_1$.

5. Consider the map $Q_{1/4}(z) = z^2 + 1/4$.

   (a) Show that $Q_{1/4}(z)$ has a single fixed point and determine its stability.

   (b) Find the repelling 2-cycles of $Q_{1/4}$.

6. Show that $Q_c(z) = z^2 + c$ has an attracting 2-cycle inside the circle with radius $1/4$ and center $(-1, 0)$.

7. Let $f(z) = e^{i\theta}z$.

   (a) Show that if θ is a rational multiple of π, then every point in $\mathbb{C}$ is periodic.

   (b) Show that if θ is not a rational multiple of π, then the orbit of $z \in \mathbb{C}$ is dense in the circle with radius $|z|$ and center at the origin.

8. Let $Q_0 : \mathbb{C} \rightarrow \mathbb{C}$ be defined by $Q_0(z) = z^2$. If S is the circle of radius 1 and center $(-1, 0)$, find and draw $Q_0^{-1}(S)$ and $Q_0^{-2}(S)$.

9. Consider $Q_2(z) = z^2 + 2$ and the unit circle $S = \{z : |z| = 1\}$. Sketch $Q_2^{-1}(S)$ and $Q_2^{-2}(S)$.

10. Prove Theorem 7.1.

11. (a) Show that the stereographic projection takes a point $(a, b, c) \in S^2 \backslash \{N\}$ to the point $z = (a + ib)/(1 - c)$ in the complex plane.

    (b) Show the converse, i.e., that a stereographic projection takes a point $z = x + iy$ in the complex plane to the point

$$\left(\frac{x}{x^2 + y^2 + 1}, \frac{y}{x^2 + y^2 + 1}, \frac{x^2 + y^2}{x^2 + y^2 + 1} \right).$$

12. Show that the map $T(z) = \frac{(1+2i)z+1}{(1-2i)z+1}$ maps the real axis in the complex plane to the unit circle.

13. Show that the map $g(z) = a_2 z^2 + 2a_1 z + a_0$, with $a_2 \neq 0$ is topologically conjugate to the map $Q_c(z) = z^2 + c$ through the conjugacy map $h(z) = a_2 z + a_1$, provided that $c = -a_1^2 + a_1 + a_2 a_0$.

14. Prove that any Möbius transformation may be written as a composition of translations (of the form $z \to z + a$), inversions (of the form $z \to \frac{1}{z}$), and homothetic transformations (of the form $z \to bz$).

15. Assume that $(a - d)^2 + 4bc = 0$ in the Möbius transformation $T = \frac{az+b}{cz+d}$.

    (a) Show that T has a unique fixed point $z^* = a - d$.

    (b) Show that T is (analytically) conjugate to a translation of the form $z \to z + a$.

16. Show that if the Möbius map T has two fixed points, then it is (analytically) conjugate to a unique linear map of the form $z \to bz$.

7.4 The Julia Set

In this section, our goal is to study the Julia set, one of the most fascinating and extensively studied objects in the theory of dynamical systems. This famous set was introduced by the French mathematician Gaston Julia (1893–1978) in his masterpiece paper, *Mémoire sur l'iteration des fonctions rationelles* (*J. Math. Pure Appl.*, **4**, 1918, 47–245). It is interesting to note that Julia was only 25 years old when he published this monumental work of 199 pages.

We begin our exposition by defining two sets: the Julia set and the filled Julia set.

DEFINITION 7.2   *Let $f : \mathbb{C} \to \mathbb{C}$. Then the **filled Julia set** $K(f)$ of the map f is defined as*

$$K(f) = \{z \in \mathbb{C} : O(z) \ \text{is bounded}\}.$$

The Julia set $J(f)$ of the map f is defined as the boundary of the filled Julia set $K(f)$. Equivalently, one may define J as the boundary of the escape set

$$E = \{z \in \mathbb{C} : |f^n(z)| \to \infty \ \text{as} \ n \to \infty\}.$$

Our main focus in this section will be on the quadratic map $Q_c(z) = z^2 + c$, where c is a complex constant. The corresponding filled Julia set, and the Julia set are denoted, respectively, by K_c and J_c.

A natural question that arises here is why restrict to polynomials of the form $Q_c(z) = z^2 + c$. The reason is simple: every quadratic map is topologically conjugate to Q_c for some real number c. In Example 3.11 we have shown that every quadratic real function of the form $P(x) = ax^2 + bx + c$ is conjugate to the logistic map $F_\mu(x) = \mu x(1 - x)$. Analogously, one may show that every quadratic complex function $P(z) = \alpha z^2 + \beta z + \gamma$ is conjugate to the quadratic complex map Q_c for some $c \in \mathbb{C}$. If $\alpha \neq 0$, $\beta, \gamma \in \mathbb{C}$, the conjugacy map h between Q_c and P is given by $h(z) = \alpha z + \beta/2$. It is easy to show that $h \circ P = Q_c \circ h$, where $c = \alpha\gamma + +\beta/2 - \beta^2/4$.

Example 7.4

1. Find K_0, and J_0.

2. Find K_{-2}, and J_{-2}. $\Box$

SOLUTION

1. From our previous analysis of $Q_0(z) = z^2$, it is easy to see that K_0 is the closed unit disk and J_0 is the unit circle.

2. To find K_{-2} and J_{-2}, we will use a conjugacy argument from Chapter 3. Consider the conjugacy map $h(z) = z + \frac{1}{z}$ defined on the complement of the closed unit disk; $A = \{z \in \mathbb{C} : |z| > 1\}$. We are going to show that the map h is a homeomorphism from A onto $B = \mathbb{C}\setminus[-2, 2]$. First, h is one-to-one. For if not, then for some points $z_1, z_2 \in A$, $h(z_1) = h(z_2)$. Hence, $z_1 + \frac{1}{z_1} = z_2 + \frac{1}{z_2}$, or $(z_2 - z_1)(z_1 z_2 - 1) = 0$. So, either $z_1 = z_2$ or $z_1 z_2 = |z_1||z_2| = 1$. The latter statement is impossible since $|z_1| > 1$ and $|z_2| > 1$. To show that h is onto, let $w = h(z)$ for some $z \in A$. Then, $z^2 - wz + 1 = 0$ has two solutions $z_{1,2} = \frac{1}{2}w \pm \frac{1}{2}\sqrt{w^2 - 4}$. Thus, if $-2 < w < 2$, then $z_{1,2} = \frac{1}{2}w \pm \frac{1}{2}i\sqrt{4 - w^2}$. This implies that $|z_{1,2}| = 1$ and consequently $z_{1,2} \notin A$. On the other hand, if $w = \pm 2$, then $z_{1,2} = \pm 1$ and thus $z_{1,2} \notin A$ also. Hence, h is onto. It is easy to check that h and h^{-1} are continuous maps and that h is a homeomorphism. Moreover, $h(Q_0(z)) = Q_{-2}(h(z))$ for all $z \in \mathbb{C}$. Consequently, Q_0 on A is conjugate to Q_{-2} on B (see Fig. 7.9).

   Since $Q_0^n(z) \to \infty$ as $n \to \infty$ for all $z \in A$, it can be shown that $Q_{-2}^n(z) \to \infty$ for all $z \in B$ (Problem 8).

   Furthermore, since the unit circle is mapped under h to the closed interval $[-2, 2]$, it follows that $J_{-2} = K_{-2} = [-2, 2]$ (see Fig. 7.10). ∎

In the previous two examples, the filled Julia sets are simple and not interesting. The question now is what happens for the filled Julia set of Q_c when

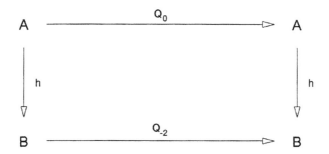

FIGURE 7.9

Conjugacy diagram : $h \circ Q_0 = Q_{-2} \circ h$.

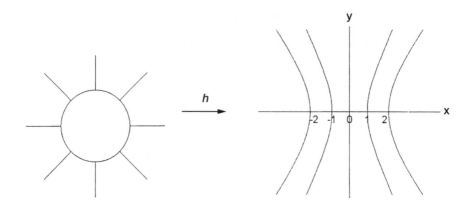

FIGURE 7.10

The map h maps lines emanating from the circle to hyperbolas with vertices on $[-2, 2]$.

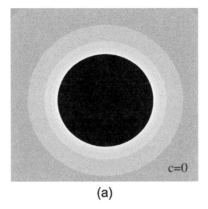

(a)

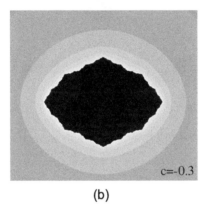

(b)

FIGURE 7.11
(a) The filled Julia set K_0. (b) The filled Julia set $K_{-0.3}$.

c varies from 0 to -2, i.e., how does the Julia set evolve from a unit circle to the closed interval $[-2, 2]$? Figs. 7.11a–7.11f show graphs of the filled Julia set K_c for real values of c. Observe that as c decreases from 0 to -2, K_c becomes more and more pinched together, until it becomes the closed interval $[-2, 2]$, when $c = -2$.

More spectacular pictures of filled Julia sets K_c are obtained when c is a complex number. But, before doing so, we will discuss some theoretical aspects that will enhance our understanding of both sets K_c and J_c. Moreover, we will develop an algorithm to generate the filled Julia set K_c, which is based on the **escape criterion** as stated below.

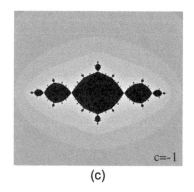

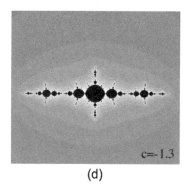

FIGURE 7.11
(c) The filled Julia set K_{-1}. (d) The filled Julia set $K_{-1.3}$.

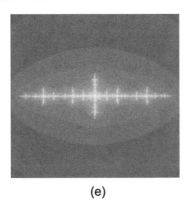

(e)

-2   2

(f)

FIGURE 7.11
(e) The filled Julia set $K_{-1.6}$. (f) The filled Julia set K_{-2}.

THEOREM 7.2 (The Escape Criterion)
Suppose that

$$|z| > \max\{2, |c|\}, \quad then \quad |Q_c^n(z)| \to \infty \quad as \quad n \to \infty.$$

PROOF   Assume that for $z \in \mathbb{C}, |z| > \max\{2, |c|\}$. Then,

$$
\begin{aligned}
|Q_c(z)| = |z^2 + c| &\geq |z^2| - |c| \\
&\geq |z|^2 - |c| \\
&> |z|^2 - |z| \quad (\text{ since } |z| > |c|) \\
&= |z| \, (|z| - 1)
\end{aligned}
\tag{7.4}
$$

Noting that $|z| > 2$, we have $|z| - 1 = 1 + \eta$, for some $\eta > 0$. Using this information in Inequality (7.4) yields

$$|Q_c(z)| > (1 + \eta)|z|. \tag{7.5}$$

By mathematical induction on n it can be shown that

$$|Q_c^n(z)| > (1 + \eta)^n |z|$$

and implies that $|Q_c^n(z)| \to \infty$ as $n \to \infty$. ∎

Algorithm for the Filled Julia Set

Set $r(c) = \max\{|c|, 2\}$. Then, $r(c)$ is called the **threshold radius** of Q_c. Observe that from Theorem 7.2, if for any $z \in \mathbb{C}, |Q^k(z)| > r(c)$, for some $k \in Z^+$, then the orbit of z escapes to ∞ and consequently $z \notin K_c$. Based on the above remark, we now introduce an algorithm to generate K_c.

Picture a square grid centered at the origin with side of length $r(c)$. We fix the maximum number of allowed iterations to be N. Then, for each z in the grid

1. If $|Q_c^k(z)| > r(c)$, for some $k \leq N$, color the point z white.

2. If $|Q_c^k(z)| \leq r(c)$, for all $k \leq N$, color the point z black.

The black points provide an approximation of the filled Julia set K_c (see Fig. 7.11).

Backward Iteration Algorithm (Encirclement of the Filled Julia Set)

Let $r(c) = \max(|c|, 2)$ be the threshold radius of Q_c. Define $L_0 = \{z \in \mathbb{C} : |z| \leq r(c)\}$. Let $L_{-1} = Q_c^{-1}(L_0) = \{z \in \mathbb{C} : Q_c(z) \in L_0\}$. Claim that $L_{-1} \subset L_0$. For if not then there exists $w \in L_{-1} \backslash L_0$ with $|w| > r(c)$. By the escape criterion (Theorem7.2), it follows that $|Q_c(w)| > |w| > r(c)$, a contradiction since $Q_c(w) \in L_0$. Define $L_{-2} = Q_c^{-1}(L_{-1}), L_{-3} = Q_c^{-1}(L_{-2}), \ldots, L_{-n-1} = Q_c^{-1}(L_{-n})$. Then, each L_{-n} is nonempty since it contains the two fixed points of Q_c (those are the roots of $z^2 - z + c = 0, \ z_{1,2} = \frac{1}{2} \pm \frac{1}{2}\sqrt{1 - 4c} \in L_0$). It is also easy to verify that

$$L_0 \supset L_{-1} \supset L_{-2} \ldots \supset L_{-n} \supset \ldots$$

and that each L_{-n} is closed and bounded. Furthermore,

$$K_c = \bigcap_{n=0}^{\infty} L_{-n} \tag{7.6}$$

$$= \lim_{n \to \infty} L_{-n} \quad \text{(in the Hausdorff metric)}$$

(see Problem 3).

Observe that $Q_c^{-1}(K_c) = K_c$ since $L_{-n-1} = Q_c^{-1}(L_{-n})$. Moreover, since Q_c is onto, $Q_c(K_c) = K_c$. Thus, K_c is positively and negatively invariant, i.e., invariant. Similarly, one may show that the Julia set J_c is a nonempty, closed, and invariant bounded subset of $\mathbb{C}$ (Problem 2).

We now summarize our findings in the following theorem.

THEOREM 7.3

The sets K_c and J_c are nonempty, closed, bounded, and invariant subsets of $\mathbb{C}$.

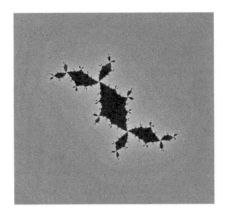

FIGURE 7.12a
(a) : $c = -0.1 + 0.8i$. The Douady Rabbit.

FIGURE 7.12b
(b) $c = -0.39 - 0.58i$. The Siegel disk.

FIGURE 7.13

$c = -0.5 + 0.5i$. Encirclement in alternating colors; in black: Q_c^0, Q_c^{-2}, Q_c^{-4}, Q_c^{-20} in white: Q_c^{-1}, Q_c^{-3}, Q_c^{-5}, Q_c^{-21}.

You may use different colors to shade the encirclements L_{-n}. In Fig. 7.13 we color L_{-2n} black while L_{-2n-1} is colored white, for $n = 0, 1, 2, \ldots$

Next, we further analyze the structure of the sets K_n. The following lemma sheds more light on K_n.

LEMMA 7.1

Let Γ be a smooth, simple, closed curve in $\mathbb{C}$. Then,

$$Q_c^{-1}(\Gamma) = \{w \in \mathbb{C} : Q_c(w) \in \Gamma\}$$

has the following properties:

1. *If c is in the interior of Γ, then $Q_c^{-1}(\Gamma)$ is also a smooth, simple, and closed curve. The interior of $Q_c^{-1}(\Gamma)$ corresponds one-to-one with the interior of Γ (see Fig. 7.14a).*

2. *If c lies on Γ, then $Q_c^{-1}(\Gamma)$ is a smooth, figure eight curve. The interior of each one of the two leaves corresponds one-to-one with the interior of Γ (see Fig. 7.14b).*

3. *If c is in the exterior of Γ, then $Q_c^{-1}(\Gamma)$ consists of two closed smooth curves, the interior of each corresponds one-to-one with the interior of Γ (see Fig. 7.14c).*

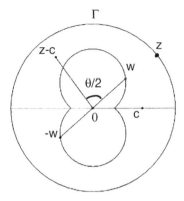

FIGURE 7.14a
$Q_c^{-1}(\Gamma)$ when c is an interior point.

PROOF   (Sketch).[1] The general proof is very technical and requires more background in complex analysis. We will provide a proof for the case when Γ is a circle.[2] Let $z \in \Gamma$ and $z - c = |z - c| \, e^{i\theta}$. Then, $w = \pm\sqrt{z - c} = \pm\sqrt{|z - c|} \, e^{i\theta/2}$.

1. If c is in the interior of Γ, then the map $z - c$ moves the center of the circle from the origin to a point inside Γ. Hence, $Q_c^{-1}(\Gamma)$ will look like the set in Fig. 7.14a.

2. If c is on Γ, then the map $z - c$ moves the center of the circle from the origin to a point on the circle, namely, $-c$, and we get a circle $\tilde{\Gamma}$ that passes through the origin. Thus, $Q_c^{-1}(\Gamma)$ will look like the set in Fig. 7.14b.

3. If c is in the exterior of Γ, then the map $z - c$ moves Γ to a circle $\tilde{\Gamma}$ that has the origin in its exterior. In this case, $Q_c^{-1}(\Gamma)$ will look like the set in Fig. 7.14c.   ▮

---

[1]The proof is optional.

[2]By virtue of the Riemann mapping theorem [1], this assumption is not restrictive. The Reimann mapping Theorem states that for any simply connected region (a region with no holes) $\mathcal{R} \neq \mathbb{C}$, there exists an analytic function mapping $\mathcal{R}$ one to one onto the disk $|z| < 1$ in $\mathbb{C}$. Now if Γ is a simple closed curve, then it encloses a region $\mathcal{R} \neq \mathbb{C}$ which is mapped one to one onto $|z| < 1$. Hence the boundary of $\mathcal{R}$, the curve Γ, is mapped analytically one to one onto the circle $|z| = 1$.

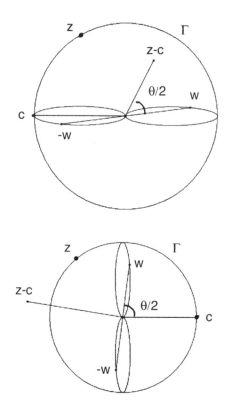

FIGURE 7.14b
$Q_c^{-1}(\Gamma)$ when c is on Γ.

7.5   Topological Properties of the Julia Set

In this section, we further analyze the structure of the Julia set and examine its topological properties. We start with the notion of connectedness.

DEFINITION 7.3   *A subset B of $\mathbb{C}$ is said to be **pathwise connected**, if for any two points z_1 and z_2 in B, there is a continuous function $f : I = [a, b] \to B$ such that $f(a) = z_1, f(b) = z_2$. Such a function f (as well as $f(I)$) is called a path from z_1 to z_2.*

We remark here that in many books about complex analysis, this definition is given for connected sets. But, here a set B is **connected** if it is not the union of two disjoint, nonempty open subsets of $\mathbb{C}$. While it is true that a pathwise

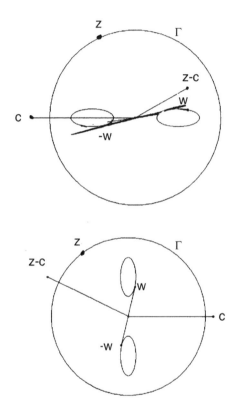

FIGURE 7.14c

$Q_c^{-1}(\Gamma)$ when c is in the exterior of Γ.

connected set is necessarily connected, the converse is false (see [108], p. 198). However, if the set B is open, then connectedness and pathwise connectedness are equivalent.

THEOREM 7.4

The following statements hold [20]:

1. *If $0 \in K_c$, then the Julia set J_c is pathwise connected.*

2. *If $0 \notin K_c$, then the Julia set J_c is totally disconnected and is actually a Cantor set.*

PROOF   (Sketch)[3]

1. Suppose $O(0)$ is bounded. Let Γ_0 be a circle that encloses $O(0)$ such that points outside Γ_0 escape to ∞. Since $Q_c(0) = c$, then c is in the interior of Γ_0. By Lemma 7.1, $\Gamma_{-1} = Q_c^{-1}(\Gamma_0)$ is a simple closed curve that lies entirely inside Γ_0. Moreover, by Lemma 7.1, Q_c maps the interior of Γ_{-1} onto the interior of Γ_0. Observe that $Q_c(c) = Q_c^2(0) \in \Gamma_0$ by assumption. Hence, $c = Q_c^{-1} Q_c(c) \in Q_c^{-1}(\Gamma_0) = \Gamma_{-1}$, and c is in the interior of Γ_{-1}. Thus, $\Gamma_{-2} = Q_c^{-1}(\Gamma_{-1})$ is a simple closed curve that lies inside Γ_{-1}. Continuing this process, we create a sequence $\{\Gamma_n\}$ of simple closed curves, where Γ_{-n-1} lies inside Γ_{-n} and $c \in \Gamma_{-n}$ for all $n \in Z^+$. Let L_{-n} be the closed region bounded by Γ_{-n}. Then, each L_{-n} is closed, bounded, and pathwise connected. It is easy to show that

$$K_c = \bigcap_{n=0}^{\infty} L_{-n} \text{ is the filled Julia set of } Q_c.$$

   From Willard [108], it follows that K_c is pathwise connected (see Fig. 7.15).

2. Suppose that the orbit $O(0)$ of 0 is unbounded. Let Γ_0 be a circle such that $\Gamma_{-1} = Q_c^{-1}(\Gamma_0)$ lies entirely inside Γ_0 and such that points outside Γ_{-1} escape to ∞. Furthermore, there exists $N \in Z^+$ such that $Q_c^N(0) \in \Gamma_0, Q_c^r(0)$ is inside Γ_0 for $0 \le r < N$ and $Q_c^r(0)$ is outside Γ_0 for $r > N$.

   As in part 1, we let $\Gamma_{-n-1} = Q_c^{-1}(\Gamma_{-n})$. Now, for $n < n_0, c$ is in the interior of Γ_{-n+1} and thus Γ_{-n} is a simple closed curve inside Γ_{-n+1}. When $n = n_0, c \in \Gamma_{-n_0+1}$ and thus by Lemma 7.1, Γ_{-n_0} is a figure eight. Now, the Julia set is divided into two subsets each of which is contained in one region of the Fig. 7.8. Hence $J(f)$ is disconnected (see Fig. 7.16). To prove that

---

[3]The proof is optional.

$J(f)$ is, in fact, a Cantor set is beyond the scope of this book. The interested reader may consult [25]. ∎

Julia Sets and the Repelling Periodic Points

Next, we show that the Julia set contains all repelling periodic points of Q_c. However, before establishing this result, we need the following inequality from **complex analysis** [1].

LEMMA 7.2 (Cauchy's Inequality)
Let Γ be a circle with center at z_0 and radius r, and let f be analytic in an open set U containing Γ and its interior. Then,

$$|f^{(n)}(z_0)| \leq \frac{Mn!}{r^n}, \quad n = 0, 1, 2, 3, \ldots.$$

where $M = \sup\{|f(z)| : z \in \Gamma\}$ and $f^{(n)}$ is the nth derivative of f.

Note that the map Q_c is analytic everywhere. A k-periodic point z_0 is called repelling if $|(Q_c^k)'(z_0)| > 1$.

THEOREM 7.5
The following statements hold true:

1. *If z_0 is a repelling periodic point of Q_c, then $z_0 \in J_c$. Furthermore, the set of repelling periodic points of Q_c is dense in J_c.*

2. *If $z_0 \in J_c$, then J_c is the closure of $\cup_{n=1}^{\infty} Q_c^{-n}(z_0)$.*

PROOF Suppose there exists a repelling N-periodic point z_0 such that $z_0 \notin J_c$. Since $O(z_0)$ is finite, it is bounded and thus $z_0 \in K_c \backslash J_c$. Hence, z_0 is an interior point of K_c and consequently there exists an open disc $B_\delta(z_0) \subset K_c$. Let Γ be a circle centered at z_0 with radius r such that $\Gamma \subset B_\delta(z_0)$.
Then, by Lemma 7.2,

$$|(Q_c^{kN})'(z_0)| \leq \frac{M}{r}, \quad \text{for all } k \in Z^+. \tag{7.7}$$

Since z_0 is a repelling periodic point, we have

$$|(Q_c^N)'(z_0)| = \eta > 1$$

and by the chain rule, we obtain

$$|(Q_c^{kN})'(z_0)| = \eta^k. \tag{7.8}$$

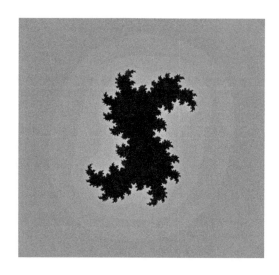

FIGURE 7.15

A connected Julia set. $c = 0.377 - 0.248i$.

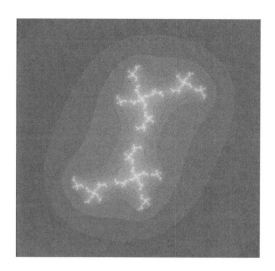

FIGURE 7.16

A disconnected Julia set. $c = -1.2i$.

Hence, $|(Q_c^{kN})'(z_0)| \to \infty$ as $k \to \infty$, which contradicts inequality (7.7).

This completes the proof of the first part of the theorem. The proofs of the second part of 1, and part 2 are beyond the scope of this book. The interested reader may consult [25]. ∎

Chaos on the Julia Set

We have seen two instances in which the Julia set is simple: $J_0 = S^1$, and $J_{-2} = [-2, 2]$. Let us examine closely the dynamics of the map $Q_0(z) = z^2$ on its Julia set, the unit circle S^1. Let $z = e^{i\theta} \in S^1$. Then, $Q_0(z) = e^{2i\theta}$, which is reminiscent of our old friend, the double-angle map, encountered in Chapter 3. As in Chapter 3, one may prove that Q_0 is chaotic on S^1. Since Q_0 and Q_{-2} are topologically conjugate, it follows by Theorem 3.9 that Q_{-2} is also chaotic on its Julia set $[-2, 2]$.

In the next result, we extend the above observations to the general quadratic map Q_c.

THEOREM 7.6
The quadratic map $Q_c(z) = z^2 + c$ is chaotic on its Julia set J_c, for every $c \in \mathbb{C}$.

PROOF   Theorem 7.4 tells us that the set of periodic points in J_c is dense in J_c. To prove transitivity of Q_c on J_c, let U and V be two nonempty subsets of $\mathbb{C}$ that intersect J_c. Let $z_0 \in V \cap J_c$. Then, by Theorem 7.5, $J_c = \cup_{n=1}^{\infty} Q_c^{-n}(z_0)$. Thus, for some $m \geq 1, Q_c^{-m}(z_0) \cap U \neq \phi$. Thus, if $u \in Q_c^{-m}(z_0) \cap U$, then $Q_c^{-m}(u) = z_0$ and consequently, $Q_c^m(U) \cap V = \phi$. This implies that Q_c is chaotic on J_c. ∎

Summary Our exposition may be extended to polynomials $P(z) = a_0 + a_1 z + \cdots + a_n z^n$ of degree n, where $P : \mathbb{C} \to \mathbb{C}$, $n \geq 2$. And indeed it can be extended further to rational functions $f(z) = u(z)/v(z)$, where $u(z)$ and $v(z)$ are polynomials on the extended complex plan $\mathbb{C} \cup \{\infty\}$. One may show that the following properties hold for the Julia set $J(P)$

(i) $J(P)$ is the closure of the repelling periodic points.

(ii) $J(P)$ is an uncountable compact set containing no isolated points and is invariant under P and P^{-1}.

(iii) If $z \in J(P)$, then $J(P) =$ the closure of the set $\cup_{k=1}^{\infty} P^{-1}(z)$.

(iv) $J(P)$ is the boundary of the basin of attraction of each attracting fixed point of P, including ∞, and $J(P) = J(P^n)$, for each positive integer n.

Exercises - (7.4 and 7.5)

1. Generate the graphs of the filled Julia sets, in Figs. 7.11a–7.11f, using the program "Julia."

2. Describe the filled Julia set for $f(z) = z^3$.

3. Prove Equation (7.6).

4. Consider the map $Q_i(z) = z^2 + i$.

   (a) Prove that 0 is an eventually periodic point.
   (b) Describe the filled Julia set K_i.

5. Show that for any $c \in \mathbb{C}$, J_c and K_c are closed and bounded invariant sets.

6. Prove that if $|z| > |c| + 1$, then the orbit $O(z)$ of z, under Q_c, is unbounded.

7. Let $F_\mu(z) = \mu z(1 - z)$ be the logistic map on the complex plane.

   (a) Show that if $|z| > \frac{1}{|\mu|} + 1$, then $|F_\mu(z)| > |z|$.
   (b) Use part (a) to develop an escape criterion for F_μ.

8. Let $B = C\setminus[-2, 2]$. Show that for every $z \in B$,
$$Q^n_{-2}(z) \to \infty \ \text{as} \ n \to \infty.$$

9. Show that for $Q_c = z^2 + c$ we have

   (a) $Q^n_c(z) \to 0$ if $|z| < \frac{1}{2} + \sqrt{\frac{1}{4} - |c|}$.
   (b) $Q^n_c(z) \to \infty$ if $|z| > \frac{1}{2} + \sqrt{\frac{1}{4} + |c|}$ as $n \to \infty$.

10. Let $f(z) = z^3 + c$. Show that if $|z| \geq |c|$ and $|z|^2 > 2$, then $|f^n(z)| \to \infty$ as $n \to \infty$.

11.* Let $f(z) = z^k + c, k > 3$. Show that if $|z| \geq |c|$ and $|z|^{k-1} > 2$, then $|f^n(z)| \to \infty$ as $n \to \infty$.

12. Let $f(z) = z^3 + az + b$. Show that if $|z| > \max\{|b|, \sqrt{|a| + 2}\}$, then $|f^n(z)| \to \infty$ as $n \to \infty$.

13. (Project) Develop an escape criterion for the map $E_\lambda = \lambda e^z$, where $e^z = e^{x+iy} = e^x(\cos y + i \sin y)$. Then sketch the filled Julia set for $\lambda = 1, \lambda = \pi i, \lambda = -2 + i, \lambda = 1 + 0.4i$.

14. (Project) Develop an escape criterion for the map $S_\lambda(z) = \lambda \sin z$. Then sketch the filled Julia set for $\lambda = 0.8, \lambda = 1, \lambda = 1.1$.

7.6 Newton's Method in the Complex Plane

In Chapter 1, we briefly discussed Newton's method on the real line. Recall that to find a real zero of a function $f(x)$, we start with an initial guess $x(0) = x_0$ and use the difference equation

$$x(n+1) = x(n) - \frac{f(x(n))}{f'(x(n))}, n = 0, 1, 2, \ldots \qquad (7.9)$$

which is equivalent to the following map:

$$N_f(x) = x - \frac{f(x)}{f'(x)}. \qquad (7.10)$$

The map N_f is commonly called the Newton map. It was shown (see Example 1.10) that zeros of $f(x)$ correspond to fixed points of $N_f(x)$. Furthermore, fixed points of $N_f(x)$ are (locally) asymptotically stable. Hence, the procedure (7.9) converges to an actual root of $f(x)$. The main goal of this section is to extend Newton's method to complex functions $f(z)$ on the complex plane $\mathbb{C}$. In this case, Formula (7.10) reads as

$$N_f(z) = z - \frac{f(z)}{f'(z)}.$$

One may show, using Corollary 7.1, that the fixed points of $N_f(z)$ are asymptotically stable (Problem 1).

Furthermore, a point z^* is a zero of $f(z)$ if and only if it is a fixed point of $N_f(z)$.

In 1879, Sir Authur Cayley [17] considered the convergence of Newton's method for complex quadratic polynomials. Let us start our exposition with the following interesting example.

Example 7.5
Let $Q_{-1}(z) = z^2 - 1$. Then $N(z) = \frac{z^2+1}{2z}$, whose fixed points are $z_1^* = 1$ and $z_2^* = -1$. Now,

$$|N'(z_i^*)| = |\frac{1}{2} - \frac{1}{2z_i^{*2}}| = 0, 1 \leq i \leq 2.$$

Hence, by Corollary 7.1, both z_1^* and z_2^* are asymptotically stable. Since the dynamics of $N(z)$ can be well understood if one can show that the maps Q_0 and N are topologically conjugate on the extended complex plane $\overline{\mathbb{C}}$. We use a Möbius transformation $T(z) = \frac{az+b}{cz+d}$ as our conjugacy map. Note that T must take the attracting fixed points of $N\{1, -1\}$ to the attracting fixed points of $Q_0\{0, \infty\}$ and the repelling fixed point ∞ of N to the repelling fixed point 1 of Q_0. Thus, $T(\infty) = 1$, which implies that $a = c = 1$. Moreover, $T(1) = 0$ and $T(-1) = \infty$, which implies that $b = -1$ and $d = -1$. Hence, $T(z) = \frac{z-1}{z+1}$. □

It is easy to verify that $T \circ N = Q_0 \circ T$ and that T is a homeomorphism on $\overline{\mathbb{C}}$. Since T maps the imaginary axis to the unit circle (Problem 2), it follows that N is chaotic on the imaginary axis. Next, we show that T maps the right half plane to the interior of the unit circle and the left half plane to the exterior of the unit circle. But, this is self-evident since for $z = a + ib$ we have

$$|T(z)| = |\frac{a + ib - 1}{a + ib + 1}| = |\frac{(a-1)^2 + b^2}{(a+1)^2 + b^2}| \begin{cases} < 1 \text{ if } a > 0 \\ > 1 \text{ if } a < 0. \end{cases}$$

Thus, the basin of attraction of $z_1^* = 1$ is given by

$$W^s(1) = T^{-1}[W^s(0)] = T^{-1}(I) = \{z = a + ib : a > 0\},$$

where I is the interior of the unit circle. Similarly, the basin of attraction of $z_2^* = -1$ is given by

$$W^s(-1) = T^{-1}[W^s(\infty)] = T^{-1}(E)$$
$$= \{z = a + ib : a < 0, \}$$

where E is the exterior of the unit circle.

Sir Authur Cayley [17] gave the following result which generalizes the above observations.

THEOREM 7.7
The following statements hold:

(a) If a complex quadratic polynomial $Q(z) = az^2 + bz + c$ has two distinct zeros, then $N_Q(z)$ is chaotic on the perpendicular bisector of the line joining the two zeros. Moreover, the basin of attraction of each root under $N_Q(z)$ is the set of points that lie on the same side of the bisector as a root (see Fig. 7.17).

(b) If $Q(z)$ has one repeated root r, then the basin of attraction of r under N_Q is the entire complex plane.

Cayley posed the question of whether or not Theorem 7.7 remains valid for complex cubic polynomials. It turns out that Cayley's problem is rather complicated and its solution is surprising. Due to the limited scope of this book, we are only going to address this problem via the following example.

Example 7.6
Consider the complex cubic map $f(z) = z^3 - 1$. Then $N_f(z) = z - \frac{z^3-1}{3z^2} = \frac{2z^3+1}{3z^2}$ has three attracting fixed points: $z_1^* = 1$, $z_2^* = e^{\frac{2\pi i}{3}}$, and $z_3^* = e^{\frac{4\pi i}{3}}$.

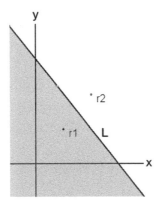

FIGURE 7.17
The roots of $Q(z) = az^2 + bz + c$ are r_1 and r_2. The basin of attraction of r_1 under $N_Q(z)$ is in gray. On the perpendicular bisector L, $N_Q(z)$ is chaotic. The remaining points are in the basin of attraction of r_2.

Using computer graphics (see Fig. 7.18), we see that $W^s(z_1^*)$ is the black region, $W^s(z_2^*)$ is the gray region, and $W^s(z_3^*)$ is the white region. This can be accomplished via a simple algorithm. Select a grid of points in the square, say 200×200, with vertices $z_1 = 2+2i$, $z_2 = 2-2i$, $z_3 = -2-2i$, and $z_4 = -2+2i$ (in the complex domain). For each point z in the grid, compute $N_f^{200}(z)$. If (a) $|N_f^{200}(z) - z_1^*| < 0.1$, then color the point black, (b) $|N_f^{200}(z) - z_2^*| < 0.1$, then color the point gray. The rest of the points are assumed to be in $W^s(z_3^*)$ and will have the white color.

In [79], it was shown that $\partial W^s(z_1^*) = \partial W^s(z_2^*) = \partial W^s(z_3^*) = J(N_f)$. In other words, the Julia set of the map $N_f(z) = z - \frac{z^3-1}{3z^2}$ is the boundary of the basins of attractions for the three attracting fixed points z_1^*, z_2^*, and z_3^*.
▯

Exercises - (7.6)

1. Let $N_f(z)$ be the Newton map of a complex function $f(z)$.

   (a) Show that r is a zero of $f(z)$ if and only if r is a fixed point of $N_f(z)$.

   (b) Show that the iterative Newtons's procedure converges to a zero of $f(z)$.

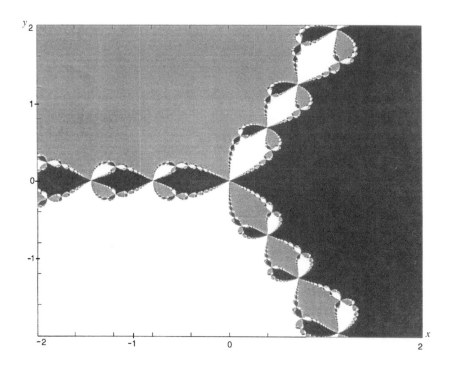

FIGURE 7.18

For $f(z) = z^3 - 1$, $N_f(z) = \frac{2z^3+1}{3z^2}$. There are three attracting fixed points: 1, $e^{2\pi i/3}$, $e^{4\pi i/3}$. Grey: Basin of attraction of $e^{2\pi i/3}$; White: Basin of attraction of $e^{4\pi i/3}$; Black: Basin of attraction of 1.

2. Let $T(z) = \frac{1-z}{1+z}$ be defined on the extended complex plane $\overline{\mathbb{C}}$. Show that T maps the imaginary axis onto the unit circle.

3. Prove Theorem 7.7 (a).

4. Prove Theorem 7.7 (b).

5. Describe the dynamics of Newton's function N_f for $f(z) = z^3$.

6. Describe the dynamics of Newton's function for any cubic polynomial with one zero of multiplicity three.

7. Show that the function $H(z) = \frac{z-i}{z+i}$ gives a conjugacy between the Newton map $N_{Q_1}(z)$ and $Q_0(z)$.

8. Describe the dynamics of Newton function N_f for $f(z) = z^3 - z^2$.

9. Describe the dynamics of Newton function N_f for $f(z) = z^3 - z$.

10.* Let $f(z) = z^3 - cz + 1$.

   (a) Show that an iterate of N_f is conjugate to $Q_{-2}(z) = z^2 - 2$ for some c.

   (b) Find a parameter value c for which N_f is chaotic on an open subset of $\mathbb{C}$.

   (c) Is there an open set of parameters c for which N_f is chaotic on a subset of $\mathbb{C}$?

7.7 The Mandelbrot Set

7.7.1 Topological Properties

The Mandelbrot set is considered by far the most complicated, yet the most fascinating fractal. It was discovered by the mathematician Benoit Mandelbrot in late 1970s [67]. For the quadratic family of maps Q_c, the Mandelbrot set, denoted by $\mathcal{M}$, is located in the parameter space and is defined as

$$\mathcal{M} = \{c \in \mathbb{C} : O(0) \text{ is bounded under } Q_c\}.$$

In other words, the Mandelbrot set $\mathcal{M}$ consists of all those c values for which the corresponding orbit of 0 under Q_c does not escape to infinity. A natural question may now be raised: why would anyone care about the fate of the orbit of 0? The answer to this question will be given later by Theorem 7.8.

As an immediate consequence of Theorem 7.4, we state the following **principle of dichotomy**.

COROLLARY 7.2

(Principle of Dichotomy). *For the quadratic map Q_c, we have either:*

1. *$c \in \mathcal{M}$ and the corresponding Julia set J_c is (pathwise) connected.*

2. *$c \notin \mathcal{M}$ and the corresponding Julia set J_c is totally disconnected (in fact, it is a Cantor set).*

PROOF

1. Observe that if $c \in \mathcal{M}$, then $O(0)$ is bounded and thus $0 \in K_c$. Hence, Theorem 7.4 is applicable.

2. The proof of the second part is analogous and will be omitted.   ∎

The above corollary provides a useful characterization of the Mandelbrot set $\mathcal{M}$. It simply says that

$$\mathcal{M} = \{c : J_c \text{ is (pathwise) connected}\}.$$

You may wonder whether or not the Mandelbrot set itself is connected. We will postpone answering this question until we generate a few pictures of the Mandelbrot set.

We now give a criterion under which a parameter value c belongs to $\mathcal{M}$. In order to accomplish this task, we need to introduce the complex version of Singer's theorem; for a proof, see [25]. Recall that a point z_0 is a critical point of a map f if $f'(z_0) = 0$. Note that 0 is the only critical point of the quadratic map $Q_c(z)$.

THEOREM 7.8

If z_0 is an attracting point of a complex map f, then z_0 attracts at least one critical point of f.

COROLLARY 7.3

If Q_c has an attracting periodic point, then $c \in \mathcal{M}$.

PROOF   This is left to the reader as Problem 1.   ∎

Our next task is to develop an algorithm to draw the Mandelbrot set and to acquire an understanding of its fascinating graph. We start our program with the following important result.

THEOREM 7.9

The Mandelbrot set $\mathcal{M}$ is contained in the closed disk of radius 2, i.e.:

$$\mathcal{M} \subset \{c : |c| \leq 2\}.$$

PROOF   Suppose that $|c| > 2$. Then if $|z| \geq |c|$ we have

$$|Q_c(z)| = |z^2 + c| = |z|\,\left|z + \frac{c}{z}\right| \geq |z|\,\left(|z| - \frac{|c|}{|z|}\right) \geq |z|(|c| - 1) = \eta|z|,$$

where $\eta = |c| - 1 > 1$.

Now, $Q_c(0) = c, |Q_c^2(0)| = |Q_c(c)| \geq |c|\eta$. We will show by mathematical induction that $|Q_c^{n+1}(0)| \geq |c|\eta^n$. So, assume that $|Q_c^n(0)| \geq |c|\eta^{n-1}$. Then,

$$|Q_c^{n+1}(0)| = |Q_c Q_c^n(0)| \geq |Q_c^n(0)|\eta$$
$$\geq |c|\eta^n.$$

It follows that $|Q_c^{n+1}(0)| \geq |c|\eta^n \to \infty$ as $n \to \infty$. Consequently, $c \notin \mathcal{M}$, and the proof is now complete.   ∎

The question arises as to how to sketch the Mandelbrot set and what algorithm can be used to draw a computer-generated $\mathcal{M}$? Well, we simply use Theorem 7.9 to write down such an algorithm.

The Pixel Algorithm for the Mandelbrot Set $\mathcal{M}$

1. Choose the number of iterations allowed N.

2. Choose a grid (in the complex plane) contained inside the square of vertices:
   $(-2, 2), (2, -2), (2, 2), (-2, -2)$.

3. For each c in the grid, compute the first N points:
   $\{Q_c(0), Q_c^2(0), \ldots, Q_c^N(0)\}$.

4. If at any iteration $k \leq N, Q_c^k(0)$ leaves our square, stop the process and color c white; otherwise color it black.

The black region corresponding to the Mandelbrot set is seen in Figure 7.19.

7.7.2   Rays and Bulbs

We are now going to analyze the Mandelbrot set in Fig. 7.20 more closely. Let us begin with the big cardioid on the right side of the picture. The interior of this cardioid corresponds to all c for which $Q_c(z)$ has an attracting fixed point. To show this, let z^* be an attracting fixed point of $Q_c(z)$. Then, by Corollary 7.3, $c \in \mathcal{M}$. Furthermore,

$$Q_c(z) = z^2 + c = z \tag{7.11}$$

and

$$|Q_c'(z)| = |2z| < 1.$$

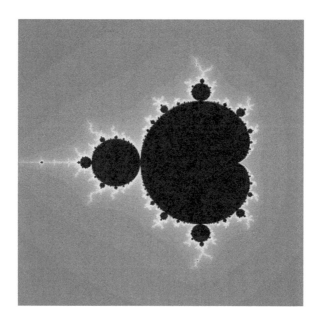

FIGURE 7.19
The Mandelbrot set $\mathcal{M}$.

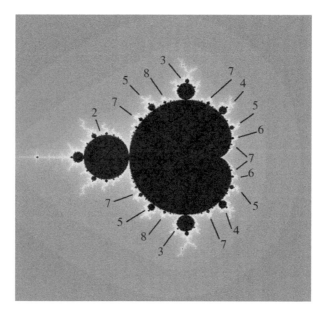

FIGURE 7.20
Periods of the primary bulbs in $\mathcal{M}$.

The boundary of such points is $|z| = \frac{1}{2}$ or $z = \frac{1}{2}e^{i\theta}, 0 \leq \theta \leq 2\pi$. Substituting into Equation (7.11) yields

$$c = \frac{1}{2}e^{i\theta} - \frac{1}{4}e^{2i\theta} \tag{7.12}$$

which traces a cardioid as θ ranges from 0 to 2π. (This corresponds to $r = \frac{1}{2} - \frac{1}{4}\cos\theta, 0 \leq \theta \leq 2\pi$, in real polar coordinates.) In summary, all the parameter values c for which $Q_c(z)$ has an attracting fixed point lie inside a cardioid that interests the x axis at $c = \frac{1}{4}$ (when $\theta = 0$) and $c = -\frac{3}{4}$ (when $\theta = \pi$).

Note that from Equation (7.11), the fixed points are

$$z_1^* = \frac{1 - \sqrt{1 - 4c}}{2}, \quad \text{and} \quad z_2^* = \frac{1 + \sqrt{1 - 4c}}{2}.$$

Hence,

$$Q_c'(z_1^*) = 1 - \sqrt{1 - 4c}, \quad \text{and} \quad Q_c'(z_2^*) = 1 + \sqrt{1 - 4c}.$$

Thus, z_2^* is always unstable, and z_1^* is asymptotically stable for all c for which $|1 - \sqrt{1 - 4c}| < 1$. This implies that for real $c, -\frac{3}{4} < c < \frac{1}{4}$. Hence, at $c = \frac{1}{4}$, we witness a saddle node bifurcation. This corresponds perfectly to the bifurcation diagram of the real map $Q_c(x) = x^2 + c, -\frac{3}{4} < c < \frac{1}{4}$, where we have a single branch (see Fig. 7.21). The scenario continues as in Chapter 1, and z_1^* loses its stability for $c < -\frac{3}{4}$ and an asymptotically stable 2-cycle is born. To find the boundary of all the attracting 2-cycles of $Q_c(z)$, we find the fixed points of $Q_c^2(z)$. This yields $(z^2 + c)^2 + c = z$ or $(z^2 + z + 1 + c)(z^2 + c - z) = 0$.

Note that $z^2 + c - z = 0$ produces the fixed points z_1^* and z_2^* of $Q_c(z)$, while

$$z^2 + z + 1 + c = 0 \tag{7.13}$$

yields the 2-periodic points $\tilde{z}_1$ and $\tilde{z}_2$. Since $\tilde{z}_1$ and $\tilde{z}_2$ are attracting, $|Q_c'(\tilde{z}_1) Q_c(\tilde{z}_2)| < 1$; thus,

$$|4\tilde{z}_1\tilde{z}_2| < 1. \tag{7.14}$$

But, from Equation (7.13), $\tilde{z}_1\tilde{z}_2 = 1 + c$. Substituting this into Inequality (7.14) yields $|1 + c| < \frac{1}{4}$.

Therefore, the parameter values c for which $Q_c(z)$ has an attracting 2-cycle lie entirely inside the circle $|1 + c| = \frac{1}{4}$, whose center is $(-1, 0)$ and radius is $\frac{1}{4}$. In Fig. 7.21, each bulb in the Mandelbrot set corresponds to c values for which $Q_c(z)$ admits an attracting periodic orbit of some period k.

As we have seen above, the main bulb is the cardioid, which contains all values of c for which $Q_c(z)$ has an attracting fixed point. This main bulb (decoration) has infinitely many smaller bulbs (decorations) and antennas. Each of these antennas, in turn, consists of a number of spokes that varies from one bulb (decoration) to another. Let us call any bulb (decoration) attached to the main cardioid as a primary bulb (decoration). For every c

that lies in the interior of a primary bulb, the orbit of zero is attracted to a certain k-cycle, where the period k is fixed for all such c. For those values of c that lie inside the other smaller bulbs (decorations) attached to the primary bulb, the orbit of zero is attracted to a cycle whose period is a multiple of k. Using the algorithm, given below, to determine the periods of the bulbs, one may check the following table which determines in which bulb in $\mathcal{M}$, the given value c is located.

$-0.12 + 0.75i$	3
$-0.5 + 0.55i$	5
$0.28 + 0.54i$	4
$0.38 + 0.333i$	5
$-0.62 + 0.43i$	7
$-0.36 + 0.62i$	8
$-0.67 + 0.34i$	9
$0.39 + 0.22i$	6

Note that the number of spokes in the largest antenna attached to a primary bulb (decoration) is equal to the period of that bulb (decoration), provided that we count the spoke emanating from the primary bulb (decoration) to the main junction point (see Figs. 7.21 to 7.24).

In the next figure (Fig. 7.25), we have imposed the bifurcation diagram for Q_c and the Mandelbrot set. The real c values that form the horizontal axis in the bifurcation diagram lie directly below the corresponding real c values in the Mandelbrot set. Note that each bifurcation corresponds to a new bulb that intersects the x axis and the period is the number of branches of the bifurcation diagram there.

An Algorithm to Determine the Period of a Bulb

The complex analogue of Singer's theorem states that if $\tilde{z}$ is a periodic point of a polynomial complex map, then it must attract a critical value. Now for our map $Q_c(z), 0$ is the only critical value. Pick a bulb in the Mandelbrot set and an approximate center $\tilde{c}$. The orbit $\{Q_{\tilde{c}}^n(0)\}$ will be asymptotic to a certain k-cycle, which may be determined computationally.

7.7.3 Rotation Numbers and Farey Addition

Recall that the main cardiod of the Mandelbrot set is given by $c = \frac{1}{2}e^{i\theta} - \frac{1}{4}e^{2i\theta}$. Let us call a "decoration" that is directly attached to the main cardiod $\mathcal{M}$ a primary bulb; all other decorations will be called secondary bulbs.

Now a primary bulb attaches to the main cardiod at an internal angle $\phi = 2\pi(m/n)$. The number m/n is called the rotation number of the bulb. The main cardiod will be called bulb $(0/1)$ ($\phi = 2\pi \cdot 0/1 = 0$). The main primary bulb is the circle $|1+c| = \frac{1}{4}$ centered at $(-1,0)$ which is attached at the angle $\phi = \pi = 2\pi(1/2)$ and hence this bulb will be called bulb $(1/2)$. It is worth

FIGURE 7.21
A bulb of period 3.

FIGURE 7.22
A bulb of period 4.

FIGURE 7.23
A bulb of period 5.

FIGURE 7.24
A bulb of period 7.

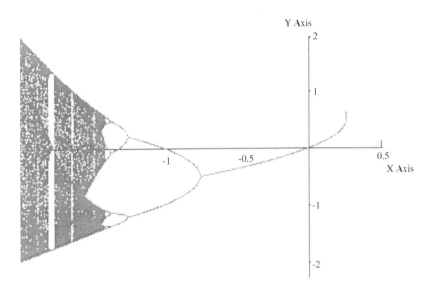

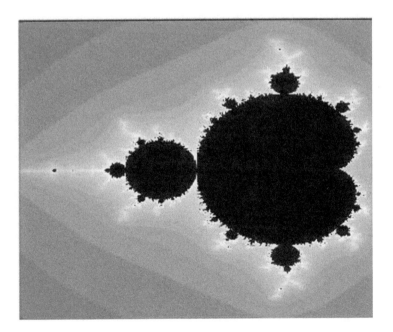

FIGURE 7.25

The bifurcation diagram of $Q_c(x) = x^2 + c$ in comparison to the Mandlebrot set $\mathcal{M}$. (This figure is a second generation computer image.)

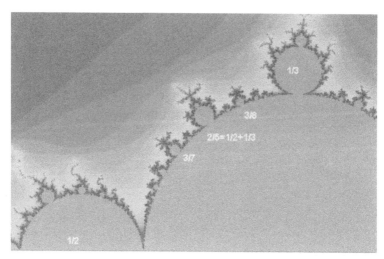

FIGURE 7.26
The primary bulbs.

noticing that the denominator n of the rotation number m/n is the period of
the attracting periodic cycle in the bulb (m/n) (Figure 7.28).

To find the rotation number m/n for the largest bulb between the $(0/1)$
and $(1/2)$ bulbs, we use the <u>Farey</u> addition: $a/b + c/d = (a+c)/(b+d)$.
Hence bulb $(1/3)$ is the largest bulb between $(0/1)$ and $(1/2)$ bulbs. Since
the Mandelbrot set is symmetric (why?), there will be a pair of bulb $(1/3)$,
one above the real axis and one below it. One of these bulbs will be located
at an angle $\theta = 2\pi/3$ and the other at an angle $\theta = 4\pi/3$. Moreover, both
will contain an attracting 3-periodic cycle. Now the largest bulb between the
$(1/2)$ and $(1/3)$ bulbs is the bulb $(2/5)$; the bulb $(2/5)$ is smaller than the
$(1/2)$ and $(1/3)$ bulbs. Next primary bulb sequences are given by

$$\frac{1}{2} + \frac{1}{3} = \frac{2}{5} + \frac{1}{2} = \frac{3}{7} + \frac{1}{2} = \frac{4}{9} + \dots \text{ and}$$
$$\frac{1}{2} + \frac{1}{3} = \frac{2}{5} + \frac{1}{3} = \frac{3}{8} + \frac{1}{3} = \frac{4}{11} + \dots \text{ and}$$
$$\frac{1}{2} + \frac{0}{1} = \frac{1}{3} + \frac{0}{1} = \frac{1}{4} + \frac{0}{1} = \frac{1}{5} + \dots$$

<u>Secondary bulbs.</u> Now we turn our attention to the secondary bulbs. Every
primary bulb (m_1/n_1) has decorations (bulbs) similar to the main bulb. Hence
the rotation number of a secondary bulb will be an ordered pair $(m_1/n_1, m_2/n_2)$,
where (m_1/n_1) is the rotation number of the primary bulb and (m_2/n_2) is
the rotation number of the attached secondary bulb. Moreover, the period of
the secondary bulb is $n_1 \times n_2$.

FIGURE 7.27
The primary bulbs.

The largest secondary bulb between $(m_1/n_1, m_2/n_2)$ and $(m_1/n_1, m_3/n_3)$ bulbs is bulb $\left(\frac{m_1}{n_1}, \frac{m_2+m_3}{n_2+n+3}\right)$. Figure 7.27 illustrates this procedure and shows few secondary bulbs in the primary bulb $(1/3)$.

7.7.4 Accuracy of Pictures

In his article [36], John Ewing questioned the accuracies of the pictures of the Mandlebrot set. He cited two reasons for his skeptisism; computers make mistakes, and so do people. By examining pictures of tiny sections of the boundary, we see disconnected pieces of $\mathcal{M}$, often resembling miniature copies of the whole set, floating nearby. One might easily conclude that $\mathcal{M}$ is not connected and that it consists of a main body with an infinite number of islands nearby. Douady and Hubbard [29] proved that the pictures are misleading and that the Mandelbrot set $\mathcal{M}$ is indeed connected.

THEOREM 7.10
The Mandelbrot set is connected.

PROOF   The proof is based on the idea that the Mandelbrot set may be viewed as a subset of the Riemann aphere $\overline{\mathbb{C}}$ using steriographic projection

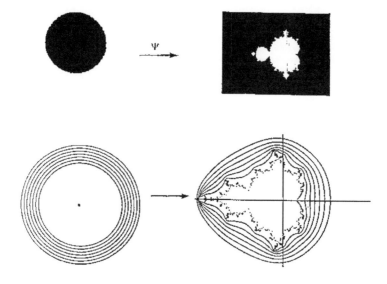

FIGURE 7.28

Mapping the disk to the outside of $\mathcal{M}$.

(see Section 7.3). Now the Mandelbrot set is connected simply when its complement (in $\mathbb{C}$) is "simply connected," that is, when its complement has no holes. The famous riemann mapping theorem says that such a set which is not the entire complex plane is homeomorphic to the unit disk.

Hubbard and Douady turned the Riemann mapping theorem around; they showed that there must exist such a map from the unit disk $\Delta = \{z \,:\, |z| \leq 1\}$ to the complement of $\mathcal{M}$ (in the sphere), and hence the complement of $\mathcal{M}$ is simply connected. They used the map

$$\Psi : \Delta \to \overline{\mathbb{C}} \quad \text{defined as}$$

$$\Psi(z) = \frac{1}{z} + b_0 + b_1 z + b_2 z^2 + \ldots$$

when δ is the unit disk.

The map Ψ gives us a new way to draw pictures of $\mathcal{M}$, different from the Pixel method. The first part of the map Ψ, $z \longmapsto \frac{1}{z}$ maps the interior of a disk Δ onto the exterior, reversing orientation and sending the origin to ∞. The power series terms $b_0 + b_1 z + b_2 z^2 + \ldots$ add a small distortion to make the image of Ψ the exterior of $\mathcal{M}$ rather than the exterior of Δ. The image under Ψ of a circle $|z| = r$, $r < 1$, is a simple closed curve bounding a region $\mathcal{M}_r$, which contains $\mathcal{M}$ as indicated in Figure 7.28. ■

John Ewing computed the area of the Mandelbrot set using the pixel method by subdividing a region of the plane containing $\mathcal{M}$ into pixels and calcualted the fraction that are inside $\mathcal{M}$. Using this method he found that the area of $\mathcal{M}$ is approximately 1.52. However, Ewing was not content with his finding. He used Green's Theorem to calcualte the area of $\mathcal{M}$ by integrating around its boundary. The map Ψ provides a change of variable that converts the problem of intcgrating around $\mathcal{M}$ to the problem of integrating around a circle (for more details see J. Ewing [36]). The estimate of the area of $\mathcal{M}$ using this method is 1.72. A challenging problem is to determine which one of these two area 1.52, 1.72 is the correct answer. It is a fascinating problem that I hope the reader will get interested in solving it.

Exercises - (7.7.1)

1. Show that if the quadratic map Q_c has an attracting periodic point, then $c \in \mathcal{M}$.

2. Generate the graph of the Mandelbrot set for $Q_{0.5}, Q_i, Q_{0.4+0.1i}$ using the program "Mandelbrot."

3. Show that if $c \in \mathcal{M}$, then $\bar{c} \in \mathcal{M}$.

4. Show that the Mandelbrot set $\mathcal{M}$ of Q_c contains all c such that $|c| \leq 1/4$. (Use mathematical induction.)

5. Consider the complex logistic map $F_\mu = \mu\, z(1-z)$, where $\mu, z \in \mathbb{C}$.

   (a) Find all fixed points of F_μ.

   (b) Find all μ for which F_μ has an attracting fixed point.

6. Let $\tilde{\mathcal{M}}$ be the set of all complex numbers μ such that $O^+(1/2)$ under F_μ is bounded. Modify the program "Mandelbrot" to obtain a graph of $\tilde{\mathcal{M}}$.

7. Let z_1^* and z_2^* be the fixed points of $Q_c(z)$. Prove:

   (a) if $c < -2$, then $Q^n(0) \geq 2 + n|c+2|$, for $n \geq 2$;

   (b) if $-2 \leq c < 0$, then Q_c maps $[-z_1^*, z_1^*]$ into itself;

   (c) if $0 \leq c \leq \frac{1}{4}$, then Q_c maps $[0, z_2^*]$ into itself;

   (d) if $c > \frac{1}{4}$, then $Q_c^n(0) \geq n(c - \frac{1}{4})$.

   (e) Conclude from the above that the intersection of the Mandelbrot set $\mathcal{M}$ with the real axis is the interval $[-2, \frac{1}{4}]$.

8. Show that $i \in \mathcal{M}$.

9. Draw a graph of the Mandelbrot set for $f_c(z) = z^3 + c$. Then, show that if $|c| > 2$, then $f_c^n(z) \to \infty$ as $n \to \infty$.

10. For the map $Q_c(z) = z^2 + c$, show that the set $\{z : Q_c^n(z) \to \infty\}$ is an open set.

In problems 11–14 find the secondary bulbs and graph them for the indicated primary bulb.

11. Primary Bulb $(1/2)$.

12. Primary Bulb $(3/7)$.

13. Primary Bulb $(2/5)$.

14. Primary Bulb $(3/8)$.

Bibliography

[1] L. V. Ahlfors, *Complex Analysis,* Third Edition, New York, 1979.

[2] K. T. Alligood, T. D. Sauer, J. A. Yorke, *Chaos: An Introduction to Dynamical Systems,* Springer-Verlag, New York, 1997.

[3] B. Aulbach and B. Kieninger, On three definitions of chaos, *Nonlinear Dyn. Syst. Theory* **1**, 2001, 23–37.

[4] J. Banks, J. Brooks, G. Cairns, G. Davis, and P. Stacey, On Devaney's definition of chaos, *Am. Math. Month.* **99**(4), 1992, 332–334.

[5] M. Barnsley, *Fractals Everywhere,* Academic Press, San Diego, 1988.

[6] M. Barnsley and L. Hurd, *Fractal Image Compression,* A. K. Peters, Ltd. Wellesley, Massachusetts, 1993.

[7] B. Basener et al., Dynamics of a discrete population model for extinction and sustainability in ancient civilizations, Preprint.

[8] M. Benedicks and L. Carleson, The dynamics of the Hénon map, *Ann. Math.* **133**, 1991, 73–169.

[9] R.J.H. Beverton and S.J. Holt, *On the Dynamics of Exploited Fish Population,* Fishery investigations, ser. 2, v. 19. H.M. Stationery Off, London, 1957.

[10] R.J.H. Beverton and S.J. Holt, *The Theory of Fishing,* In Sea Fisheries: Their Investigation in the United Kingdom, M. Graham, ed., 372–441, Edward Arnold, London, 1956.

[11] George D. Birkhoff, *Collected Mathematical Papers,* Dover Publication, New York, 1968.

[12] L. S. Block and W. L. Coppel, *Dynamics in One Dimension,* Lecture Notes in Mathematics No. 1513, Springer-Verlag, New York, 1992.

[13] K.M. Brooks and H. Bruin, *Topics from one-dimensional dynamics,* London Mathematical Society, Student Texts 62, Cambridge University Press, Cambridge, 2004.

[14] R. Brown, Horseshoes in the measure preserving Hénon map, *Ergod. Th. Dynam. Syst.* **15**, 1995, 1045–1059.

[15] L. Carleson and T.W. Gamelin, *Complex dynamics,* Springer–Verlag, New York, 1995.

[16] J. Carr, *Applications of Center Manifold Theory,* Springer-Verlag, New York, 1981.

[17] Arthur Cayley, The Newton-Fourier imagery problem, *Am. J. Math.* **2**, 1879, 97.

[18] K.G. Coffman, W.D. McCormic, R.H. Simoyi, and H.L. Swinney, Universality, multiplicity and the effect of iron impurities in the Belousov–Zhabotinskii reaction, *J. Chem. Phys.* **86**, 1987, 119.

[19] K. Cooke and L. Ladeira, Applying Carvalho's method to find periodic solutions of difference equations, *J. Difference Eq. Appl.* **2**, 1996, 105–115.

[20] R. Crownover, *Introduction to Fractals and Chaos,* Jones and Bartlett, Boston, 1995.

[21] P. Cull, Stability of discrete one-dimensional population models, *Bull. Math. Biol.* **50(1)**, 1988, 67–75.

[22] F. Dannan, S. Elaydi, and V. Ponomarenko, Stability of hyperbolic and nonhyperbolic fixed points of one-dimensional maps, *J. Difference Equ. Appl.* **9** (2003), 449–457.

[23] J. D. Delahaye, Fonctions admettant des cycles d'ordre n'importe quelle puissance de 2 et aucun autre cycle, *C. R. Acad. Sci. Paris* Ser. A-B **291**, 1980, 323–325; Addendum, 671.

[24] W. de Melo and S. Van Strien, *One-Dimensional Dynamics,* Springer-Verlag, New York, 1992.

[25] R. L. Devaney, *An Introduction to Chaotic Dynamical Systems,* Second Edition, Addison-Wesley, Reading, Massachusetts, 1989.

[26] R. L. Devaney, *A First Course in Chaotic Systems: Theory and Experiments,* Addison-Wesley, Reading, Massachusetts, 1992.

[27] R. Devaney and Z. Nitecki, Shift automorphism in the Hénon mapping, *Comm. Math. Phys.* **67**, 1979, 137–148.

[28] R. DeVault, E. Grove, G. Ladas, R. Levins, and C. Puccia, Oscillation and stability in a delay model of a perennial grass, *J. Difference Eq. Appl.* **1**, 1995, 173–185.

[29] A. Douady and J. Hubbard, Itération des polynômes quadratique complexes, *C. R. Acad. Sci. Paris* **29**, Ser. **I-1982**, 123–126.

[30] J. Dugundji, *Topology,* Allyn and Bacon, Boston, 1966.

[31] S. Elaydi, A converse of Sharkovsky's theorem, *Am. Math. Month.* **103**(5), 1996, 386–392.

[32] S. Elaydi, *An Introduction to Difference Equations,* Third Edition, Springer-Verlag, New York, 2005.

[33] S. Elaydi and W. Harris, On the computation of A^n, *SIAM Rev.* **40**(4), 1998, 965–971.

[34] S. Elaydi and A. Yakubu, Global stability of cycles: Lotka-Volterra competition model with stocking, *J. Diff. Equations and Appl.* **8**(**6**) (2002), 537-549.

[35] R.H. Enns and G.C. McGuire, *Nonlinear Physics with Maple for Scientists and Engineers,* Second Edition, Birkhauser, Boston, 2000.

[36] J. Ewing, Can we see the Mandelbrot set?, *The College Math. J.,* Vol 26, No. 2 (1995), pp. 90–99.

[37] K. Falconer, *Fractal Geometry: Mathematical Foundations and Applications,* John Wiley & Sons, New York, 1990.

[38] P. Fatou, Sur les equations fouctionneelles, *Bulletin Societé Math. France,* vol. 47, 1919, pp. 161–271.

[39] M. Feigenbaum, Quantitative universality for a class of nonlinear transformations, *J. Stat. Phys.* **19**, 1978, 25–52.

[40] J. Franks, Homology and dynamical systems, *Am. Math. Soc.,* CBMS 49, Providence, Rhode Island, 1982.

[41] J. Fröyland, *Introduction to Chaos and Coherence,* Institute of Physics Publishing, Philadelphia, 1992.

[42] S. Grossmann and S. Thomae, Invariant Distributions and Stationary Correlation Functions of One-Dimensional Discrete Maps, *Z. Naturforsch,* 32a, 1977, 1353–1363.

[43] J. Guckenheimer, On bifurcation of maps of the interval, *Invent. Math.,* vol. 39, 1977, pp. 165–178.

[44] D. Gulick, *Encounters with Chaos,* McGraw-Hill, New York, 1992.

[45] J. Guckenheimer, On bifurcation of maps of the interval, *Invent. Math.,* vol. 39, 1977, pp. 165–178.

[46] R. G. Harrison and D. J. Biswas, Chaos in light, *Nature* **321**, 1986, 504.

[47] F. Hausdorff, Dimension und Äusseres Mass, *Math. Ann.* **79**, 1919, 157–179.

[48] M. Hénon, A two-dimensional mapping with a strange attractor, *Comm. Math. Phys.,* **50**, 1976, 69–77.

[49] R.A. Horn and C.R. Johnson, *Matrix Analysis,* Cambridge University Press, Cambridge, 1999.

[50] T. Hüls, Bifurcation of connecting orbits with one nonhyperbolic fixed point for maps, *SIAM J. Appl. Dyn. Sys.,* **4(4)**, 2005, 985–1007.

[51] J. Hutchinson, Fractals and self-similarity, *Indiana Univ. J. Math.* **30**, 1981, 713–747.

[52] G. Julia, Memoire sur L'iteration des fonctions Rationnelles, *Journel des mathematiques pures et appliquées,* vol. 4, 1918, pp. 47–245.

[53] E. I. Jury, *Theory and Applications of the z-Transform Method,* John Wiley & Sons, New York, 1964.

[54] L. Edelstein-Keshet, *Mathematical Models in Biology,* Random House, New York, 1988.

[55] U. Kirchgraber and D. Stoffer, On intervals: transitivity → chaos, *Am. Math. Month.* **101**(4), 1994, 353–355.

[56] V. L. Kocic and G. Ladas, *Global Behavior of Nonlinear Difference Equations of Higher Order with Applications,* Kluwer Academic, Boston, 1993.

[57] A. N. Kolmogrov, Local structure of turbulence in an incompressible liquid for very large Reynolds numbers, *C. R. Acad. Sci. USSR* **30**, 1941, 299–303.

[58] M. Kot, *Elements of Mathematical Ecology,* Cambridge University Press, Cambridge, 2001.

[59] Y. A. Kuznetsov, *Elements of Applied Bifurcation Theory,* Springer-Verlag, New York, 1995.

[60] J. P. LaSalle, *The Stability of Dynamical Systems,* SIAM, Philadelphia, 1976.

[61] H. Lauwerier, *Fractals,* Princeton University Press, Princeton, New Jersey, 1991.

[62] T. Y. Li and J. A. Yorke, Period three implies chaos, *Am. Math. Month.* **82**, 1975, 985–992.

[63] A. M. Liapunov, Problème général de la stabilité du mouvement. *Ann. Math. Studies 17,* Princeton University Press, Princeton, New Jersey, 1949.

[64] A. Libchaber, C. Larouche, and S. Fauva, Period doubling cascade in Mercury, a quantitative measurement, *J. Physique Lett.,* **43:L211**, 1982.

[65] E. N. Lorenz, Deterministic nonperiodic flow, *J. Atm. Sci.* **20**, 1963, 130–141.

[66] S. N. MacEachern and L. M. Berliner, Aperiodic chaotic orbits, *Am. Math. Month.* **100**(3), 1993, 237–241.

[67] B. Mandelbrot, *The Fractal Geometry of Nature,* W. H. Freeman, New York, 1982.

[68] R. May, Simple mathematical models with very complicated dynamics, *Nature* **261**, 1976, 459–467.

[69] R. M. May, Course 8: Nonlinear problems in ecology and resource management, in G. Ioos, R.H.G. Helleman, and R. Stora, eds., *Chaotic Behavior of Deterministic Systems,* North-Holland, Amsterdam, 1983.

[70] F. C. Moon, *Chaotic Vibrations,* John Wiley & Sons, New York, 1987.

[71] R. D. Neidinger and R. J. Annen, The road to chaos is filled with polynomial curves, *Am. Math. Month.* **103**(8), 1996, 640–653.

[72] Y. Neimark, On some cases of periodic motions depending on parameters, *Dokl. Acad. Nauk. SSSR* **129**, 1959, 736–739.

[73] R. Nillsen, Chaos and one-to-oneness, *Math. Magazine* **72**(1), 1999, 14–21.

[74] J. Ortega, *Matrix Theory,* Plenum Press, New York, 1987.

[75] E. Ott, *Chaos in Dynamical Systems,* Cambridge University Press, New York, 1993.

[76] A. Pakes and R. A. Maller, *Mathematical Ecology of Plant Species Competition,* Cambridge University Press, Cambridge, Massachusetts, 1990.

[77] E. C. Pielou, *An Introduction to Mathematical Ecology,* John Wiley & Sons, New York, 1969.

[78] H. O. Peitgen, H. Jürgens, and D. Saupe, *Chaos and Fractals,* Springer-Verlag, New York, 1992.

[79] H. O. Peitgen and P. H. Richter, *The Beauty of Fractals: Images of Complex Dynamical Systems,* Springer-Verlag, Berlin, 1986.

[80] Henri Poincaré, *Science and Method,* Dover Publication, New York, 1952.

[81] Y. Pomeau and P. Manneville, Intermittent transition to turbulence in dissipative dynamical systems, *Comm. Math. Phys.,* 74(1980), 189.

[82] M. H. Protter and C. B. Morrey, *A First Course in Real Analysis,* Second Edition, Springer-Verlag, New York, 1991.

[83] P. Prusinkiewicz and J. Hanan, *Lindenmayer Systems, Fractals, and Plants,* iLecture Notes in Biomathematics, **No. 79**, Springer–Verlag, New York, 1989.

[84] W.E. Ricker, Stock and Recruitment, *J. Fisheries Res. Board Canada,* **11**, 1954, 559–623.

[85] C. Robinson, *Dynamical Systems,* CRC Press, Boca Raton, 1995.

[86] C. Robinson, *An Introduction to Dynamical Systems: Continuous and Discrete,* Pearson Prentice Hall, Upper Saddle River, 2004.

[87] O.E. Rössler, Horseshoe-map chaos in the Lorenz equation, *Phys. Lett.* **60A**, 1977, 392–394.

[88] J.C. Roux, R.H. Simoyi, and H.L. Swinney, Observation of a strange attractor, *Physica D* **8**, 1983, 257.

[89] R. S. Sacker, A new approach to the perturbation theory of invariant surfaces, *Comm. Pure Appl. Math.* **18**, 1965, 717–732.

[90] P. Saha and S. H. Strogatz, The birth of period 3, *Math. Mag.* **68**(1), 1995, 42–47.

[91] H. Sedaghat, The impossibility of unstable, globally attracting fixed points for continuous mappings of the line, *Am. Math. Month.* **104**, 1997, 356–358.

[92] H. Sedaghat, *Nonlinear Difference Equations: Theory with Application to Social Science Models,* Springer, 1999.

[93] A. N. Sharkovsky, Co-existence of cycles of a continuous mapping of a line into itself, *Ukranian Math. Z.* **16**, 1964, 61–71.

[94] A. N. Sharkovsky, Yu. L. Maistrenko, and E. Yu. Romanenko, *Difference Equations and Their Applications,* Kluwer Academic, Dodrecht, 1993.

[95] W. Sierpinski, Sur une Corbe Cantorienue qui contient une image biunivoquet et continué detoute Corbe donée, *C. R. Acad. Paris* **162**, 1916, 629–632.

[96] R.H. Simoyi, A. Wolf, and H.L. Swinney, One-dimensional dynamics in a multicomponent chemical reaction, *Phys. Rev. Lett.* **49**, 1982, 245.

[97] D. Singer, Stable orbits and bifurcation of maps of the interval, *SIAM J. Appl. Math. 2* **35**, 1978, 260–267.

[98] S. Smale, Differentiable Dynamical Systems, *Bull. Am. Math. Soc.* **73**, 1967, 747–817.

[99] J. Maynard Smith, *Mathematical Ideas in Biology,* Cambridge University Press, 1968.

[100] D. Sprows, Digitally determined periodic points, *Math. Magazine* **71**(4) (1998), 304–305.

[101] P. Štefan, A theorem of Sharkovsky on the existence of periodic orbits of continuous endomorphisms of the real line, *Comm. Math. Phys.* **54**, 1977, 237–248.

[102] P. Touhey, Persistent properties of chaos, *J. Difference Equ. Appl.* **9** (2000), 249–256.

[103] P. Touhey, Yet another definition of chaos, *Am. Math. Month.* **104**(6) (1997), 411–414.

[104] M. Vellekoop and R. Berglund, On intervals: transitivity→ chaos, *Am. Math. Month.* **101**(4), 1994, 353–355.

[105] H. Von Koch, Sur une corbe continué sans tangente, obtenue par une construction géometrique élémentaire, *Arkiv für Mathematik* **1**, 1904, 681–704.

[106] John Von Neumann, The character of the equation, *Recent Theories of Turbulence* (1949), in Collected Works, ed. A.H. Taub, Oxford: Pergamon Press, 1963.

[107] S. Wiggins, *Introduction to Applied Nonlinear Dynamical Systems and Chaos, Texts in Applied Mathematics 2,* Springer–Verlag, 1990.

[108] S. Willard, *General Topology,* Addison-Wesley, Reading, Massachusetts, 1970.

Answers to Selected Problems

Exercises - (1.2–1.4)

1. $x(n) = 4 + \left(\frac{1}{2}\right)^n (c - 4)$

2. $x(n) = 1$

5. $x(n) = \sin^2(2^n \sin^{-1} \sqrt{x_0})$

7. (a) $a(n+1) = \frac{9}{8}a(n)$

   (b) $a(1) = 9$, $a(2) = 1125$

9. (a) $a(n+1) = rq(1-m)a(n) - 2$

11. (a) $y(n+1) = \frac{3}{4}y(n) + \frac{1}{4}$

    (b)

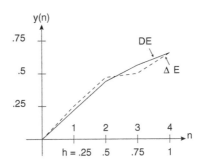

    (c) $y(t) = 1 - e^{-t}$

DE

t	0	.25	.5	.75	1
$y(t)$	0	.22	.39	.53	.63

ΔE

n	0	1	2	3	4
$y(n)$	0	.25	.44	.5	.63

Exercises - (1.5 and 1.6)

1. (a)

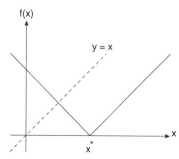

$x^* = \frac{1}{2}$ is a fixed point

(b) The eventually fixed points are $\pm\frac{3}{2}, \pm\frac{5}{2}, \pm\frac{2k+1}{2}, \ldots$

3. (a) Let $f(x) = 2x(x-1)(x-2)(x+1) + x$. Then f has four unstable fixed points $x_1^* = 0$, $x_2^* = 1$, $x_3^* = 2$, $x_4^* = -1$.

   (b) $f(x) = x^2 + 1$ has no fixed points

   (c) Let $f(x) = x^2 - \frac{1}{2}$. The fixed point $x_1^* = \frac{1+\sqrt{3}}{2}$, $x_2^* = \frac{1-\sqrt{3}}{2}$ is asymptotically stable.

5. (a) $x^* = \frac{\alpha - 1}{\beta}$

11. There are three fixed points.

$$x_1^* = \frac{1}{2} \quad : \quad \text{asymptotically stable}$$

$$x_2^* = -\frac{1}{2} \quad : \quad \text{asymptotically stable}$$

$$x_3^* = 0 \quad : \quad \text{unstable}$$

13. (a) $x_1^* = 0$ is unstable

    (b) $x_2^* = \ln 2$ is asymptotically stable

Exercises - (1.7)

1. Fixed Points: $x_1^* = 0$, $x_2^* = 1$

   x_1^* is asymptotically stable, x_2^* is unstable

3. $x_1^* = 0$ is unstable

$x_2^* = \dfrac{2}{3}$ is asymptotically stable

5. $r_1^* = 0$ is unstable

$x_2^* = 1$ is asymptotically stable

7. $x_1^* = 0$ is unstable

9. $x_1^* = 0$ is unstable

$x_2^* = \dfrac{\alpha - 1}{\beta}$ is asymptotically stable

13. $x_1^* = 0$ is asymptotically stable if $-2 \le \alpha < 0$

$x_1^* = 0$ is unstable

$x_2^* = 1$ is asymptotically stable if $0 < \alpha \le 2$

19. $x_1^* = 0$ is semiasymptotically stable from the right

x_2^* is unstable

Exercises - (1.8)

1. $\bar{x}_1 = \dfrac{-1 + \sqrt{.4}}{2} \approx -.1838$

$\bar{x}_2 = \dfrac{-1 - \sqrt{.4}}{2} \approx -.8162$ The 2-cycle is stable.

3. There are no 2-periodic cycles.

5. Every point is a 2-periodic cycle under h, since $h^2(x) = x$. Moreover, every point is stable but not asymptotically stable.

7. (b) The periodic 2-cycle is asymptotically stable.

9. (a) $\bar{x}_1 = \dfrac{1}{3}, \bar{x} = \dfrac{2}{3}$ is a 2-cycle

The 2-cycle is unstable.

(b) There are (2^k) k-periodic points.

11. $x(n) = \dfrac{1}{2\mu} \left[\mu + 1 + (-1)^n \sqrt{(\mu + 1)(\mu - 3)} \right]$, if $\mu > 3$

13. If $\alpha \neq -1$, then $c_1 = 0$ which leads to the fixed point $x(n) = c_0$. In this case the equation has no nontrivial 2-periodic solution.

If $\alpha = -1$, then we have two cases:

(i) If $\beta = 0$, then $c_0 = 0$ and $x(n) = (-1)^n c_1$, c_1 arbitrary, is a 2-cycle.

(ii) If $\beta \neq 0$, then $c_1^2 = \dfrac{c_0(2 + \beta c_0)}{\beta}$.

For any c_0 for which c_1 is real, the solution is

$$x(n) = c_0 + (-1)^n \sqrt{\frac{c_0(2 + \beta c_0)}{\beta}}. \tag{15}$$

If $\beta < 0$, there is a family of 2-periodic solutions: for any $c_0 < 0$ or any $c_0 > -\frac{2}{\beta}$. However, for $\beta > 0$ there is a family of 2-periodic solutions: for any $c_0 > 0$ or any $c_0 < -\frac{2}{\beta}$. The solution is given by equation (15).

Exercises - (1.9)

3. (a) $x_1^* = \dfrac{1 + \sqrt{1 - 4\mu}}{2}$, $x_2^* = \dfrac{1 - \sqrt{1 - 4\mu}}{2}$, x_1^* is unstable, x_2^* is asymptotically stable if $-\dfrac{3}{4} < \mu < 0$.

(b) $\left\{ \overline{x_1} = \dfrac{-1 + \sqrt{-3 - 4\mu}}{2}, \overline{x_2} = \dfrac{-1 - \sqrt{-3 - 4\mu}}{2} \right\}$ is a 2-cycle which is asymptotically stable if $-\dfrac{5}{4} < \mu < -\dfrac{3}{4}$.

10. $k = 2m$, $k = 4$, $m = 2$
   4-periodic points are given by

$$x(n) = c_0 + (-1)^n c_2 + c_1 \cos\left(\frac{n\pi}{2}\right) + d_1 \sin\left(\frac{n\pi}{2}\right)$$

$$\begin{aligned}
n = 0, \quad & x(0) = c_0 + c_2 + c_1 \\
n = 1, \quad & x(1) = c_0 - c_2 + d_1 \\
n = 2, \quad & x(2) = c_0 + c_2 - c_1 \\
n = 3, \quad & x(3) = c_0 - c_2 - d_1
\end{aligned}$$

since $x(4) = x(0)$, we get by substitution in the equation

$$x(4) = x(0) = \mu x(3)[1 - x(3)]$$
$$c_0 + c_2 + c_1 = \mu(c_0 - c_2 - d_1)[1 - c_0 + c_2 + d_1] \tag{16}$$
$$x(5) = x(1) = \mu x(4)[1 - x(4)] = \mu x_0(1 - x_0)$$
$$c_0 - c_2 + d_1 = \mu(c_0 + c_2 + c_1)[1 - c_0 - c_2 - c_1] \tag{17}$$
$$c_0 + c_2 - c_1 = \mu(c_0 - c_2 + d_1)[1 - c_0 + c_2 - d_1] \tag{18}$$
$$c_0 - c_2 - d_1 = \mu(c_0 + c_2 - c_1)[1 - c_0 - c_2 + c_1]. \tag{19}$$

We have 4 equations in 4 unknowns; using Maple we obtain c_0, c_1, c_2, d_1.

Exercise - (2.4)

3. Consider the function $f(x) = 2\cot^{-1} x$. There are two asymptoti-
   cally stable fixed points $x_1^* \approx -1.8$, $x_2^* \approx 1.8$ where $2\cot^{-1} x_1^* = x_1^*$,
   $\mu \cot^{-1} x_2^* = x_2^*$.

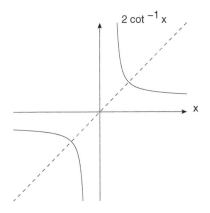

$$L_\mu'(x) = \frac{-2}{1+x^2}$$

$$|L_\mu'(x)| = \frac{2}{1+x^2} < 1 \Rightarrow 1 < x^2 \Rightarrow \text{either } x > 1 \text{ or } x < -1.$$

So we have two attracting fixed points with basins of attraction $(-\infty, -1)$
and $(1, \infty)$, respectively. Another example: $f(x) = 2\sin x$ on $[-\pi, \pi]$.

Exercise - (2.6)

1. $0.\overline{0010111010}$

5. $f : [1, 7] \to [1, 7]$, $f(1) = 4$, $f(2) = 7$, $f(3) = 6$, $f(4) = 5$, $f(5) = 3$,
   $f(6) = 2$, $f(7) = 1$.

6. Consider the map $f : [1, 9] \to [1, 9]$ defined by $f(1) = 5$, $f(2) = 9$
   $f(3) = 8$, $f(4) = 7$, $f(5) = 6$, $f(6) = 4$, $f(7) = 3$, $f(8) = 2$, $f(9) = 1$.

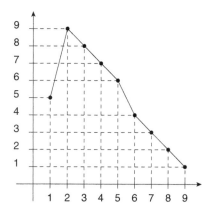

7. Let $f : [1, 2k + 1] \rightarrow [1, 2k + 1]$ be defined as follows:

$$f(1) = k+1, f(2) = 2k+1, f(3) = 2k, \ldots, f(k+1) = k+2, f(k+2) = k,$$
$$f(k + 3) = k - 1, f(k + 4) = k - 2, \ldots, f(2k) = 2, f(2k + 1) = 1.$$

Then connect the points in the graph linearly.

9. $\tilde{f}(x) = \begin{cases} f(x) + 12, & 1 \leq x \leq 7 \\ x - 12, & 13 \leq x \leq 19 \end{cases}$

12. To construct a map with a prime period 8 but no points of prime period 16, we start with the map $f : [1, 4] \rightarrow [1, 4]$ defined as follows:

$$f(1) = 3, \quad f(2) = 4, \quad f(3) = 2, \quad f(4) = 1,$$

and on each interval $[n, n + 1]$ we assume f to be linear.

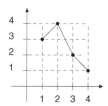

Now we use the double map

$$\tilde{f}(x) = \begin{cases} f(x) + 6 & \text{for } 1 \leq x \leq 4, \\ x - 6 & \text{for } 7 \leq x \leq 10 \end{cases}$$

$$\tilde{f} : [1, 10] \rightarrow [1, 10]$$

Then $\tilde{f}$ has points of period 8 but no points of period 16.

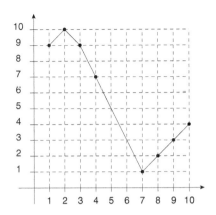

16. Let $f : I \rightarrow I$ defined as follows: $f(1) = 1$, $f(x) = f_k(x)$ if $x \in I_k = \left[1 - \frac{1}{3^k}, 1 - \frac{2}{3^{k+1}}\right]$, $k \in \mathbb{Z}^+$, where $f_0(x) : \left[0, \frac{1}{3}\right] \rightarrow \left[0, \frac{1}{3}\right]$ defined by $f_0(x) = x$, $f_1(x) = \left[\frac{2}{3}, \frac{7}{9}\right] \rightarrow \left[\frac{2}{3}, \frac{7}{9}\right]$ defined by $f_1(x) = -x + \frac{13}{9}$; $f_2(x) = \left[\frac{8}{9}, \frac{25}{27}\right] \rightarrow \left[\frac{8}{9}, \frac{25}{27}\right]$ defined by $f_2(x) = -x + \frac{49}{27}$ for $x \in \left[\frac{8}{9}, \frac{73}{81}\right]$, $f_2(x) = x - \frac{2}{81}$ for $x \in \left[\frac{74}{81}, \frac{25}{27}\right]$ and then connect linearly. In general, we have $f_k(x) = -x + \frac{2(3^{k+1}) - 5}{3^{k+1}}$ for $x \in \left[1 - \frac{1}{3^k}, 1 - \frac{1}{3^k} + \frac{1}{3^{k+2}}\right]$ and $f_k(x) = x - \frac{2}{3^{k+2}}$ and then connect linearly.

Exercises - (3.4)

1. (a) $n = 5$, $f^5(0.1 + 0.01) = f^5(0.11) = 0.97$, $f^5(0.1) = 0.59$
   $\lambda(0.1) \approx \frac{1}{5} \ln\left(\frac{0.97 - 0.59}{0.01}\right) \approx 0.73$

   (b) $n = 6$, $f^6(0.11) = 0.12$, $f^6(0.1) = 0.97$
   $\lambda(0.1) \approx \frac{1}{6} \ln\left|\frac{0.12 - 0.97}{0.01}\right| = 0.74$

   (c) $n = 7$, $f^7(0.11) = 0.42$, $f^7(0.1) = 0.11$
   $\lambda(0.1) \approx \frac{1}{7} \ln\left(\frac{0.42 - 0.11}{0.01}\right) = 0.49$

3. (a) $n = 5$, $|f^5(0.31) - f^5(0.3)| = 0.0080057346$, $\lambda(x_0) \approx -0.044$

   (b) $n = 6$, $\lambda(x_0) = -0.044$

   (c) $n = 7$, $\lambda(x_0) \approx -0.043$

5. $f(x) = \begin{cases} 3x, & 0 \leq x \leq \frac{1}{3} \\ 2 - 3x, & \frac{1}{3} < x \leq \frac{2}{3} \\ 3 - 3x, & \frac{2}{3} < x \leq 1 \end{cases}$

   $|f'(x)| = 3$

   (a) $\log 3 \approx 1.0986$

   (b) f possesses sensitive dependence on initial conditions.

Exercises - (3.5)

7. $f(x) = \begin{cases} \frac{3}{2}x, & 0 \le x \le \frac{1}{2} \\ \frac{3}{2}(1-x), & \frac{1}{2} < x \le \frac{3}{4} \end{cases}$ on $\left[0, \frac{3}{4}\right]$

(a) $f(x) = \begin{cases} \frac{3}{4} - \frac{3}{2}\left(\frac{1}{2} - x\right) = \frac{3}{2}x & \text{if } 0 \le x \le \frac{1}{2} \\ \frac{3}{4} - \frac{3}{2}\left(x - \frac{1}{2}\right) = \frac{3}{2}(1-x) & \text{if } \frac{1}{2} < x \le 1 \end{cases}$

(b)

$$|f'(x)| = \frac{3}{2}$$

$$\lambda(x) = \lim_{n \to \infty} \frac{1}{n} \sum_{k=0}^{n-1} \log \frac{3}{2}$$

$$= \lim_{n \to \infty} \frac{1}{n} \cdot n \log \frac{3}{2}$$

$$= \log \frac{3}{2} > 0$$

Hence f possesses sensitive dependence on initial conditions.

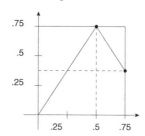

Let $x \in \left(0, \frac{3}{8}\right)$. Then there exists $r > 0$ such that $f^r(x) \ge \frac{1}{2}$. Claim that if $z \ge \frac{1}{2}$, then $f(z) \ge \frac{3}{8}$. Assume the contrary, that is, $\frac{3}{2}(1-z) < \frac{3}{8}$. Then $\frac{3}{2} - \frac{3}{2}z < \frac{3}{8}$ which implies that $z > \frac{3}{4}$, which is not possible since $f\left[0, \frac{3}{4}\right] = \left[0, \frac{3}{4}\right]$. Hence f has no periodic points in the interval $\left(0, \frac{3}{8}\right)$.

Exercises - (3.6 and 3.7)

1. (a) $x_1^* = 0$, $x_2^* = \frac{3}{4} = .75$

    (b) $x_1^* = 0.\overline{0}$ $x_2^* = 0.20\overline{20}$

2. $\left(\dfrac{1}{9},\dfrac{2}{9}\right) \xrightarrow{h} \left(\dfrac{1}{3},\dfrac{2}{3}\right) \xrightarrow{h} \left(1,\dfrac{3}{2}\right) \xrightarrow{h} \left(-\dfrac{3}{2},0\right) \xrightarrow{h} \left(-\dfrac{9}{2},0\right) \cdots (-\infty,0)$

3. (a) Let $x = \cdot x_1 x_2 x_3 \cdots \in E$ be in a ternary expansion. If $x \notin K$, then for some i, $x_i = 1$. Case (i): $x_1 = 2$. Then

$$h(x) = 3 - \left[x_1 + \frac{x_2}{3} + \frac{x_3}{3^2} + \cdots + \frac{x_i}{3^{i-1}} + \cdots\right]$$

$$= \cdot\bar{2} - \cdot x_2 x_3 x_4 \cdots$$

$$= \cdot y_1 y_2 y_3 \cdots$$

where $x_i = y_{i-1} = 1$.

4. Perfect. Let $p \in \wedge$. Then $p \in A_n$ for each n. Hence $p \in [a_{n-1}, b_{n-1}] \subset A_n$, with $b_{n-1} - a_{n-1} < \frac{1}{(1+\varepsilon)^n}$. It follows that $|a_{n-1} - p| < \frac{1}{(1+\varepsilon)^n}$. Thus $\lim\limits_{n\to\infty} |a_{n-1} - p| = 0$ which implies that $\lim\limits_{n\to\infty} a_n = p$.

5. (i) Clearly $\tilde{K} = \cap_{n=1}^{\infty} S_n$ is closed, being the intersection of closed sets.

(ii) Totally disconnected. Notice that the length of each subinterval in S_n is $\left(\frac{2}{5}\right)^n$. If $\tilde{K}$ contains an interval of length d, then $d > \left(\frac{2}{5}\right)^n$ for all large n, which is absurd.

(iii) Perfect. Assume that $p \in \tilde{K}$. Then $p \in \tilde{S}_n$ for all n. Hence $p \in [a_n, b_n] \subset \tilde{S}_n$, with $b_n - a_n = \left(\frac{2}{5}\right)^n$. Thus $|a_n - p| < \left(\frac{2}{5}\right)^n$. For $\varepsilon > 0$, let N be large enough such that $\left(\frac{2}{5}\right)^N < \varepsilon$. Then for $n > N$, $|a_n - p| < \varepsilon$. This implies that $a_n \to p$ as $n \to \infty$.

Exercises - (3.8)

3. (a) $h\left(\dfrac{\sqrt{5}-1}{2\sqrt{5}}\right) = \cdot 01\bar{0}$

(b) $h\left(\dfrac{\sqrt{5}+1}{2\sqrt{5}}\right) = \cdot 11\bar{0}$

7. (a) $h(x) = \dfrac{\mu}{\lambda}x - \dfrac{\mu}{2\lambda}$, $c = \dfrac{\mu^2 - 2\mu}{4\lambda} = 1$, $\lambda = \dfrac{\mu^2 - 2\mu}{4}$

(b) Now $\lambda = 2$ corresponds to $\mu = 4$. Since $F_4 \approx Q_2$ and F_4 is chaotic on $[0,1]$, Q_2 is chaotic on $[-1,1]$.

9. $T(x) = \begin{cases} 2x, & 0 \le x \le \frac{1}{2} \\ 2(1-x), & \frac{1}{2} < x \le 1 \end{cases}$

(a) Let $h(x) = \sin^2\left(\dfrac{\pi x}{2}\right)$

Exercises - (4.1 and 4.2)

1. eigenvector $\begin{pmatrix} 1 \\ \frac{3}{2} \end{pmatrix}$ is associated with $\lambda_1 = 3$, $\begin{pmatrix} 1 \\ 1 \end{pmatrix}$ is associated with $\lambda_2 = \frac{1}{2}$, $A^n = \begin{pmatrix} -2 \cdot 3^n + \frac{3}{2^n} & 2 \cdot 3^n - \frac{1}{2^{n-1}} \\ -3^{n+1} + \frac{3}{2^n} & 3^{n+1} - \frac{1}{2^{n-1}} \end{pmatrix}$

3. $\lambda_1 = \lambda_2 = 2$

$V_1 = \begin{pmatrix} 1 \\ -2 \end{pmatrix}$

$V_2 = \begin{pmatrix} 1 \\ 1 \end{pmatrix}$

$A^n = \begin{pmatrix} \frac{2^n(n+3)}{3} & \frac{n \cdot 2^{n-1}}{3} \\ -\frac{n2^{n+1}}{3} & \frac{2n(3n-1)}{3} \end{pmatrix}$

5. $\lambda_1 = \frac{-1+\sqrt{3}i}{2}$, $\lambda_2 = \frac{-1-\sqrt{3}i}{2}$

$v = \begin{pmatrix} -\frac{3}{2} + \frac{\sqrt{3}}{2}i \\ 1 \end{pmatrix}$

$A_n = \begin{pmatrix} -\sqrt{3}\sin\frac{2n\pi}{3} + \cos\frac{2n\pi}{3} & -2\sqrt{3}\sin\frac{2n\pi}{3} \\ \frac{2\sqrt{3}}{3}\sin\frac{2n\pi}{3} & \cos\frac{2n\pi}{3} + \sqrt{3}\sin\frac{2n\pi}{3} \end{pmatrix}$

7. $X(n) = \begin{pmatrix} 2^{n-1}(n+2) \\ 2^n(1-n) \end{pmatrix}$

9. $f^n \begin{pmatrix} 0 \\ 1 \end{pmatrix} = A^n \begin{pmatrix} 0 \\ 1 \end{pmatrix} = \begin{pmatrix} -2\sqrt{3}\sin\frac{2n\pi}{3} \\ \cos\frac{2n\pi}{3} + \sqrt{3}\sin\frac{2n\pi}{3} \end{pmatrix}$

Exercises - (4.3 and 4.4)

1. $X(n) = \frac{1}{3}\begin{pmatrix} (-1)^n + 2^{n+1} \\ 2^{n+1} \end{pmatrix}$

3. $X(n) == \begin{pmatrix} k_1 2^n + k_2 3^n \\ k_1 2^n + 2k_2 3^n \end{pmatrix}$

5. $X(n) == \begin{pmatrix} 3^n x_1(0) + 3^{n-1}x_2(0) \\ 3^n x_2(0) \end{pmatrix}$

7. $X(n) = k_1 \cos n\theta + k_2 \sin n\theta$ where $\theta = \tan^{-1}(4) \approx 76°$

15. $Y(n) == \begin{pmatrix} 2^n + n2^{n-1} - \frac{3}{4}n \\ 2^n - 1 \end{pmatrix}$

Exercises - $(4.5 - 4.7)$

1. The origin is asymptotically stable since $\rho(A) = \frac{1}{2} < 1$ (Theorem 4.13).

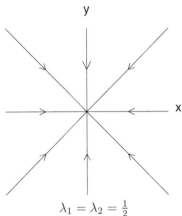

$\lambda_1 = \lambda_2 = \frac{1}{2}$
Phase portrait: origin is asymptotically stable.

3. Since $\rho(A) = 2 > 1$, it follows by Theorem 4.13 that the origin is asymptotically stable.

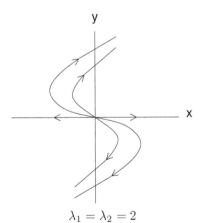

$\lambda_1 = \lambda_2 = 2$
Phase portrait: origin is unstable.

5. Eigenvalues of A are $\lambda_1 = \frac{1}{2} + \frac{1}{4}i$, $\lambda_2 = \frac{1}{2} - \frac{1}{4}i$.

$$|\lambda_1| = |\lambda_2| + \sqrt{\frac{1}{4} + \frac{1}{16}} = \frac{\sqrt{5}}{4} < 1$$

Hence by Theorem 4.12, the origin is asymptotically stable.

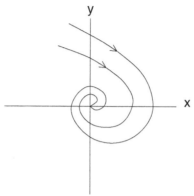

$\lambda_1 = \frac{1}{2} + \frac{1}{4}i, \quad \lambda_2 = \frac{1}{2} - \frac{1}{4}i$
A stable focus.

7. The eigenvalues of A are $\lambda_1 = 2$, $\lambda_2 = 3$. The corresponding eigenvectors are $V_1 = \begin{pmatrix} 4 \\ -5 \end{pmatrix}$, $V_2 = \begin{pmatrix} 2 \\ -3 \end{pmatrix}$, straight-line solutions are

$$Y_1 = \begin{pmatrix} 4 \\ -5 \end{pmatrix} 2^n, \quad Y_2 = \begin{pmatrix} 2 \\ -3 \end{pmatrix} 3^n.$$

The origin is unstable since $\rho(A) = 3 > 1$.

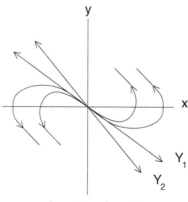

$\lambda_1 = 2, \quad \lambda_2 = 3$
Unstable node.

9. The eigenvalues of A are $\lambda_1 = i$, $\lambda_2 = -i$. The origin is stable but not asymptotically stable.

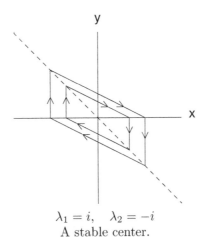

$$\lambda_1 = i, \quad \lambda_2 = -i$$
A stable center.

Exercises - 4.8

1. (i) $a > 4$: oscillatory source

   (ii) $a = 4$: unstable origin (bifurcation value)

   (iii) $a < 4$: a saddle

   (iv) $a = -\dfrac{4}{3}$: unstable origin

   (v) $a < -\dfrac{4}{3}$: spiral sources

3. Bifurcation value: $a = -3$

   $a < -3$: a saddle

   $a = -3$: unstable oscillatory

5. (i) $a \geq -\dfrac{7}{6}$ where the eigenvalues are $\lambda_2 = -1$ and $\lambda_1 = -\det A = \dfrac{2}{3}$ and we have an "oscillatory" stable origin, for $a < -\dfrac{7}{6}$, we have an oscillatory saddle.

   (ii) $a = -1$, the eigenvalues are $\lambda_1 = \lambda_2 = 0$ and every point is a fixed point, for $-\dfrac{7}{6} < a < -1$, we have a sink, and for $-1 < a < 0$, we have a spiral sink.

(iii) $a = 0$, where the eigenvalues are $\lambda_1 = 1$, $\lambda_2 = 1$, where we have eigher a stable or unstable origin, for $a > 0$, we have a saddle.

7. Region 5: $a < b$ and $3a + 6 > b$

   Region 6: $a < b$ and $3a + 6 < b$

   Region 7: $a > b$ and $3a + 6 < b$

   Region 1: $b > -\dfrac{1}{4}(a-1)^2$

   Region 8: $b < 3a + 6$ and $b < 2a + 1$

   Region 4: $b < a$ and $b > -\dfrac{1}{4}(a-1)^2$ and $b < 2a + 1$

Exercises - (4.9)

1. Let $V(x_1, x_2) = x_1^* + x_2^*$.

3. Let $V = x_1 x_2$.

5. Let $V(x, y) = x^2 + 4y^2$.

7. Let $V(x, y) = x^2 + y^2$.

11. Let $V(x, y) = xy$.

Exercises - (4.10 and 4.11)

3. $|\alpha| < 1$

5. $\alpha\beta < 1$

6. (a) $X_1^* = \begin{pmatrix} 0 \\ 0 \end{pmatrix}$ is asymptotically stable

   (b) $X_2^* = \begin{pmatrix} 6 \\ 3 \end{pmatrix}$ is unstable

7. $|b| < 1 - a < 2$

Exercises - (4.12)

1. $A = Df \begin{pmatrix} x_1^* \\ x_2^* \end{pmatrix} \begin{pmatrix} 1 - b_1 - a_1 \left(\frac{1}{1+\gamma_2}\gamma_1\gamma_2\right) & u_1 - u_1 \left(\frac{1-\gamma_1\gamma_2}{1+\gamma_2}\right) \\ a_2 - a_2 \left(\frac{1-\gamma_1\gamma_2}{1+\gamma_2}\right) & 1 - b_2 - a_2 \left(\frac{1-\gamma_1\gamma_2}{1+\gamma_2}\right) \end{pmatrix}$

$0 < a_i < 1, \ 0 < b_i < 1, \ i = 1, 2$

3. (a) $N^* = \dfrac{r}{a(r-1)} \ln r, \ r > 1, \ a > 0$

   (b) N^* is asymptotically stable

5. (c)
$$\operatorname{tr} A_1 = 1 - b - \sqrt{(1-b)^2 + 4a}, \quad \det A_1 = -b$$
$$1 - 1 + b + \sqrt{(1-b)^2 + 4a} - b > 0$$
$$1 + 1 - b - \sqrt{(1-b)^2 + 4a} - b = 2(1-b) - \sqrt{(1-b)^2 + 4a} > 0$$
$$4(1-b)^2 > (1-b)^2 + 4a$$

7. Use the Liapunov function $V(p,q) = (1-\alpha)^{-1}[p - p^* - p^* \ln\left(\frac{p}{p^*}\right)] + (c(1-\beta))^{-1}[q - q^* - q^* \ln\left(\frac{q}{q^*}\right)]$

Exercises - (6.1, 6.2 and 6.3)

1. (b) $D_t = 1$

   (c) $D_f = \lim\limits_{h \to 0} \dfrac{\ln N(h)}{\ln\left(\frac{1}{h}\right)} = \lim\limits_{n \to 0} \dfrac{\ln 8^n}{\ln 4^n} = 1.5$

3. (b) $D_t = 1$

   (c) $D_f = \lim\limits_{h \to 0} \dfrac{\ln 5^n}{\ln 3^n} = \dfrac{\ln 5}{\ln 3}$

5. (b) $D_t = 1$

   (c) $D_f = \lim\limits_{n \to \infty} \dfrac{\ln(3^{n+1})}{\ln(2^n)} = \dfrac{\ln 3}{\ln 2}$

7. (b) $D_t = 2$

   (c) $D_f = \lim\limits_{n \to \infty} \dfrac{\ln 20^n}{\ln 3^n} = \dfrac{\ln 20}{\ln 3} \approx 2.73$

9. (b) $D_t = 1$

19. $D_t(C_5) = 0, \ D_f(C_5) = \lim\limits_{n\to\infty} \dfrac{\ln 2^n}{\ln \left(\frac{5}{2}\right)^n} = \dfrac{\ln 2}{\ln \left(\frac{5}{2}\right)}$

20. $D_5(C_{2n+1}) = 0, \ D_f(C_{2n+1}) = \dfrac{\ln 2}{\ln \left(\frac{2n+1}{n}\right)}$

Exercises - (6.4)

1. (a) $F\begin{pmatrix} x \\ y \end{pmatrix} = \begin{pmatrix} -x \\ y \end{pmatrix}$

    (b) $F\begin{pmatrix} x \\ y \end{pmatrix} = \begin{pmatrix} -\frac{1}{2}x \\ 2y \end{pmatrix}$

    (c) $F\begin{pmatrix} x \\ y \end{pmatrix} = \begin{pmatrix} \frac{\sqrt{2}}{2} & -\frac{1}{2} \\ \frac{\sqrt{2}}{2} & \frac{\sqrt{3}}{2} \end{pmatrix}\begin{pmatrix} x \\ y \end{pmatrix}$

3.

$$F_1\begin{pmatrix} x \\ y \end{pmatrix} = \begin{pmatrix} \frac{1}{3} & 0 \\ 0 & \frac{1}{3} \end{pmatrix}\begin{pmatrix} x \\ y \end{pmatrix}$$

$$F_2\begin{pmatrix} x \\ y \end{pmatrix} = \begin{pmatrix} \frac{1}{3} & 0 \\ 0 & \frac{1}{3} \end{pmatrix}\begin{pmatrix} x \\ y \end{pmatrix} + \begin{pmatrix} 0 \\ \frac{2}{3} \end{pmatrix}$$

$$F_3\begin{pmatrix} x \\ y \end{pmatrix} = \begin{pmatrix} \frac{1}{3} & 0 \\ 0 & \frac{1}{3} \end{pmatrix}\begin{pmatrix} x \\ y \end{pmatrix} + \begin{pmatrix} \frac{1}{3} \\ \frac{1}{3} \end{pmatrix}$$

$$F_4\begin{pmatrix} x \\ y \end{pmatrix} = \begin{pmatrix} \frac{1}{3} & 0 \\ 0 & \frac{1}{3} \end{pmatrix}\begin{pmatrix} x \\ y \end{pmatrix} + \begin{pmatrix} \frac{2}{3} \\ 0 \end{pmatrix}$$

$$F_5\begin{pmatrix} x \\ y \end{pmatrix} = \begin{pmatrix} \frac{1}{3} & 0 \\ 0 & \frac{1}{3} \end{pmatrix}\begin{pmatrix} x \\ y \end{pmatrix} + \begin{pmatrix} \frac{2}{3} \\ \frac{2}{3} \end{pmatrix}$$

5.

$$F_1\begin{pmatrix} x \\ y \end{pmatrix} = \begin{pmatrix} \frac{1}{3} & 0 \\ 0 & \frac{1}{3} \end{pmatrix}\begin{pmatrix} x \\ y \end{pmatrix}$$

$$F_2\begin{pmatrix} x \\ y \end{pmatrix} = \begin{pmatrix} \frac{1}{3} & 0 \\ 0 & \frac{1}{3} \end{pmatrix}\begin{pmatrix} x \\ y \end{pmatrix} + \begin{pmatrix} \frac{2}{3} \\ 0 \end{pmatrix}$$

7.

$$F_1\begin{pmatrix}x\\y\end{pmatrix}=\begin{pmatrix}\frac12&0\\0&\frac12\end{pmatrix}\begin{pmatrix}x\\y\end{pmatrix}$$

$$F_2\begin{pmatrix}x\\y\end{pmatrix}=\begin{pmatrix}\frac12&0\\0&\frac12\end{pmatrix}\begin{pmatrix}x\\y\end{pmatrix}+\begin{pmatrix}\frac12\\0\end{pmatrix}$$

$$F_2\begin{pmatrix}x\\y\end{pmatrix}=\begin{pmatrix}\frac12&0\\0&\frac12\end{pmatrix}\begin{pmatrix}x\\y\end{pmatrix}+\begin{pmatrix}0\\\frac12\end{pmatrix}$$

Exercises - (6.5)

1. (a) $D(A,B)=\sqrt5$
    (b) $D(A,B)=2$

5. By Lemma 8.16, $F_1, F_2, \ldots, F_N$ are contractions on H. This implies by Lemma 8.17 that $F = \cup_{i=1}^N F_i$ is also a contraction on H. The conclusion now follows by applying the contraction mapping principle (Theorem 8.15).

7. Let $F : X \to X$ be a contraction with a contraction factor $\alpha \in (0,1)$. Now

$$d(x(n+1),x(n)) = d(F(x(n)),F(x(n-1)))$$
$$\leq \alpha d(x(n),x(n-1))$$
$$\vdots$$
$$\leq \alpha^n d(x(1),x(0)).$$

Index

Phaser Scientific Software, LLC.

PHASER 3.0 Installation Instructions: On Microsoft Windows Vista: After inserting the enclosed CD-ROM in the system's drive, an *"AutoPlay"* dialog box will appear with the following options: *"Install or run program" Run phaser3.0-win32-setup.exe* **or** *"General options" Open folder to view files.* Click on *Run phaser3.0-win32-setup.exe.* On Microsoft Windows XP: The software installer should automatically run without user intervention after inserting the enclosed CD-ROM. For other platforms, please consult the README.txt file on the CD-ROM for installation instructions.

PHASER 3.0 Supplemental Modules: On Microsoft Windows Vista: After inserting the enclosed CD-ROM in the system's drive, an *"AutoPlay"* dialog box will appear with the following options: *"Install or run program" Run phaser3.0-win32-setup.exe* **or** *"General options" Open folder to view files.* Click on *Open folder to view files* option. To access the supplemental modules, navigate to the *modules/Discrete_Chaos* directory and double-click on the *index.html* file. On Microsoft Windows XP and assuming PHASER software is already installed, insert the CD-ROM while holding down the *Shift* key for 15 seconds (this temporarily prevents the OS from autorunning the installer on the CD-ROM). Open the *"My Computer"* folder and right-click on the *CD Drive* with the Phaser logo and select *"Explore"*. Navigate to *modules/Discrete_Chaos* directory and double-click on the *index.html* file. For other platforms, the supplemental material for *Discrete Chaos* is located at *modules/Discrete_Chaos/index.html* on the CD-ROM. Alternatively, one can copy the *modules/Discrete_Chaos.zip* archived file from the CD-ROM to the local hard drive, unpack it, and then double-click on the *index.html* file.

PHASER 3.0 Notes: This fully functional copy of PHASER is licensed for individual use by owners of *Discrete Chaos, 2nd Edition* for 9 months from the date of first installation. After this period, PHASER becomes a Reader which can still run the simulations in the supplemental modules or any other modules. For further information on permanent individual or site licenses and special discounts to the owners of *Discrete Chaos*, please visit our website or write to us, respectively, at:

http://www.phaser.com   *info@phaser.com*